Swidden Agriculture in Indonesia

New Babylon

Studies in the Social Sciences

43

MOUTON PUBLISHERS · BERLIN · NEW YORK · AMSTERDAM

Swidden Agriculture in Indonesia

The Subsistence Strategies of the Kalimantan Kantu'

Michael Roger Dove

MOUTON PUBLISHERS · BERLIN · NEW YORK · AMSTERDAM

Michael Roger Dove
USAID/Islamabad
Agency for International Development
Washington, D.C. 20520
U.S.A

Library of Congress Cataloging in Publication Data

Dove, Michael.
Swidden agriculture in Indonesia.

(New Babylon, studies in the social sciences ; 43)
Bibliography: p.
Includes index.
1. Kantu' (Indonesian people)—Economic conditions.
2. Shifting cultivation—Indonesia—Kalimantan Barat.
I. Title. II. Series.
DS646.32.K36D68 1985 338.1'09598'3 85-11497

ISBN 0-89925-036-X (alk. paper)

CIP-Kurztitelaufnahme der Deutschen Bibliothek

Dove, Michael Roger:
Swidden agriculture in Indonesia : the subsistence strategies of the Kalimantan Kantu' / Michael Roger Dove. - Berlin ; New York ; Amsterdam : Mouton, 1985.
(New Babylon ; 43)
ISBN 3-11-009592-0 (Berlin)
ISBN 0-89925-036-X (New York)

NE: GT

Printed on acid free paper.

Printing: Druckerei Hildebrand, Berlin. — Binding: Dieter Mikolai, Berlin. Printed in Germany.

To the memory of Michelle Z. Rosaldo

Table of Contents

List of Illustrations

List of Tables

Preface

This book represents the first half of my study of the economy and ecology of the swidden Kantu' of West Kalimantan. The present volume focuses on the swidden system itself. The planned second volume will relate the swidden system to broader issues of Kantu' social structure, politics, religion and history.

The field research on which this study is principally based was carried out during a two-year stay in Indonesia, from June 1974 to June 1976. After returning to the United States from Indonesia, I devoted the ensuing three years to an initial analysis of the results of this research, which I submitted to the Department of Anthropology at Stanford University in partial fulfillment of the requirements for the Ph.D. (awarded in 1981). In 1979 I returned to Indonesia, where I worked periodically on this manuscript for an additional five years. Although I was based in Java, I was able to make five separate return trips to Kalimantan during these years, in each of which I carried out further studies of swidden ecology and economy. Thus, this monograph is the result of almost ten years of research and writing on the topic, an investment of time and energy that was made possible only by the cumulative assistance of a number of different institutions and individuals.

The initial field work was supported by a training grant from the (U.S.) National Institute of General Medical Sciences, and by research funds from the Center for Research in International Studies (Stanford University) and the (U.S.) National Science Foundation (grant # GS-42605). The initial analysis and write up of the field data was supported by a fellowship from the Richard D. Irwin Foundation. Subsequent analysis and writing was supported by a postdoctoral fellowship and succession of grants from the Rockefeller Foundation, as well as by a fellowship from the East-West Environment and Policy Institute and the Ford Foundation. The actual publication of the monograph also was subsidized by grants from the Rockefeller and Ford Foundations and the East-West Environment and Policy Institute. However, the opinions expressed in this publication do not necessarily reflect those of any of these institutions.

The initial two years of field research was sponsored by the Indonesian Institute of Science (LIPI). I also received considerable support from the West Kalimantan provincial government, in particular from the former governor, Pak Kadarusno, and the former subdistrict officer in the Empanang river valley, Pak Yakub. Upon returning to Indonesia in 1979 I became affiliated with Gadjah Mada University in Yogyakarta, in the capacity of visiting professor and research fellow. I am grateful for the institutional support that I received during this period from the Faculty of Arts and Letters, the Population Studies Center, the Environmental Studies Center and, especially, the past and present directors of the latter, Ir. Soemantri Sastrosoeharjo and Dr. Sugeng Martopo.

There are many more individuals to whom I am indebted for advice and encouragement at one stage or another in the course of this project. These include George N. Appell, Carol Carpenter, Loy V. Crowder, Charles Frake, Derek Freeman, Thomas G. Kessinger, Colin MacAndrews, A. Terry Rambo, Renato Rosaldo and Laurence D. Stifel. For their continued support in seeking publication of this work I am especially indebted to Joseph E. Black and Harold C. Conklin. In preparing the manuscript for publication, I owe thanks to H.J. Koesoemanto, Bambang Agus Suripto, and Dwi Susanto.

The individuals most directly responsible for this work are the Kantu' themselves, especially the one-hundred odd members of the longhouse Tikul Batu. They did something for me that is still, in a way, beyond my comprehension. They took me into their lives; they worked with me, ate, drank, sang and danced with me; taught me, humored me, cared for me - all in the absence of any personal advantage or profit. I think I can say that no anthropologist has ever had, in this regard, a warmer, more enjoyable (and easier) field situation. This development, the development of close friendships (or, more precisely, fictive kin relations), was totally unexpected, but it came to be the most satisfying aspect of my life in the field. Sometimes, the shortcomings of anthropological research are attributed to recalcitrant informants. That charge does not hold here. Whatever are the shortcomings of this study, they are decidedly my own.

Finally, I wish to acknowledge the greatest debt to my parents, Roger and Betty Dove, who have always been a source of support and inspiration to me.

Michael R. Dove
June 1984
Yogyakarta

Units of Measure

I. CURRENCY (June 1974 - June 1976)

415 rupiah = one dollar U.S.

II. AREA

1 *belah* = one swidden-sized section of forest, averaging 2.9 hectares in area.

III. TIME

Sunrise = 6:00.
Sunset = 18:00.
One workday = 350 minutes.
2/3/75 = 2nd day of 3rd month of 1975 (viz., 3 March 1975).

IV. VOLUME

1 *mok* = .295 liter.
1 *kolak* = 20 *mok* = 5.9 liters.
1 *raga' pemenih* 'sowing basket' = 5 liters.
1 *takin pemenih* 'sowing basket' = 34 liters.
1 *takin pengetau* 'harvesting basket' = 15.5 liters.
1 *lanyi'* 'carrying basket' = 88.5 liters.

V. RICE

tangkai = unthreshed, unwinnowed, unhusked grain.
renas = threshed, unwinnowed, unhusked grain.
padi = threshed, winnowed, unhusked grain.
beras = threshed, winnowed, husked grain.
nasi = cooked *beras*.

1 liter *tangkai* = .75 liter *renas* (approximate).
1 liter *renas* = .75 liter *padi* (approximate).
1 liter *padi* = .41 liter *beras*.
1 liter *padi* = .62 kilogram *padi*.
1 ton (metric) = 1000 kilograms.

VI. STATISTICAL TESTS

X^2 = chi square test.
X_c^2 = chi square test with Yates' correction.
(P=) = Fisher's exact test.
r = Pearson's product moment correlation coefficient.
r_s = Spearman's rank order correlation coefficient.

Men that undertake only one district are much more likely to advance natural knowledge than those that grasp at more than they can possibly be acquainted with: every kingdom, every province, should have its own monographer.

Gilbert White, The Natural History of Selbourne (1789)

Introduction

THE STUDY

Theory

The study of traditional, non-Western systems of agriculture is a relatively new field. Even within anthropology, where such studies have attained their greatest prominence, for long there was little enthusiasm for dealing with this subject. Such pioneer figures as Boas and Radcliffe-Brown considered the study of agriculture to be at best unimportant and at worst an improper subject for the anthropologist[1]. A change in this attitude was heralded by the publication, in the 1950s, of monograph-length studies of Southeast Asian systems of swidden agriculture by Condominas, Conklin, Freeman, and others[2]. These were followed in short order by a proliferation of studies on non-Western agriculture by all manner of anthropologists in pursuit of all sorts of theoretical goals. Indeed, agriculture has provided much of the substantive content for most of the anthropological subfields and issues developed since World War II, at least in North America. This includes the subfields of economic anthropology, cognitive anthropology and, especially, ecological anthropology. It includes the theoretical debates concerning the prior determinacy of production systems, on the one hand, and social structure, political structure, and religion, on the other[3]. It also includes the extended debates over the theories of scholars such as Boserup and Chayanov, as well as work in the areas of risk taking and decision making[4]. Finally, studies of agriculture have been central to the recent efflorescence of applied and developmental anthropology[5].

There are several possible reasons for this historically increasing prominence of agriculture as a subject matter of anthropoloy. One is the sheer complexity of the subject matter. The early practitioners of anthropology may have found the study of agriculture unrewarding in part because it was beyond their comprehension - and indeed there is little to suggest that the early ethnographers understood very much about the systems of agriculture of their subjects. Only as the field of anthropology has matured, perhaps, have anthropologists been given the methodological and theoretical tools prerequisite to the study of these systems. Another possible explanation is based on a view of agriculture as not merely complex but also as crucial. Some anthropologists believe that the character of agriculture and other systems of production is a primary determinant of the character of other aspects of society. If this is indeed the case, then it follows that as anthropology has matured as a science, and as anthropologists have come closer to a true understanding of the nature of human society, so inevitably they have come to deal more with agriculture as one of its primary determinants. A final reason for the increasing prominence of agriculture in anthropological studies involves the extraordinary postwar boom in rural development and studies of rural development in the Third World. A great deal of this development has focused on agriculture, and anthropologists - although they often know little about Western, mechanized, capitalized, agriculture - often know a great deal about traditional agriculture in the Third World. It can be said without fear of hyperbole that there are in the world today but two groups of experts on the nature of traditional, non-Western agriculture: the agriculturalists themselves, and those anthropologists (and anthropologically minded in other disciplines) who have taken the time and effort to live, work, and study with them. With rare exceptions, the government officers and the university agronomists and agricultural economists have sought to improve or more often replace the traditional systems of agriculture, but they have not sought to understand them. Because of this, anthropological studies of non-Western systems of agriculture have much to offer not only to the field of anthropology itself, but also to those other fields that work with these systems but do not know them.

It is in this context of past and present involvement of anthropologists in the study of agriculture that I offer the current study, an analysis of the system of swidden agriculture of the Kantu' of West Kalimantan (fig. 1). Swidden studies have occupied a prominent role in the history of anthropological studies of agriculture. As I noted earlier, the pioneering studies in the 1950s all were concerned with systems of swidden agriculture. These studies were followed by a plethora of others[6]. In some of these studies, swidden agriculture was just

FIGURE 1
Kalimantan And The Research Site

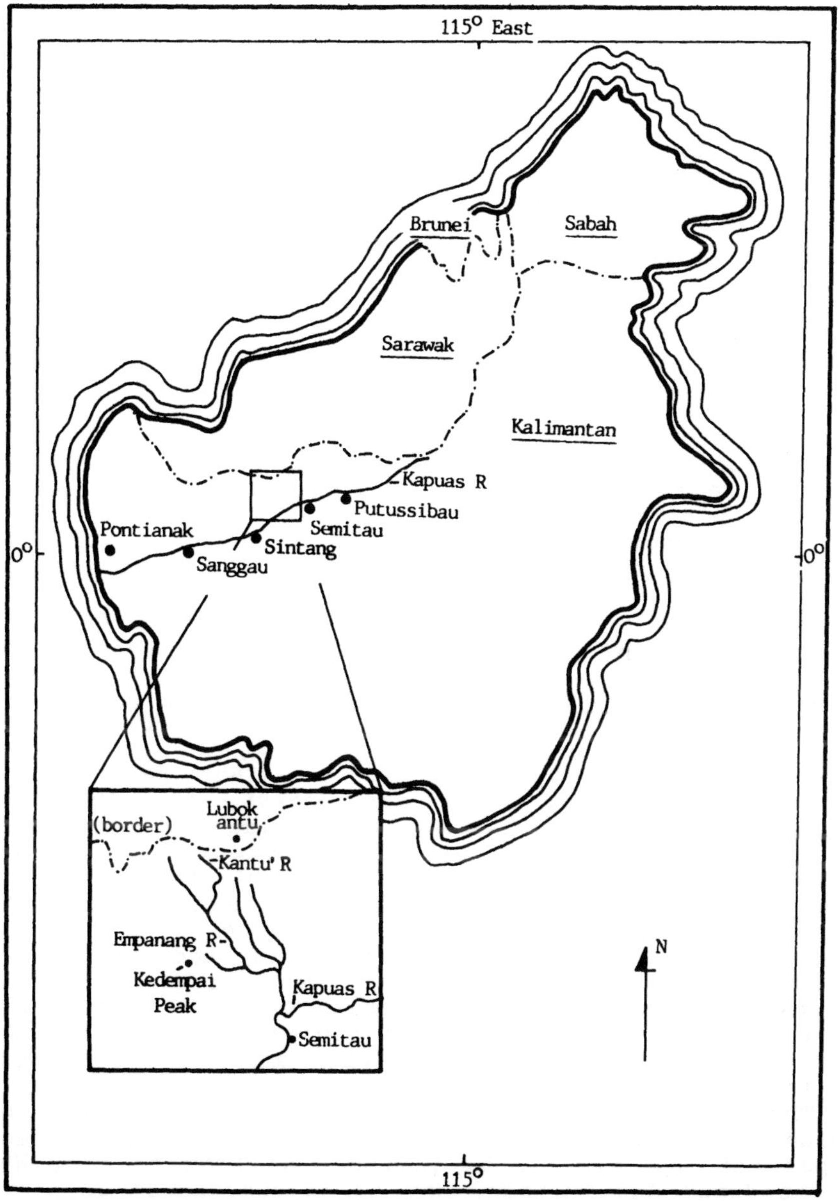

a convenient medium in which to examine problematic aspects of theory - some of which were mentioned earlier - while for others swidden agriculture was itself the object of study. Studies of the latter sort have focused on such areas as ethnobotanical knowledge, carrying capacity, migration, warfare, and environmental degradation - referring especially to grassland succession[7]. More recent studies have focused on such issues as energy tranfer, ecological patch theory, and forest mimicry[8]. Some of the most interesting research on swidden agriculture has involved studying it not in isolation, but in relation to other systems of more intensive agriculture. A number of anthropologists have emphasized the various distinctions between the swidden cultivation of dry rice and the irrigated cultivation of wet rice and have related this to differences in economy and society[9]. Others have focused on the development of systems of swidden agriculture into more intensive systems of cultivation (and on the reverse process as well)[10]. Finally, related studies have been carried out by anthropologists looking at the role of swidden agriculture in rural development[11]. A promising direction for future studies may involve looking at the 'problem' of developing swidden agriculture in terms of the political and economic implications of extensive systems of cultivation for the policies and interests of central governments[12].

In this study of the Kantu' system of swidden agriculture, I have two purposes. The first is to present the swidden system[13]. This consists in translating the Kantu' swidden terminology into English and, more generally, in describing the swidden system[14]. This description includes three aspects of the system, in particular, that have been slighted in most extant descriptions of swidden systems. These are first, the diversity of swidden strategies within the village or longhouse, which most studies treat as a homogenous unit[15]; second, the diversity of swidden strategies within the household, based on the cultivation of two or more separate swiddens, a phenomenon that also has gone unreported in other studies; and third, the diversity and uncertainty of the local environment, and the influence of this on the swidden system. The overall emphasis in this description is on production as opposed to distribution, the latter of which is dealt with only in the context of labor's role in production[16].

The second purpose of this study is to explain the swidden system of the Kantu'. After describing each aspect of the system, my purpose has been to ask, Why is it done? And further, Why is it done as it is done?

I posed and pursued these queries based on the theory that the swidden system is purposeful and each aspect of it is best explained in terms of how it relates and contributes to this purpose. This inquiry lies within an intellectual tradition stretching directly back to the functionalism of Hempel and Nagel[17]. In viewing the swidden system as central to the lives of the Kantu', and in seeking to explain this system largely in terms of material needs and constraints, as I do, this inquiry lies within the contemporary tradition of cultural materialism, as practised by Harris and others[18].

My adoption of this approach was motivated in part by the character of the swidden system itself. Early in my study I observed that this system seemed ill able to tolerate nonfunctional or dysfunctional behaviour. In an average year, a majority of Kantu' households fail to reap a harvest sufficient for one year's consumption requrements. Given this genuine threat of system failure and hunger, I hypothesized that most behaviour is likely to contribute to the success of the swiddens. This success, it became clear to me, depends upon overcoming two principal difficulties: first, the abovementioned diversity and uncertainty of the local environment and, second, related time constaints on the intensification of human labor within the system[19]. I hypothesized, then, that the principal goals of Kantu' swidden strategy are to minimize the effects of this uncertainty, to maximize exploitation of this diversity, and to maximize labor intensity.

I regard the swidden system, in addressing these needs, as the major mechanism of human adaptation to the local environment. I do not regard this adaptation as static[20]. The Kantu' environment is constantly changing, as are the Kantu' themselves; and in order to deal with these changing realities the Kantu' are continually modifying their swidden system, sometimes at a very conscious level, and sometimes at a less conscious level. I illustrate this process, wherever possible, through the use of historical data on changes in the swidden system.

I attempt to explain the swidden system largely but by no means solely with respect to the local physical environment. I recognize that the character of Kantu' society - their social and political structure, their belief system and technoloby, as well as their history - is not merely some function of this environment, but is at least in part itself an independent and constraining variable, constraining the way that they have adapted to this environment through their swidden system. For example, the very fact that the Kantu' make swiddens is not determined solely by the character of the local environment. The environment of the Empanang valley is a sufficient but not necessary condition for the practice of swidden agriculture. Nonswidden subsistence economies, for example hunting and gathering, could also have been practised in this valley, as they are elsewhere in Kalimantan. Their absence in this case probably is due more to the distance from trading centers than to any peculiarity of the local environment[21].

In attempting to explain the swidden system, I present explanations from two different perspectives. The first perspective is that of the Kantu' themselves. The Kantu' knowledge of their environment, and of the optimal response to it through their swiddens, is impressive. A small fraction of that knowledge is presented in this study in explanation of their swidden behavior. The second perspective is external to the Kantu'. It consists in quantification, categorization and interpretation of Kantu' behaviour according to the external systems of knowledge of anthropology, cultural ecology, and agricultural economics. As Rappaport has written:

> The ethnographer must prepare two accounts of his subject matter. One, which I have called 'cognized model', is a description of a people's knowledge of their environment and of their beliefs concerning it. The second, the 'operational model', describes the same ecological system (including the people and their activities) in accordance with the assumptions and methods of the objective sciences, in particular the science of ecology[22].

Methodology

I base this study largely upon data on two sets of swiddens: the 34 swiddens planted by the fourteen households of Tikul Batu in the fall of 1974 and harvested in the spring of 1975, hereafter referred to as the 1975 swiddens; and the 35 swiddens planted by sixteen households in the fall of 1975 and harvested in the spring of 1976, hereafter referred to as the 1976 swiddens (figure 2). A small portion of the study is based on data on the 1974 swiddens, which were harvested four to six months before I arrived at Tikul Batu, and the 1977 swiddens, which were still being cleared when I departed from Tikul Batu in 1976. Some points regarding the evolution and adaptation of the swidden system are based on data from the more distant past, gathered using oral historical techniques. Individual swiddens will be referred to either by their location, year of harvest,

FIGURE 2
The Territory Of Tikul Batu And The
1975 And 1976 Swiddens

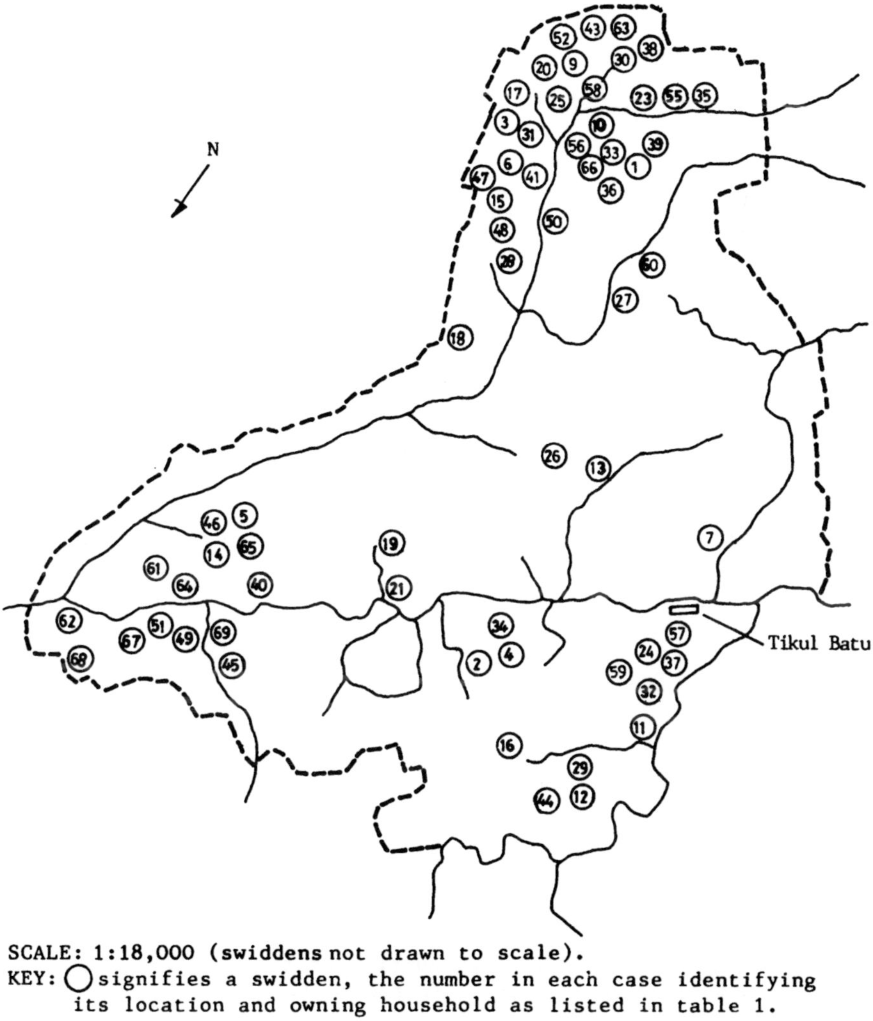

SCALE: 1:18,000 (swiddens not drawn to scale).
KEY: ◯ signifies a swidden, the number in each case identifying its location and owning household as listed in table 1.

TABLE 1
Location and Ownership Of The 1975 And 1976 Swiddens

1975 Swidden Year			1976 Swidden Year		
Swidden number	House-hold	Location	Swidden number	House-hold	Location
1	1	Kenua'	35	1	Kenua' upstream
2	1	Serempa'	36	1	Kenua' downstream
3	2	Kenua'	37	1	Seluyang
4	2	Serempa'	38	2	Kenua' upstream
5	2	Setungkus	39	2	Kenua' downstream
6	3	Kenua'	40	2	Lubuk Rabing
7	3	Tingi' Tebing	41	3	Kenua'
8*	3	Lubuk Kepayang	42*	3	Lubuk Kepayang
9	4	Kenua' upstream	43	4	Kenua'
10	4	Kenua' downstream	44	4	Serangir
11	4	Kantu'	45	5	Sempinih
12	4	Seluyang	46	5	Setungkus
13	5	Sengkadai	47	6	Kenua' upstream
14	5	Setungkus	48	6	Kenua' downstream
15	6	Kenua'	49	6	Sempinih
16	6	Kenua'	50	7	Kenua'
17	7	Kenua' upstream	51	7	Lubuk Menserai
18	7	Kenua' downstream	52	8	Kenua'
19	7	Sebeliya'	53*	9	Kantu' upriver
20	8	Kenua'	54*	9	Kantu' downriver
21	8	Sebeliya'	55	11	Kenua' upstream
22*	9	Kantu'	56	11	Kenua' downstream
23	11	Kenua'	57	11	Seluyang
24	11	Seluyang	58	12	Kenua'
25	12	Kenua'	59	12	Seluyang
26	12	Sengkadai'	60	13	Kenua'
27	13	Kenua' upstream	61	13	Lubuk Rabing
28	13	Kenua' downstream	62	13	Sepering
29	13	Seluyang	63	14	Kenua'
30	14	Kenua' upstream	64	14	Lubuk Rabing
31	14	Kenua' downstream	65	14	Setungkus
32	14	Seluyang	66	15	Kenua'
33	15	Kenua'	67	16	Lubuk Menserai
34	15	Serempa'	68	16	Sepering
			69	17	Sempinih

* These swiddens were located outside of the longhouse territory proper and hence are not shown in figure 1.

and owning household, or by their number in the overall sequence of sixty-nine swiddens (table 1). Individual households will be referred to by the sequential numbers that the household members themselves painted on each pintu 'door' or household in the longhouse.

I gathered data on the swiddens from three different sources. The first was participant observation. I participated in virtually every major swidden activity and most minor ones as well. During breaks in my own swidden work, I studied the work of others - timing, measuring, and photographing it. Some sampling and generalization was necessary in these studies. It was not possible to study the performance of every member of the longhouse during every activity in the swidden cycle. On many days during this cycle, the population of the longhouse is scattered among three dozen different swiddens, spread over the entire ten square kilometers of the longhouse territory.

While I could not be exhaustive in my participant observation, I was exhaustive in my interviewing, which was the second source of data for his study. I established data files for each of the sixty-nine swiddens cleared during 1975 and 1976. During each phase of the swidden cycle, I interviewed each household regarding the activities in its swiddens. My rule, whenever I asked a question, was to ask it of all households and of all swiddens. I deviated from this rule in one respect, only. I pursued some matters further (if not exclusively) when questioning households in which a daily journal was being kept for me.

These daily journals were the third souce of data for this study. They were kept at my behest by three young, married, Kantu' males: Angkol of household #3, Suring of household #4 and Ayung of household #8 (later of household #16). Each of these gentlemen was partially literate in Indonesian. Given that skill and given a Kantu' orthography that we jointly created, they were able to write in the Kantu' language a daily record of their households' activities. At my direction, this record included (1) work performed by members of the household; (2) transfers of labor between households; (3) ritual work proscriptions (e.g., as the result of omens); (4) harvesting of comestibles; (5) hunting, fishing and gathering; (6) household meals; (7) meals of domestic pigs and fowl; (8) rice threshing, drying, winnowing, and husking; (9) ritual activities; (10) the presence of guests in the household; and (11) the giving, receiving, buying, selling and trading of goods[23].

The analysis in this monograph will follow the structure of the swidden cycle itself. I commence with a chapter on the selection of the prospective swidden site, following with chapters on slashing, felling, burning, planting, weeding, guarding, harvesting and then the carrying in of the harvest. Then follow three chapters on activities that are less clearly sequential: the harvesting of nonrice cultigens, the building of swidden houses, and the making of swidden tools. The discussion in most chapters is divided into separate analyses of, first, technology, with separate discussion of material, behavioral, spatial, and temporal factors; second, labor, with separate discussions of rates of labor and sources of labor; and third, the chronology of the work.

THE PEOPLE

Borneo, the third largest island in the world, is partitioned among three modern nations, the Sultanate of Brunei, the Federated States of Malaysia, and the Republic of Indonesia. The areas belonging to Malaysia are known as Sarawak and Sabah, while the largest part of the island, belonging to Indonesia, is known as Kalimantan. The interior of the island, representing territory of all three nations, is inhabited largely by tribal peoples. In the ethnographic literature dating back over one century, these peoples have been referred to, collectively, as 'Dayak'. The Dayak often are subdivided into three main groups, based on cultural, linguistic and geographic affinities. These are, first, a northern grouping, including the Dusun and Murut; second, a southern grouping, including all of the so-called 'Ngadju' peoples; and third, a central grouping, including the Kenyah, Kayan and Kayang, as well as all of the Ibanic peoples. In their comprehensive survey of Bornean languages, Cense and Uhlenbeck defined 'Iban' as a Malay dialect, spoken by the Iban of Sarawak and by related groups living along both the eastern and western tributaries of the upper Kapuas river in West Kalimantan[24]. The term 'Iban' or 'Ibanic' was clarified further in Hudson's survey of the Dayak languages of Western Borneo. He called Ibanic one of the Malayic Dayak languages, as spoken by the Iban, Seberuang, Mualang, Ketungau, and Kantu' groups[25].

The Kantu' are one of the least known of these Ibanic groups. They inhabit the Kapuas river and its tributaries between Semitau in the South to above Putussibau in the north (figure 1). They are mentioned in passing in several Dutch works, in recent articles by the Englishman, King (who worked among a proximate but different group, the Maloh), and in the Master's thesis of an American agricultural extension worker, Bucher[26]. These Kantu' are part of the original complex of Ibanic peoples in the Kapuas river valley, from which the present-day, well studied Iban of Sarawak sprang, sometime in the sixteenth century[27]. The contemporary Kantu' trace their own origin to an area far to the south of their current distribution, along the Sanggau river, closer to the mouth of the Kapuas. They say that their ancestors, while still living in the Sanggau region, explored the upper reaches of the Kapuas on expeditions to gather wild resins. They found the Empanang valley uninhabited, they say, and so they gave to the tributaries of the Empanang the names that they bear to this day. They named the principal tributary of the Empanang after themselves, 'Kantu''. These early travelers eventually removed their families from their Sanggau homeland and migrated to the Empanang as well as to other rivers in the upper Kapuas. By the last quarter of the nineteenth century Kantu' settlement of the Empanang had reached its peak, with over thirty longhouses scattered the length of the valley. Kantu' settlement was halted and then reversed by increasing warfare with Iban from across the watershed in Sarawak. In 1882, the Kantu' in the Empanang were forced to flee downriver, and (with the assistance of the then Dutch authorities in Kalimantan) they relocated along the banks of the middle and upper Kapuas, where they mostly are found today[28]. The conclusion of peace agreements between the Kantu' and Iban early this century permitted the two former antagonists to resettle the Empanang. Large numbers of Iban left Sarawak to occupy the western reaches of the valley, while a few Kantu' returned from the Kapuas to resettle the eastern reaches. In 1975, there were 5434 tribesmen in the Empanang valley, comprising 4775 Iban living in several dozen longhouses, and 659 Kantu' living in five longhouses and one village of detached dwellings. This study is primarily concerned with the inhabitants of one of these longhouses, Tikul Batu Ulu (hereafter referred to as Tikul Batu), who numbered 115 persons as of September 1974 (figure 3)[29].

Political Structure

The Kantu' are subject to the authority of the various non-Kantu' civil and military officials who occupy local positions in the administrative hierarchy of the national government. On a day to day basis, however, the Kantu'

FIGURE 3
The Longhouse Tikul Batu

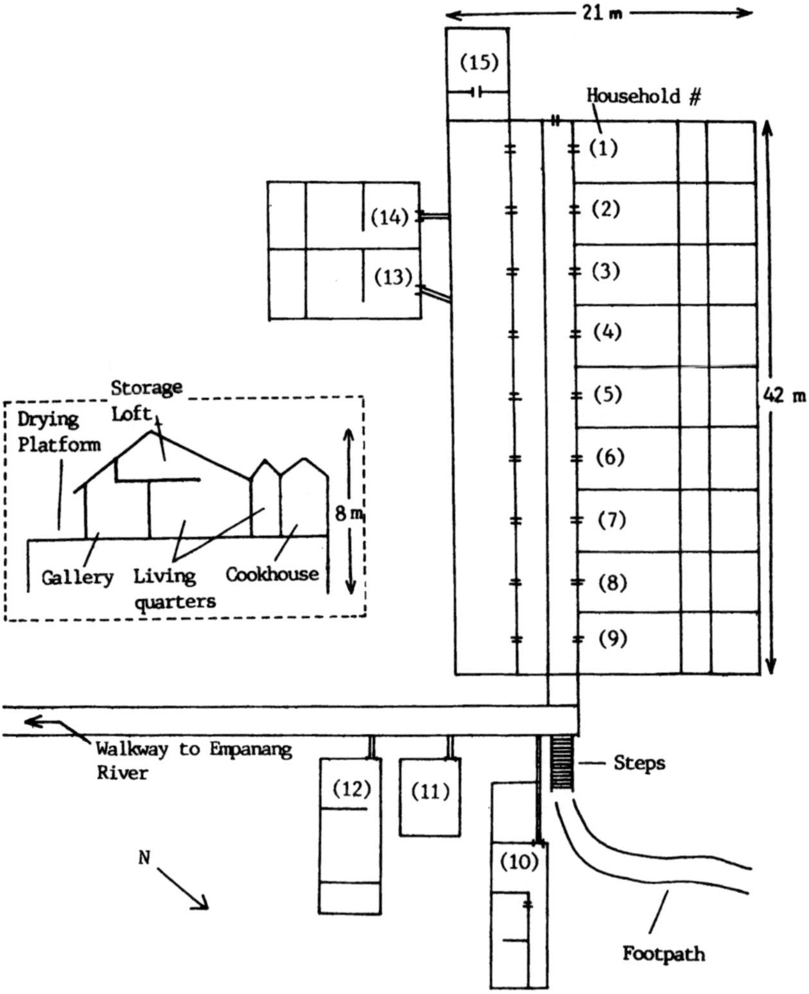

are most influenced by the authority of several of their own people, whose authority either has been granted or recognized by the national government. The Kantu' (and Iban) who occupy these several positions or ranks are chosen (whether by the people or the government) in part on the basis of their knowledge of tribal custom and law. Nevertheless, the positions themselves are not traditional but were created by the government. Before the Kantu' came under the aegis of a national government, they recognized the traditional Tungul and Manok Sabong, leaders in war, and Tua', leaders in ritual. Both are now but memories to the Kantu'. However, some other types of traditional authority have survived to the present day.

Within each longhouse there still are certain households that play a central role in longhouse activities. One is the pun rumah 'longhouse stem' or pun burong 'bird stem'. This is the household that holds the kayu burong 'omen sticks' that were uprooted following the initial construction of the longhouse. The second notable household in the longhouse is the bilek tuai 'senior household'. This is the one household whose origin, in an act of partition from another household, occurred the greatest number of generations ago. Tangible evidence of this antiquity is the amount of ritual, heirloom property (including trophy heads and ancient Chinese jars) that belong to such a household. The eldest male and female in both the longhouse stem and the senior household (households #2 and 4, respectively, in Tikul Batu) tend to carry special weight in longhouse affairs. More generally, this is true of all tuai 'elders'. However, any adult male can participate in the village moots held in the gallery of the longhouse, at which all longhouse wide decisions (concerning legal disputes, ceremonies, etc.) are made. Indeed, participation in these moots is obligatory for adult males, in the sense that failure to participate is regarded as uncooperative and antisocial, and brings down the censure of the longhouse. Adult females also may participate in the moots, but it is rare for them to do so. There are other areas of authority, primarily involving intrahousehold as opposed to interhousehold affairs, in which females play a greater role than males. A final focus of authority in the longhouse is the manang 'shaman'. At the time of this fieldwork, there happened to be no shamans in the longhouse Tikul Batu. When an inhabitant of Tikul Batu became ill, any one of four shamans living in the two nearest Kantu' longhouses was called in.

Within each longhouse household, there usually are just two determinants of authority. One is age.

Younger siblings defer to older siblings, and younger generations defer to older generations. The second determinant has to do with postmarital residence. Whichever spouse does not move upon marriage is known as the pun bilek 'household stem' or kuasa bilek 'household authority'. The authority of this person is especially marked with regards to any devolution of the household's property (e.g., land or heirlooms)[30].

Social Structure

Kantu' society is bilaterally structured. Their kinship terminology stresses the difference between generations but does not distinguish, for the most part, between kin on the father's side and kin on the mother's side, nor between consanguines and affines. There are two discrete social groupings in Kantu' society: the bilek 'household', and the rumah 'longhouse'. The households of Tikul Batu contained an average of 8.5 persons and 7.2 persons (spanning 2.9 and 2.6 generations) during 1974-1975, and 1975-1976, respectively[31].

The Household

The term bilek refers to each separate, walled off living unit within the longhouse (figure 3), as well as to the separate families that inhabit each unit. Both the living quarters and the family also may be called, although with less frequency, pintu 'door', which refers to the door that opens out from each enclosed living quarters onto the open gallery of the longhouse. Each household builds and maintains its own living quarters as well as the other portions of the longhouse that are aligned with these quarters, including the storage loft, and a section of the longhouse roof, gallery and drying platform. The household is partially defined, therefore, as the group that makes common residence in one of the longhouse's living units. One household may nyarang 'share' the longhouse quarters of another, but this is permitted only as a short term expedient since it is thought to threaten the ritual health of the households involved, if not of the entire longhouse.

In addition to being a residential unit, the household is a unit of production. Indeed, it is the most important unit of production in Kantu' society. Of the nine major, sequential swidden activities (viz., selecting, slashing, felling, burning, planting, weeding, guarding, harvesting and carrying), only two - planting and carrying - are carried out largely using extrahousehold workers. In large part, each household works on its own in its own swiddens. Each swidden is identified exclusively and without exception with a single household. Each household refers to its swiddens as umai kami 'our swiddens' (kami, the first person, plural, exclusive pronoun, signifies in this context the household members)[32].

The sharing of production activities and goals is the most important defining characteristic of the household. From this characteristic, others follow. Thus, the household also is the sole unit of consumption. No unit smaller than the household can function as a unit of consumption. It is a tenet of Kantu' law that goods must be shared and consumed within the household with absolute equality, regardless of which members of the household were directly responsible for producing them. The Kantu' say that the partitioning of households usually is due to internal quarrels over the unequal consumption of goods. (Often, two married couples of the same generation within one household will quarrel over some inequity in the distribution of goods among their respective offspring[33].) Nor does any unit larger than the household, such as the longhouse, function as a unit of consumption[34]. Even in cases in which two households temporarily share the same longhouse quarters (as noted above), each will cook its own rice. Indeed, this is the test of whether two households are nyarang 'sharing' quarters or actually are sebilek 'one household'[35].

The household also is an important ritual unit. Almost all swidden rituals are carried out by the households, individually, and the attendant prescriptions and proscriptions usually are limited to the individual households as well. Similarly, the proscriptions (e.g., against swidden labor) arising from an omen encountered by any given individual usually apply only to that individual's own household and not to the whole longhouse. With respect to illness the household is also a ritual unit, since it is always the household that stages curing ceremonies. The proscriptions that arise from such ceremonies apply to the longhouse as a whole more often than not, but they apply in their severest form only to the household that stages the ceremony.

Finally, the household is the major unit of appropriation, especially with regard to forest land. In opening up the forest and making swiddens therein, the individual household both creates and exercises rights to forest. In order to understand the devolution of these rights, some prior discussion of the developmental cycle of the household is warranted. Kantu' males and females usually marry in their middle to late teens. The Kantu' have no prescriptive rule as to postmarital residence, and in practice it is ambilocal. Of the twenty-four couples living in Tikul Batu in 1974-1976, twelve practiced uxorilocal residence upon marriage and twelve practiced virilocal residence. (The initial decision on residence is usually permanent, except possibly in the case of the early divorce and remarriage of the same two parties.) Since the frequencies of virilocal and uxorilocal residence tend to be equal, and since the average couple produces more than four offspring, the average household experiences two or more of its members (from the same sibling set) marrying out and two or more marrying in. The former cases are bemoaned by the household as a loss of dear kinsmen and valued fellow workers. The latter cases, while hailed as victories for the household, are problematic in their own way.

Two siblings and their respective spouses cannot get along in the same household for more than a short period of time. There is a capacity for conflict between couples of the same generation that does not exist for couples of adjacent or alternate generations. As a result, when an older sibling in the household already has married in and a younger sibling then marries in as well, a source of tension is created that can be relieved only by the departure from the household - by means of its pedua' 'partition' - of one of the couples. In such circumstances it is usually the later married couple that stays in the parent household. In a newly married couple, the female often is occupied with the care of newly born infants, leaving only the one male available for work in the swiddens. When a couple has been married for some time, however, they usually have older children (e.g., five or six years of age or more) who can look after younger children and infants. This frees their mother for swidden work, effectively doubling the work force[36]. It is couples at this stage of their domestic cycle - that are capable of establishing an economically viable household on their own - that will leave a parent household through partition. Since elder siblings marry before younger ones as a norm, it is they who will leave the household through partition or outmarriage, eventually leaving just the youngest of the sibling set in the household along with his/her spouse and children, along with the parents of that sibling set so long as they live. This last remaining member of the sibling set perpetuates that household, which thus is a perduring corporate group. The longhouse Tikul Batu is composed of several such long established households, plus a larger number of younger households created from them by partition. In addition to these relations of partition, the households are interrelated by marriage (figure 4).

FIGURE 4
Ties Of Consanguinity And Affinality Among The Households Of Tikul Batu

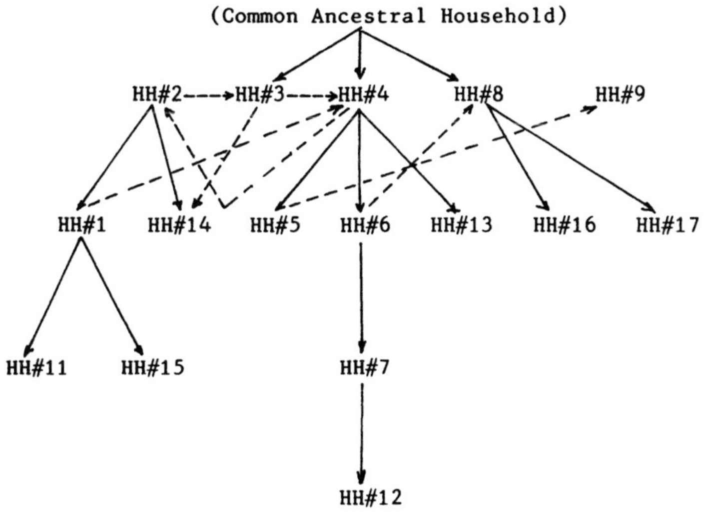

NOTE: (←——) = Direct conganguineal link (viz., by household partition).
(←---) = Direct affinal link (viz., by marriage).
The movement of people, whether during partition or marriage, is indicated by the direction of the arrows.

Based on this outline of the developmental cycle, the devolution of property rights now can be explained. When an infant is born into a household, she/he immediately acquires rights to all property to which the household as a whole holds sole rights, as opposed to rights shared with one or more other households. Thenceforth, this new household member automatically acquires rights to any property to which the rest of the household acquires rights, whether sole or shared. This continues and these rights accrue until this household member marries. If the member marries natolocally, these rights remain intact. But if the member marries out and into another household, called <u>nguang k'urang</u> 'join [other] people', all accrued rights to the property of the natal household are forfeited. The Kantu' say (and often it is literally true) that the outmarrying child takes nothing from the natal household but the clothes that she/he wears. The Kantu' justify this forfeiture of rights by saying that the outmarrying child enters the *hak* 'rights' or 'claims' of the household into which she/he marries and, consequently, has no further use for rights in the natal household. Indeed, whoever marries into a household immediately acquires rights equal to those of his/her spouse, whose natal household it is (although the latter, who eventually will become the 'household stem' or 'household authority', has somewhat greater say in the ultimate use or disposition of these rights). However, in the event of divorce and the attendant division of property, the outsider who married into the household is recognized to have a claim only on the property acquired by the household since the marriage, and not on any property to which his/her former spouse holds rights but which was acquired before the marriage. This division of property is justified, the Kantu' say, by the fact that the outsider can return to his/her natal bilek and reacquire all of the property rights that were forfeited at the time of the marriage.

After birth and marriage, the next occasion in the developmental cycle upon which property rights devolve is household partition. A newly married couple lives for some time in one of their natal households and then (unless that one is the youngest of his/her sibling set) they establish their own household, which is called *angkat kediri'* 'leave alone'. This involves *pedua' bilek* 'household division', which consists in division not only of the household's members, but also its propety. All property to which the departing couple holds rights is subject to this division, on a per person basis. This includes all of the household's rubber groves, but none of its secondary forest. Rights to this forest thereafter must be *kuntsi* 'shared' by the two households, the senior one and the junior one. These forest rights must be shared further with any additional households created by further partitioning of the senior household, but they cannot be shared with any households created from partitioning of the junior one. The Kantu' say that forest rights can be shared only with *simpang suti'* 'one branch'. Essentially, this means that the forest rights held by any given generation can be shared, through the partitioning of its households, only with the resulting first descending generation. They cannot be shared further, in the partitioning of this first descending generation's households, with the second descending generation. However, once these second generation households begin to come into existence, it is possible for the two older generations to divide up their forest rights so that they are held once again by single households. The households of the first descending generation households then can share these rights with the households of the second descending generation without violating the 'one branch' rule.

For example, figure 4 shows that households #5, 6, and 13 were created through the partition of household #4. According to the Kantu' law just discussed, the forest rights of the original household #4 had to be shared with each of the junior households #5, 6, and 13. However, these rights could not be further shared with (e.g.) household #7, which was created through the partition of household #6. Subsequently, however, some of the forest rights shared by households #4, 5, 6 and 13 were divided up amongst them. The fourth portion of these rights given to household #6 in this division then could be shared by it with household #7.

Death does not occasion any devolution of property rights, so long as at least one child of the household's senior couple remains in the household. Sometimes, if the elders are particularly difficult to live with, all of their children, including the youngest, may move out of the household. In that event, the household becomes punas 'heirless' upon the death of both parents. Claims then would be made against the household's property by kinsmen in other households[37].

The Longhouse

Since 1888, when the Kantu' began to return to the Empanang valley from their refuges along the Kapuas, one group of Kantu' and their descendants have lived there in a succession of eleven different longhouses, culminating in the contemporary longhouse Tikul Batu. Construction on Tikul Batu was begun in 1955, and it finally was occupied in 1957. Some of the work on the longhouse was done by the households individually, each working on its own section; and some work was done by the longhouse as a whole, working first on one household's section, and then on another's. The longhouse consists of the following: an open, drying platform running its entire length; a covered but uninterrupted gallery, also running the length of the longhouse and divided lengthwise into a walkway and a work area; nine living apartments that also run the length of the longhouse, but are walled off from one another; a series of cook houses appended to each apartment; and a series of storage lofts built over the walkway and part of each apartment (figure 3). Nine households participated in the initial construction of the longhouse. Five households, created by partition in the years since, have built less elaborate quarters attached to the longhouse proper by means of catwalks. Also forming a part of the longhouse is a walkway running from it to the river bank, and two sets of steps running from it to the ground.

The longhouse is first and most obviously a unit of residence, and the basis for this is not _de facto_ but _de jure_. A household can neither enter into nor terminate residence in a given longhouse at will. Residence is subject to certain constraints. If all members of one household are away from their longhouse apartment for more than nine nights (meaning that no fire is lit in their cook house for more than nine nights), the errant household must sacrifice one chicken for the ritual well-being of the rest of the longhouse. If that same household does this three times in a row (viz., is away from its apartment for thirty days), then the rest of longhouse can demand that it abandon its apartment for a period of not less than three years. The Kantu' say that the purpose of this proscription is _Awak urang ngidup api kabut_ 'So that people light [their] fires a lot', and _Awak urang nyaga rumah_ 'So that people look after the longhouse'. This proscription is important enough to define membership in the longhouse. The Kantu' say that whoever is susceptible to this proscription therefore is susceptible to all other longhouse proscriptions.

The Kantu' value not only the frequency of the household's residence in the longhouse, but also its continuity. Frequent moving by households from one longhouse to another is discouraged[38]. During the first nineteen years (1957-1976) after the construction of Tikul Batu, no households from other longhouses joined it and only two of its own households moved out. One of these latter, household #15, moved out for two years and then returned. The second, the former household #10 (created by a partition of household #5), moved away permanently. The Kantu' say that someone who moves about a lot is just searching for an easier way of life (and this is said disparagingly) and is unable to get along with people in any given place: _Ia nadai tau' begulai ngau urang_ 'He/she cannot mix with people'. In addition, they say that anyone who moves a lot runs the risk of encountering ill omens (the move being seen as in effect an invitation to the spirits to pass judgement on the household and its actions) that will jeopardize the subsequent well-being of the household. Indeed, the Kantu' believe that the move of a household threatens not only its own well-being, but the well-being of the entire longhouse from which it moves. In order to forestall any ill consequences, the longhouse therefore demands a pemali 'fine' (usually the sacrifice of a chicken) from the departing household. In addition, the departing household is forbidden to disassemble its section of the longhouse, until such time as the remaining households also choose to abandon the longhouse. Nor is the departing household permitted to sell or otherwise dispose of its rights to forest in that longhouse's territory. Upon the move of the household, its rights become the joint rights of the remaining households, under the administration of the longhouse headman.

This ability of the longhouse as a whole to exercise rights over the dwelling and property of a departed household reflects the fact that the longhouse is not only a residential body, but also, to a limited extent, a corporate body, a unit of appropriation. Thus, the forest rights of a departed household are said to be kuntsi sekayu rumah 'shared by the whole longhouse'. This is not the only property so shared. The walkway from the longhouse to the river, the steps to both this walkway and to the longhouse, and the walls enclosing the open gallery at each end of the longhouse, all are similarly shared[39]. So too are all footpaths in the longhouse territory shared. Finally, and of most importance, this longhouse territory is shared. All primary forest within this territory, which is forest to which no one household as yet holds unquestioned rights, comes under the hak rumah 'territorial rights of the longhouse'. These rights consist in the ability to prohibit anyone from another longhouse from clearing and farming such forest. The boundaries of these longhouse territories (in the Empanang valley) are less clear between any two Kantu' longhouses, which will share a common origin, than between one Kantu' longhouse and one Iban longhouse, which will share only a history of warfare and mistrust. Thus, the boundary between the territory of Tikul Batu and the nearby Iban longhouse, Empakan, is the clearly defined ridge between two southerly tributaries of the Empanang, the Kenua' Yepit 'Little Kenua' and the Kenua' Besai 'Big Kenua'[40].

The longhouse also is a unit of production. As a unit it does not make any swiddens, nor does it engage in any other productive enterprise. However, the longhouse does represent a pool of workers that is drawn upon in the making of each individual household's swiddens[41]. The extent to which extrahousehold labor is used in the swiddens of Tikul Batu is summarized in table 2. This shows that of the nine sequential swidden

TABLE 2
Use Of Extrahousehold Labor In The Swiddens

Swidden Phase	Percentage of 1975 Swiddens in Which Extrahousehold Labor Was		Percentage of 1976 Swiddens in Which Extrahousehold Labor Was	
	Used	Not used	Used	Not used
Selecting	0 %	100 %	0 %	100 %
Slashing	32	68	57	43
Felling	32	68	60	40
Burning	0	100	0	100
Planting	65	35	80	20
Weeding	23	77	71	29
Guarding	0	100	0	100
Harvesting	71	29	57	43
Carrying	41	59	52	48

work phases, extrahousehold labor was used in no swiddens during three of these phases (viz., selecting, burning and guarding), in some swiddens during four phases (viz., slashing, felling, weeding and carrying), and in a clear majority of the swiddens in two phases (viz., planting and harvesting). In some instances this labor is extra-longhouse as well as extrahousehold, but in most instances it is recruited from other households within the same longhouse.

Extrahousehold labor can be recruited under any one of several arrangements, which differ principally according to the general versus specific nature of the way in which it is reciprocated. Least specific and most general is betolong 'help'. This is unsolicited labor, given freely - hence I will call it 'gift labor' - by one household to another closely related household that finds its own labor resources inadequate for a particular swidden activity. The household donating this labor expects in return only that the recipient household will return the favor in a similar situation. In addition, however, the recipient household is expected to feed the donors in the middle of the working day and perhaps at its beginning as well.

Less generalized reciprocity is involved in berimpoh, which I translate as 'cooperative labor'. The Kantu' distinguish this labor arrangement from gift labor by noting that the donor household selects the one or two households in the longhouse with whom it will work in gift labor, whereas all of the longhouse's households work together in cooperative labor. In addition, the Kantu' say that gift labor is not dibalas 'repaid' or 'reciprocated', whereas cooperative labor is. Under a cooperative labor arrangement, all of the households in the longhouse work for one household on one day (and this household feeds all the others on that day), and then they work for another household the next day, and so on. The only major stage of the swidden cycle in which cooperative labor is used is the carrying in of the rice harvest. Cooperative labor also is used in the construction of the longhouse and any separate dwellings that later are added on to it. By contrast, gift labor is used throughout the swidden cycle, albeit in small amounts.

More common than the gift labor arrangements, and less generalized than cooperative labor arrangements, is bedurok or 'reciprocal labor'. Labor that is given under a reciprocal labor arrangement is repaid, as in a cooperative labor arrangement, but there is a difference. In cooperative labor one household's labor throughout one work phase is reciprocated by another household's labor throughout that same phase, whether the latter represents the same number of workdays as the former or not. In contrast, in reciprocal labor one man-day of work is reciprocated by one man-day of work. As the Kantu' say they Rikin tuboh-ari 'Reckon body-days'. Careful track is kept of the creation and discharge of reciprocal labor. The household receiving reciprocal laborers may provide them with side dishes for their midday meal, but the laborers themselves must provide the main part of their meal, the rice dish.

Slightly less common than reciprocal labor arrangements, and requiring even stricter reciprocation, are arrangements to bekuli 'work for wages'. In wage labor arrangements, labor performed by one party to the arrangement is reciprocated by the other party not in kind, but by a wage payment, calculated on a strict per person per day basis. The standard wage usually is reckoned either in unhusked rice, at two kolak per worker per day, or in currency, at 300 rupiah per worker per day, plus one or two meals (or 400 rupiah, without the meals). Wage labor is used in all phases of swidden work in which reciprocal labor is used except for planting, in which reciprocal labor arrangements are the rule. The differences among these several types of labor arrangements are outlined in table 3[42].

The longhouse also is a ritual body. This is seen most clearly in the existence of the begela' rumah 'longhouse offering'. This offering is made whenever some

TABLE 3
Methods Of Recruiting Extrahousehold Labor

Type of Labor Arrangement	Strictness of Reciprocation	Frequency of Use	Provision of Meals by Host Household
Betolong 'Gift labor'	4th (least)	3rd	Yes
Berimpoh 'Cooperative labor'	3rd	4th	Yes
Bedurok 'Reciprocal labor'	2nd	1st	No (side dish only)
Bekuli 'Wage labor'	1st (most)	2nd	Variable (depending upon wage)

misfortune that would affect the whole longhouse is foretold, whether in someone's dream or from the behavior of an omen bird in or near the longhouse. It consists of the propitiatory sacrifice of chickens (ideally, one from each household in the longhouse) and sometimes pigs. In addition, small offering baskets of cooked rice, one basket for each member of the longhouse, are hung from lianas strung around the longhouse. The unity of the longhouse as a ritual body also is seen in the existence of a second category of longhouse offerings or sacrifices, the pemali rumah 'longhouse fines'. Any one of a number of different proscribed actions, carried out by a single individual or household in the longhouse, can place the entire longhouse in a state in which it is susceptible to retributive actions by offended spirits. These proscribed actions include (in increasing order of severity) theft, quarrels, the move of a household out of the longhouse, the failure of a household to light a fire in its hearth for more than nine nights, adultery, the birth of a fatherless child, and incestuous sexual relations or marriage. Any longhouse member or household that violates one of these proscriptions is forced, by the rest of the longhouse, to pay the longhouse a fine of a stipulated amount (ranging from one chicken to several pigs). The longhouse as a whole then sacrifices the animals involved, whereupon its ritual health is restored. The longhouse is defined as a unit, therefore, both in the danger from broken proscriptions and in the means for allaying it. The proscription itself also may define the longhouse as a unit. For example, the day following a major belian 'curing ceremony', swidden work is proscribed for the entire longhouse, not just for the single household that staged the ceremony. Finally, while the unit in most swidden ceremonies is the individual household, one or two major ceremonies are staged by the longhouse as a whole. These include the annual postharvest gawa' batu' 'whetstone feast', as well as the occasional gawa' ngalih tanah 'earth turning feast' that precedes the initial omen taking and slashing of the new swiddens, and the postplanting gawa' nutup lubang tugal 'feast of covering the dibble holes'.

The features that define the longhouse that have been discussed so far - residence, appropriation, production, and ritual - also define the household. There are two additional features of the longhouse for which this is not true. First, the longhouse alone is a jural body. Disputes, whether between members of one household or between two or more households, are resolved in longhouse wide moots, convened in the open gallery of the longhouse and presided over by the kepala kampung 'village head'. Second, and again unlike the household, the longhouse is a grouping within which marriage can take place. It is at least partially endogamous, whereas the household is completely exogamous. The Kantu' say that marriage is forbidden to one's aya' or ibo' 'male or female of parents' generation', and also to one's aki or ini' 'male or female of parents' parents' generation'. This limits the choice of marital partners to one's own generation, all members of which are possible partners with the possible exception of petunggal sekali' 'first cousins' and with the implicit exclusion of biological siblings. Given these marital proscriptions, the number of eligible partners in one's own longhouse may be almost as great as in another longhouse. And indeed, the frequency of longhouse endogamous marriages is not too much less than the frequency of exogamous marriages. The twenty-four couples living in Tikul Batu during 1974-1976 represented ten endogamous unions and fourteen exogamous unions. The endogamy versus exogamy of a union may be one determinant of the pattern of postmarital residence: residence was uxorilocal in nine of the fourteen exogamous unions, compared with just three of the ten endogamous ones. The Kantu' say that it is easier to persuade girls to reside natolocally upon marriage than it is to persuade boys. This principle finds its greatest expression in longhouse exogamous marriages, in which the move of the outmarrying spouse is especially traumatic for the household concerned (viz., compared to the move of an outmarrying child in a longhouse endogamous marriage). Thus, in an exogamous marriage postmarital residence is likely to be uxorilocal. This pattern is made possible by the converse preponderance of virilocal residence in longhouse endogamous marriages, which comes about because a newly married couple will choose not to reside (cet.par.) in that one of their two households in which there already resides a married sibling and spouse.

The unity of the longhouse as a social body is summed up in the manner in which membership in it is taken away. The Kantu' of Tikul Batu say that they, as a longhouse, will order a constituent household to leave the longhouse if its members (1) Nadai ngau urang bepekat 'Do not join in the longhouse moots', (2) Nadai ngau urang ngeranang jalai 'Do not join the others in clearing the [longhouse] trails', (3) Nadai diau ke pantang begela' rumah 'Do not observe the proscriptions [against swidden labor] following longhouse offerings', and/or (4) Nadai turun k'umai ngau urang 'Do not work in the swiddens with the others [e.g., when the others have proposed an interhousehold work party]'.

Supralonghouse Groupings

Some of the features that define the longhouse and the household also define larger groupings that may include other longhouses, other Kantu' subgroups, and even other tribal groupings. For example, two or more longhouses often are drawn together into productive activities. As has been noted, a minority of interhousehold labor arrangements are made between households in different longhouses. These include gift labor, reciprocal labor and especially wage labor arrangements, excluding only cooperative labor. Such arrangements are occasioned by the closeness of the kin ties between the households involved and/or the proximity of their swiddens. Supralonghouse groupings also may constitute ritual bodies. Thus, all of the Kantu' at Tikul Batu observe a one-day work proscription for the death of any Kantu' in any of the surrounding Kantu' communities. The members of Tikul Batu also observe work proscriptions for the sacrifice of pigs by these communities. The ritual body is seen at its broadest in the belief that the violation of proscribed actions (especially those involving sexual congress) by other Kantu', even those in other river valleys, can adversely affect the lives of the Kantu' at Tikul Batu, by causing crop failures, famine, flood, and so on. A third feature of supralonghouse groupings is jural. Longhouse members sometime seek assistance outside the longhouse, in the good offices of tribal leaders with supralonghouse jurisdictions, to settle disputes within the longhouse. Just as frequently, these officials are called in to adjudicate disputes involving two or more longhouses. A fourth and final feature of supralonghouse groupings is kinship. Every Kantu' acknowledges some degree of kinship with all other Kantu', based on geneologies and myths that affirm a common ancestry for them all. This affirmation is given added weight by the fact that marriages between Kantu' and non-Kantu' are, and have been, rare (at least within the Empanang valley). Thus, the Kantu' constitute a practically endogamous group. Within the endogamous, putative, descent group termed Kantu', or Kitai Kantu' 'We Kantu', they recognize only one major subgrouping of kinsmen (which still crosses longhouse [and household] boundaries, however), the kaban 'kindred'. Unlike the longhouse or the household, the kindred is not a discrete body but a relative one. It is ego based. According to the Kantu', one's kindred includes everyone to whom one can establish a kin tie, whether consanguineal or affinal, whether in one's own generation or in one of the ascending generations, up to and including the seventh.

Belief System

Much of the ritual life of the Kantu' is bound up with their swidden cycle. This ritual is designed to promote the health and abundance of the rice crop and to ward off any threat to it from animal pests, malevolent spirits, and human sorcerers (see table 4). The postharvest feast that ends the ritual cycle is performed jointly by all of the households at the longhouse. All of the other phases of swidden ritual are performed by the households individually, in their individual swiddens. Not all phases of ritual are performed in all swiddens. Performance of the complete ritual sequence is restricted to that minority of swiddens in which padi pun 'stem rice' is planted.

A major ritual activity of the Kantu' involves the observation of omens. Such omens consist chiefly in the flight and calls of seven, named varieties of burong antu 'spirit birds', although certain other animals (especially deer and bear) are important, as are the contents of dreams. The observance of an omen typically entails the proscription of work - especially work in the swiddens - for one day, for the whole household. The observance of omens is not uniform throughout the swidden cycle. Omens are keenly observed during selecting, slashing, felling, harvesting, carrying and perhaps guarding, but they tend not to be observed during burn-

TABLE 4
The Ritual Cycle In The Swiddens

Swidden Phase	Associated Ritual Phases
Selecting	Beburong 'Using the birds'.
Slashing	Nganyung batu umai 'Bringing the ritual whetstone' to the swidden. Ngami' batu umai 'Bringing the ritual whetstone back' from the swidden.
Felling	
Burning	
Planting	A complex of planting rituals.
Weeding	Nyingkelan umai 'Swidden offering'.
Guarding	
Harvesting	Matah padi 'Rice breaking'.
Carrying	A complex of rituals. Gawa' batu Postharvest, year's end 'Feast of the whetstone'.

ing, planting and weeding. The observance of lunar phases results in similar work proscriptions. The full moon is the most important phase in this respect, while the first waxing moon and the last waning moon are also important. The manner of observance varies through the lunar year (table 5).

TABLE 5
Proscriptions On Swidden Labor Due To Lunar Phases

Stage in Kantu' Lunar Calendar	Maximum Number of Proscribed Workdays
4th moon (11/5 - 5/6/76)	5
5th moon (9/6 - 5/7/76)	8
6th moon (9/7 - 3/8/76)	5
7th moon (7/8 - 1/9/76)	1

NOTE: The dates of the Kantu' moons are given above for the year 1976 and will vary somewhat in other years.

The Kantu' perform a number of rituals to guard their health and well-being. These rituals may focus either on the longhouse as a whole or on just one of its members. The former consists in the already mentioned bege-la' rumah 'longhouse offering', which is performed upon perception of a clear and present danger to the longhouse. In 1975, such offerings were made on four separate occasions (19/3, 29/5, 12/7, and 3/8). The first was provoked by widespread illness in the longhouse; the second by the coming of a chicken plague to a nearby longhouse and by the flight into the longhouse of a tinting lema', a bird that is believed to be the servant of one of the major omen birds; and the fourth by the occurrence of a rainbow, one end of which touched the longhouse's drying platform. The third longhouse offering was performed just for good measure, after the staging of the year's end harvest festival. While the longhouse offering focuses on the health of the longhouse, the belian 'curing ceremony' focuses on the health of the individual. Curing ceremonies are performed either upon the perception (e.g., in a dream) of a future threat to someone's health by a malevolent spirit, or upon the evidence (in illness) of a present threat. There are six different ranked varieties of ceremonies, among which the manang 'shaman' can choose, according to the perceived severity of the threat [43]. In 1975, thirty-two separate curing ceremonies, each one focused on one or more individuals in one or more households, were performed in Tikul Batu (table 6).

TABLE 6
Curing Ceremonies Performed At Tikul Batu In 1975

Month	Dates of Curing Ceremonies (and Households Involved)
January	-
February	24/2(#4), 25/2(#4), 18/2(#8), 26/2(#8).
March	13/3(#4).
April	8/4(#4).
May	11/5(#12), 13/5(#8), 15/5(#14), 16/5(#4).
June	5/6(#6,7,8,9,13), 9/6(#11), 16/6(#14), 18/6(#4).
July	1/7(#2), 8/7(#5), 9/7(#15), 20/7(#4), 21/7(#3).
August	3/8(all households), 4/8(#9), 6/8(#8), 12/8(#17), 13/8(#3), 21/8(#9).
September	5/9(#4), 6/9(#2).
October	6/10(#3,6,7,8,12,16), 15/10(#8), 22/10(#11).
November	-
December	3/12(#14), 15/12(#7).

A third and final focus of ritual pertains to life crises. The Kantu' currently stage three major life crisis ceremonies. These are the gawa' gunting 'feast of cutting', staged when a boy's hair is first cut; gawa' nusok 'feast of piercing', staged when a girl's ears are pierced; and the gawu' antu 'feast of the spirits', staged at the end of the mourning period that follows every death. These ceremonies are called gawa' because, like the postharvest gawa' batu that terminates each swidden cycle, they are major feasts involving extensive preparations and the participation of not only the entire longhouse, but also - by formal invitation - several of the surrounding longhouses. A longhouse that is given an invitation to such a feast must, if it accepts, reciprocate at a later date. The longhouse Tikul Batu is tied to a number of other longhouses in the valley (and even in neighboring valleys) in an ongoing, longhouse-based web of such feasting relationships[44].

Economy

The Kantu' use of currency is limited but by no means nonexistent. As they say, they need cash to meli tembako-garam 'buy tobacco [and] salt'. In addition to tobacco and salt, the Kantu' also buy from traders most of their clothing, as well as kerosene for the tiny lamps made from sardine cans that provide their light at night. In order to obtain the money or tradable commodities needed to buy these four staples as well as some miscellaneous goods, the Kantu' engage in various nonsubsistence activities. During otherwise slack times of the year, members of the poorest households make the two-day march across the watershed to Sarawak to work in the pepper plantations there, for a period of a few days to at most one week. The wages paid there equalled about one thousand Indonesian rupiah per day in 1974-1976, which compared favorably to the standard wage of just three hundred rupiah per day in the Empanang valley for work in the swiddens.

Most Kantu' disdain such wage labor, however, preferring to grow their own lada 'black pepper' as a cash crop. The pepper is grown in gardens cleared from the forest near the longhouse, each containing perhaps two hundred pepper plants and covering 250 to 500 square meters. The Kantu' learned how to raise pepper from Iban living in the Empanang valley, who in turn obtained this technology from fellow Iban living across the border in Sarawak. Not until the early 1950s did any significant number of Kantu' experiment with pepper gardens, but some older Kantu' claim that their fathers first planted pepper at a far earlier time (estimated at 1910 to 1920). In 1975, only three households in Tikul Batu (households #3, 4 and 13) had producing pepper gardens; but all but one of the remaining households had made gardens at some time in the recent past. The garden is started with seedlings. Two years after these are planted, the garden produces a marketable crop, and it continues to produce one crop per year for the next three to four years. The crop usually is carried on foot over the watershed, to be sold in the border towns of Sarawak. As the price in those markets fluctuates, so too does the Kantu' commitment to pepper cultivation. Even when the market price is high, some households choose not to cultivate pepper, first, because it requires some capital investment in the form of chemical fertilizers, which is always problematic for the cash poor Kantu'; and second, because pepper is labor intensive, and this demand on labor can conflict with demands from the swiddens.

Usually more desirable as a source of cash income than pepper is getah 'rubber'. The Kantu' grow rubber trees in groves containing from one hundred to three hundred trees, and covering one-quarter to three-quarters of a hectare. Most of the land immediately surrounding Tikul Batu, and also along the banks of the waterways navigable by canoe, has been cleared and planted with rubber trees (figure 5)[45]. The Kantu' obtained their first rubber seedlings from a Catholic mission downriver on the Kapuas, and from Iban across the border in Sarawak. A few Kantu' households had rubber groves prior to World War II, but the growing and tapping of rubber did not become widespread until after the war. By 1975, the households of Tikul Batu owned a total of sixty-six rubber groves (all households owning at least one grove, with the exception of household #15). The rubber trees are tapped for latex beginning about fifteen years after planting. The individual trees die after a productive period of perhaps twenty-five years, but the life of the grove as a whole is potentially open-ended, because the older trees replace themselves from the fruit that they drop to the ground. The trees can be tapped or not, at will, throughout the year. For this reason, rubber complements well the swidden cycle of the Kantu'. It can be tapped when there is no work in the swiddens and not tapped when there is. The product of this tapping, the cured slabs of rubber, is taken downriver by boat to the mouth of the Kapuas, for sale in the port city of Pontianak. The value-to-weight ratio of

FIGURE 5
Location Of Rubber Groves In The Territory Of Tikul Batu

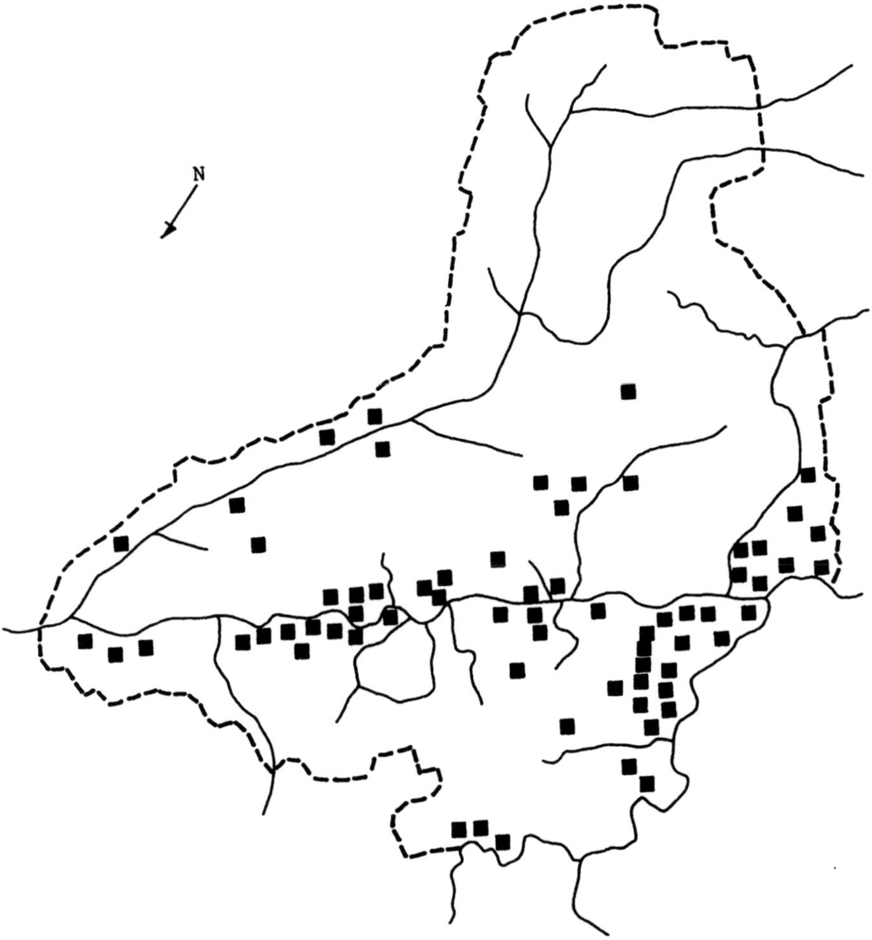

SCALE: 1:18,000 (rubber groves not drawn to scale).
KEY: ■ = one rubber grove.
NOTE: This figure shows more than the sixty-six rubber groves belonging to the households of Tikul Batu, because some groves belonging to households in other longhouses are located in the territory of Tikul Batu.

rubber is too low to justify carrying it on foot over the watershed to the markets in Sarawak, as is done with pepper. Most Kantu' do not take their rubber downriver themselves. Rather, they sell or trade it to Kantu' traders who make periodic trips downriver.

All Kantu' participate in trade to some extent. The Empanang river valley offers the best access in this region to the watershed between Kalimantan and Sarawak and the markets beyond. To these markets the Kantu' bring pepper, trade arak, and sometimes pigs and dried fish; which they exchange for clothing, pepper fertilizer, cookware and trade goods. The Kantu' take some of these trade goods across the hills into the Ketunggau river valley to the south of them, and exchange them for chickens. They take nonrice cultigens from their swiddens (especially tubers) downriver to the fishing villages of Islamicized Dayak, to trade for dried fish and live turtles. One Iban longhouse supplies the Kantu' with canoes, usually for cash. Some Kantu' have a more extensive involvement in trading. Since 1971, households #2, 4, 5 and 9 in Tikul Batu have carried on a year-round trading venture. They acquire the local output of rubber through trade, and take it downriver to Pontianak; trading it there for tobacco, salt, kerosene, cloth, clothing and other trade goods, which they bring back upriver to be sold or more often traded for more rubber.

Trade and cash earning are important aspects of the Kantu' economy, but they are less important than its subsistence aspects. Fishing is one important part of Kantu' subsistence. Most households eat at least a little fish every day, taken from the main rivers (the Empanang and the Kantu'), which flow near Tikul Batu, or from the myriad of streams that flow through the forest. The drainage of this area is such that every average-sized swidden (viz., two hectares) is bisected by at least one stream. They take fish - and also frogs, turtles, water-snakes, and freshwater crabs, shrimp and mussels - by stalking them, and by casting net, gill net, hoist net, funnel net, purse seine, funnel trap, box trap, basket, hand line, trot line, spring line, spoon, spear, and poison[46]. The Empanang valley yields a significant amount of fish to these various techniques. Indeed, the fishing opportunities in the Empanang were cited as a partial reason for the recent move to the Empanang of several households of a related Kantu' group, previously living in a hillier valley to the south[47]. Also important to this subsistence economy are hunting and the gathering of wild comestibles. (These activities are discussed in greater detail in chapters seven and ten, respectively.) A fourth aspect of the subsistence economy involves the raising of pigs and chickens. These animals are slaughtered and eaten (or not) on ritual occasions alone. A few roosters also are raised for fighting. In 1976 there were forty-one pigs and seventy-eight assorted fowl distributed among the households of Tikul Batu. Seven or eight of the households kept a total of nine domesticated dogs and fifteen cats. The former are used in hunting and the latter are valued for their control of rodents in the longhouse.

The most important aspect of the subsistence economy of the Kantu', by far, is the swidden cultivation of rice and, to a lesser extent, certain nonrice cultigens. Each year the average Kantu' household makes an average of 2.3 new swiddens, each averaging about two hectares in area. Bumai betaun 'Making yearly swiddens' is more than just one of many different economic activities in which the Kantu' may or may not engage. They say that whatever other activities they may have in hand, Kitai nadai tau' keleka' bumai 'We must not forsake the making of swiddens'. Swidden farming is more than the central economic activity in the life of the Kantu'; in a real sense it is the life of the Kantu'.

The Kantu' begin a new set of swiddens each year. The swidden cycle largely consists of a squence of nine discrete activities. These are (1) beburong 'taking omens' and selecting the swidden site, (2) nebas 'slashing' the underbrush and saplings, (3) nebang 'felling' the larger trees, (4) nunu 'burning' the slashed, felled and dried material, (5) nugal 'planting' in the ashes from this burn, (6) mantun 'weeding' (in some but not all cases) the new swidden, (7) nginang 'guarding' the growing crops from predators, (8) ngetau 'harvesting' the rice crop, and (9) ngangkut 'carrying in' the harvest. There is only one rice crop per swidden and, hence, only one rice crop per year. In addition to this sequence of nine activities, there are three swidden activities that are less sequential in nature. These are (1) ngega' ngkayu 'harvesting the nonrice cultigens', (2) ngawai langkau umai 'building swidden houses', and (3) nganyam 'plaiting', nempa' 'forging', and ngawai 'hewing' swidden tools.

All swiddens are not alike. The Kantu' make five types of distinctions (all overlapping) among their swiddens. One distinction is based upon the presence in or absence from the swidden of the ritually preeminent rice variety, the padi pun 'stem rice'. Swiddens in which it is planted are called umai pun 'stem swiddens', while all other swiddens are called umai dijap 'opposed swiddens'. A second distinction among swiddens is based on the age of the trees in the forest cut for them. If the forest either has never been cut on that spot before, or not cut within living memory, the forest is called kampung and a swidden made there is called an umai kampung. If the forest is regrowth, following one (or more) prior cuttings of the forest on that spot, the forest is called memudai and a swidden made there is called an umai memudai. A third distinction is based upon the drainage characteristics of the tanah 'land' lying beneath the forest. If that land is waterlogged or is covered with standing water, it is called tanah paya' 'swampland' and a swidden made there is called an umai paya' 'swamp swidden'. Swampland is found chiefly in low-lying areas along the main rivers, the Empanang and the Kantu', but small patches of it also occur at the sources and along the banks of small, meandering streams. Land that is not swampland is called tanah darat 'dryland' and a swidden made there is called an umai darat 'dryland swidden'. The fourth distinction among the swiddens involves the elevation and to a lesser degree the slope of the land. Low-lying land along the main rivers, which is inundated when these rivers flood, is called tanah lempa' 'floodzone land'. A swidden made on such land is called an umai lempa' 'floodzone swidden'. All land that lies above this floodzone, stretching from the edge of this zone to the hills that form the valley's walls, is called tanah mungo' 'highland' and a swidden made on such land is called an umai mungo' 'highland swidden'. All land that lies above this last zone, on the very wall of the valley, is called tanah bukit 'mountain land'. The Kantu' distinguish bukit from mungo' not only by the greater elevation of the former, but also by its greater slope. A swidden made on such land is called an umai bukit 'mountain swidden'. A fifth distinction among swiddens (made much less often than the previous four) is based on timing. The Kantu' say that each year's swiddens ideally should be planted (with rice) during the seventh or eighth moons of their lunar calendar. (In 1975, the seventh and eighth moons roughly corresponded to the months of August and September.) Most years, all swiddens are planted in or near this ideal period. Such swiddens are called umai taun 'yearly swiddens' or umai benar 'true swiddens'. Occasionally, if the harvest from this set of swiddens is very poor, the Kantu' immediately thereafter (in April or May) plant a new set of swiddens (in addition to planting a set of 'true swidden' later in the year, as usual, in August and September). Such swiddens are called umai dulau or umai mpukap 'early swiddens'. Finally, there is a distinction between all of the swidden types just discussed, on the one hand, and any other type of cultivation on the other, based upon the presence or absence of rice. Any cultivated land in which rice is present is called an umai, while any other cultivated land - even though the process of cultivation otherwise may be identical - is called tempalai. Some of these tempalai are in effect swiddens, but they rarely are made by the Kantu'. Accordingly, the focus of this study is the umai 'rice swiddens' of the Kantu'.

The Kantu' classification of their swiddens is summarized in table 7[48]. The eleven swidden types produced by this set of distinctions are not mutually exclusive. There is considerable (although not total) overlap between them[49]. The relative prevalence of each of the eleven swidden types is presented in table 8.

TABLE 7
Kantu' Classification Of Swidden Types

Swidden Types	Basis of Distinction
(1) Umai Pun vs. Umai Dijap	Presence or absence in the swidden of stem rice.
(2) Umai Kampung vs. Umai Memudai	Swidden cut from old, virgin forest versus younger regrowth.
(3) Umai Paya' vs. Umai Darat	Swidden made on swampland versus dryland.
(4) Umai Lempa' vs. Umai Mungo' vs. Umai Bukit	Swidden made in riverine floodzone versus highland versus mountain land.
(5) Umai Taun vs. Umai Mpukap	Swidden planted at the normal time of year (August & September) versus early (April & May).

The Kantu' make varying use of the terms for the eleven types of swiddens. Given that each household usually farms two or more swiddens each year, terms are likely to be used in reference only if they can differentiate amongst these and, though of lesser importance, if they can differentiate the household's swiddens from the swiddens of other households in the longhouse. Thus, a swidden never is referred to as umai taun 'yearly

TABLE 8
Incidence Of The Different Swidden Types At Tikul Batu

Swidden Type	1975 Swiddens (34 Total)	1976 Swiddens (35 Total)
Pun vs. Dijap	29 % vs. 71 %	31 % vs. 69 %
Kampung vs. Memudai	38 % vs. 62 %	24 % vs. 76 %
Paya' vs. Darat	40 % vs. 60 %	36 % vs. 64 %
Lempa' vs. Mungo' vs. Bukit	? ? ?	16 % vs. 37 % vs. 11 %
Taun vs. Mpukap	100 % vs. 0 %	100 % vs. 0 %

swidden', umai dijap 'opposed swidden', or umai darat 'dryland swidden', because these terms are relatively so undifferentiating in reference. Somewhat more common is the use of umai mungo 'highland swidden' in reference, and more common yet are the terms umai lempa' 'floodzone swidden', and umai bukit 'mountain swidden'. Very common is the use of umai paya' 'swamp swidden' and umai pun 'stem swidden', because one household rarely has more than one swidden of the former type and cannot have more than one of the latter. Equally common is the use of umai kampung 'primary forest swidden' and umai memudai 'secondary forest swidden', because a household often has one swidden of each type in a single year, but rarely has two of either type. Umai mpukap 'early swiddens' are rarely made, but when made they would - for that very reason - most likely be called by this same term.

Reference terms for swiddens are not confined to the terms for swidden types. At least as common is the use of geographical terms that locate a swidden near some known geographical feature of the Kantu' territory. All bukit 'mountains' in the Kantu' territory are individually named (e.g., Bukit Bo'). A swidden located on such a mountain may be referred to by the mountain name, as in umai di Bo' 'swidden at Bo' mountain'. Mungo 'highlands' are not individually named, but are referred to by the name of the nearest stream or other salient geographical feature. For example, mungo' penyaut is the term of reference for a stretch of highland noted for its stand of the hardwood tree, penyaut. A swidden located in that highland thus may be called after the latter's term of reference, as in umai mungo' penyaut 'penyaut highland swidden'. Swiddens may be named similarly after other geographical features, such as umai sepan (sepan referring to a deer wallow), or umai kerapa kesinap (kerapa referring to stretches of marshy heath forest, and kesinap referring to a species of tree that predominates in and hence identifies one such stretch in particular). Common geographical referents are the jalai 'footpaths' that run between different longhouses. A path between two longhouses is called by the members of each longhouse, respectively, by the name of the other longhouse. Thus, the footpath that runs between the longhouses Tikul Batu and Empakan is called jalai Empakan by the inhabitants of the former and jalai Tikul Batu by the inhabitants of the latter. A swidden located close to this path might be called umai jalai Empakan 'Empakan trail swidden'.

Most often, the geographical referent in a swidden name is the river or stream to which it is nearest. If a single household has two swiddens on the same river or stream, it distinguishes between the two by referring to one as ulu 'upriver' or 'upstream' and to the other as ili' 'downriver' or 'downstream', as in umai Kenua' ulu 'upstream Kenua' swidden' versus umai Kenua' ili' 'downstream Kenua' swidden'. The single exception to this involves swiddens that are located on or near the Empanang river. Since this is the major river in the territory, and since many swiddens are located on or near it, the term of reference umai Empanang 'Empanang swidden' likely would not enable the listener to locate or identify the swidden in question. Instead, swiddens along the Empanang are referred to most often by the name of the lubuk 'river hole' in the Empanang to which they are nearest. The river holes themselves are named after other salient features of the surrounding environment. Thus, lubuk rabing is named after a notorious rabing 'crocodile' that long made that particular river hole its home (see figure 6)[50].

After the rice harvest has been completed in a swidden, the Kantu' replace umai 'swidden' with memudai 'regrowth' or 'secondary forest' in its term of reference. Thus, instead of saying umai Kenua' 'Kenua' swidden', they then would say memudai Kenua' 'Kenua' regrowth'. In order to distinguish among swiddens made not just last year, but two, three, four or more years ago, the Kantu'

FIGURE 6
The Streams, Rivers, And River Holes
Of The Territory Of Tikul Batu

SCALE: 1:18,000
KEY: S = <u>Sungai</u> 'River' or 'Stream'
L = <u>Lubuk</u> 'River Hole'
P = <u>Pampang</u> 'Fork'

make reference to the taun 'year' in which the swidden was made. The Kantu' reckon years not as a numerical progression, however, but as a spatial progression. Thus, they might refer to a swidden as umai taun bumai Kenua' 'swidden of the year [we] farmed the Kenua' [stream]'.

THE PLACE

The longhouse Tikul Batu is situated on the right or north bank of the Empanang river, just below its juncture with the Kantu'. The territory of Tikul Batu lies mostly south of the longhouse, on the left bank of the Empanang. Within this territory are approximately 7.9 square kilometers of farmable forest and rubber groves. If consideration also is taken of land given over to scattered groves of fruit trees, nonfarmable heath forest, cemeteries, the longhouse site, and to the surface area of the numerous trails, streams, and rivers, the total extent of Tikul Batu's territory equals approximately ten square kilometers[51]. The territory can be traversed from north to south on foot (covering 4.6 statute kilometers) in approximately two hours, and it can be traversed from east to west by canoe, paddling with the current on the Empanang river (and covering 4.2 statute kilometers), in approximately one hour.

It is within this territory that the people of Tikul Batu practice their swidden system year after year. The character of this swidden system and the way in which it is practiced is markedly attuned to the character of this territory. There are a number of features of this local environment that act as real constraints on the swidden system. These include climatic, edaphic and biotic factors[52].

Climatic Factors

During twenty-one months in which I made thrice-daily readings of the air temperature at Tikul Batu (in the shade), the lowest and highest readings obtained (during the day or night) were 21.7°C and 35.6°C, respectively. This range of fourteen degrees is deceptive, however. On an average day, the range is no more than nine degrees (table 9). The air temperature is relatively high not only throughout the day (and night), but also throughout the year. The difference in temperature between the hottest and coolest months (if indeed they even merit this

TABLE 9
Mean Hourly And Daily Air Temperatures At Tikul Batu

6:00- 7:00 - 22.8° C	January - 26.8° C
7:00- 8:00 - 23.3	February - 26.9
8:00- 9:00 - 24.4	March - 26.9
9:00-10:00 - 25.3	April - 26.9
10:00-11:00 - 26.7	May - 27.5
11:00-12:00 - 27.9	June - 27.5
12:00-13:00 - 30.1	July - 26.1
13:00-14:00 - 31.1	August - 27.3
14:00-15:00 - 31.6	September - 26.8
15:00-16:00 - 31.7	October - 26.6
16:00-17:00 - 28.7	November - 26.2
17:00-18:00 - 29.4	December - 26.1
18:00-19:00 - 28.3	
19:00-20:00 - 26.4	Mean 26.8
20:00-21:00 - 24.6	
21:00-22:00 - 24.4	
22:00-23:00 - 24.3	
23:00-24:00 - 23.9	
Mean 26.9	

NOTE: The mean daily temperature is calculated as the mean of the eighteen hourly temparature from 6:00 to 24:00. These mean hourly temperatures are based on readings made during the twelve-month period from January to December 1975.

distinction) is approximately 1.8 C. Thus, there is no predictable, seasonal variation in temperature at Tikul Batu.

When possible the Kantu' do not work in the open, under a clear sky, during the hottest hours of the day. When they must work in the sun, which they call mata panas 'eye of the heat', they clothe all parts of their bodies (except face, hands and feet) against exposure to it. During some seasons they also fear exposure of their crops to too much sun. An immature rice crop, they say, can be devastated by intense and prolonged heat from the sun. On the other hand, a nearly mature rice crop actually is hastened to maturity by exposure to the sun's heat. In addition, the Kantu' actively avail themselves of the sun's heat to dry cleared swiddens before burning them and to dry grain (as well as fish, meat, reeds for plaiting, and so on) before use or storage.

The ribut 'winds' in the Empanang valley seem to draw neither strength, direction, nor seasonality from the broader equatorial weather patterns. Robequain has suggested that the trade winds and monsoons, themselves weak in the region of Borneo, are replaced in the interior of this island by equally weak valley or mountain winds[53]. The Kantu' rarely feel the wind on the forest floor. Only in open swiddens located in the highlands or on the valley wall can mild breezes be felt during the day. Stronger gusts of wind precede and sometimes accompany rainstorms. Once or twice a year, these winds blow strong enough to topple rubber trees across the paths leading away from the longhouse. The toppling of the larger, buttressed forest trees is even more infrequent. Whether strong or mild, there appears to be no seasonality to the wind regime in the Empanang valley. The Kantu' do speak of musim ribut 'wind seasons'. However these are not true seasons, occurring at fixed times of the year. Rather, they are merely periods of days or weeks in which the velocity of the wind is relatively high and constant, which coincide with periods of drought. Just as the Kantu' do not recognize any true seasonality to wind activity, so they do not recognize any prevailing wind direction[54]. The strength and direction of the wind are critical to the success or failure of each swidden burn, and in this regard the relative unpredictability of the winds is a matter of great concern to the Kantu'.

I measured the ujan 'rainfall' at Tikul Batu every twenty-four hours (in the morning) over a period of twenty-one months. The daily totals ranged from 0-144 millimeters of precipitation. The total for one year (January-December 1975) amounted to 4290 millimeters. This rainfall was distributed fairly evenly throughout the year. Table 10 shows that rainfall, like air temperature and movement, does not vary through the year according to a seasonal pattern. Again, this is not to say that there is no variation at all. Sometimes the rain comes down every day or every other day for weeks on end, while at other times there is no rain at all for a week or more (as in June and August of 1975). Such variation exists, but it is short-term, nonseasonal, and unpredictable. Marked yearly variation also is not uncommon. For example, the rainfall during the first five months of 1976 amounted to only 60 percent of that for the first five months of 1975. The Kantu' have preserved the memory of even more erratic variations in rainfall. At least once since the construction of Tikul Batu in 1957, the rains have been so heavy that the Empanang river completely submerged the longhouse, which is otherwise 4.6 meters above the mean level of the river. Most middle-aged and older Kantu' also can remember times when the rains were so scarce that the Empanang dried up completely.

TABLE 10
Monthly Pattern Of Rainfall At Tikul Batu In 1975

Month	Total Rainfall (in millimeters)	Number of Rain Days (>1 mm/day)	Most Consecutive Drought Days (<1 mm/day)
January	312	20	4
February	391	15	3
March	269	20	3
April	310	?	?
May	333	14	4
June	221	10	9
July	353	20	4
August	262	12	7
September	678	17	3
October	452	22	3
November	376	23	2
December	333	21	2
Means	358 mm	16.2 days	3.7 days

The timing and extent of rainfall affects the lives of the Kantu' in a variety of ways. Two are particularly important. While the rice crop is in the ground and maturing, continuous rainfall is undesirable because it inhibits this maturation. But the least desired and most feared rains are those that fall just before a dried swidden is going to be burned or during the burn itself. In the former case the burn is posponed, which may in turn delay the planting beyond the optimal time of year. In the latter case the burn is extinguished while in process, with disastrous results for the swidden.

The Empanang river is the largest river in the territory of Tikul Batu. Only it and the second largest river, the Kantu', are navigable by canoe. Near the longhouse Tikul Batu (which is situated about 85 meters from its right bank), the Empanang is approximately 14 meters wide and 2.3 meters deep at mean water. The waters of the Empanang are derived primarily from rains further up in the valley and in the mountainous watershed in which it has its source. Therefore the volume of water in the river varies, sometimes severely, according to the rainfall or lack of rainfall further upstream. As a result, the depth of the river also varies, which over time has cut a deep channel for the Empanang. At mean water (viz., 2.3 meters deep) it flows 2.7 meters below the top of its banks. After a drought the water drops to as much as 3.8 meters below its banks, but after rains it can crest to as much as 1.2 meters (or more) above them, giving a range of variation in the level of the river of at least 5 meters. These figures are based on my once-daily measurements at Tikul Batu of the level of the river, vis-à-vis its banks, over a period of twenty-one months.

When the river rises above its banks at Tikul Batu, it floods all of the low-lying land in the area, especially land lying within bends of the river. When the Empanang rises to within 1-1.5 meters of its banks at Tikul Batu, it only floods low-lying land in the downriver reaches of Tikul Batu's territory. The frequency with which the Empanang reaches these flood levels is analyzed in table 11.

TABLE 11
Flood Pattern Of The Empanang River In 1975

Month	Mean River Level (Below its Banks)	Number of Days Per Month on Which River Level Was			
		More than 2 m below its banks	1-2 m below its banks	0-1 m below its banks	At or above its banks
January	-2.6 m	29 days	0 days	2 days	0 days
February	-2.5	20	3	0	5
March	-2.5	28	1	1	1
April	-3.3	30	0	0	0
May	-2.8	26	1	4	0
June	-3.1	29	0	0	1
July	-2.9	30	1	0	0
August	-3.2	29	1	1	0
September	-2.2	22	1	4	3
October	-2.6	23	6	2	0
November	-2.5	26	2	1	1
December	-2.2	21	8	2	0
Means	-2.7 m	26.1 days	2 days	1.4 days	0.9 day

The existence and the character of the Empanang's floods is important to the Kantu' for several reasons. The Empanang's flood waters can drown fowl and pigs, make traveling difficult by water (and, if the trails are covered, by land) and, very rarely, threaten the longhouse itself - the floor of which is about two meters above the river banks. But the greatest concern of the

Kantu' is for the possibility of damage to their swiddens. For example, if a dried swidden is flooded just prior to being burned, the burn will be hindered. If a swidden is flooded just after being planted, the floodwaters will wash the seed rice out of the dibble holes. If the swidden is flooded after the seed has produced plants, the plants may be damaged by floating timber or more likely just by the inundation. The Kantu' say that their rices, except for the swamp rice varieties, can live in standing water for one to three days only. After three to four days of this, the rice plants die. Most nonrice cultigens succumb even quicker to inundation.

Given the vulnerability of their cultigens, the Kantu' are fortunate in that severe flooding of the Empanang is a short-lived phenomenon. The speed with which flood waters subside is a direct function of their severity. When the Empanang is raised by rains to a point not far above its lowest level (3.8 meters below its banks), it subsides gradually from its crest. But when the Empanang is raised far up, to flood levels, its subsidence is precipitous (just as the initial rise is precipitous). This is illustrated in the rise and fall of the river on the two most severe instances of flooding that I recorded (table 12). During the twenty-one month period in which records were kept, the Empanang maintained a

TABLE 12
The Most Severe Floods During 1974-1976

Date	River Level During Previous 24 Hours	Flood Level	River Level 24 Hours Afterwards	River Level 48 Hours Afterwards
23 Sept. 1974	- 2.7 m	+ 1.2 m	- 2.4 m	- 4.3 m
4 Sept. 1975	- 2.9 m	+ 1.2 m	- 0.8 m	- 2.4 m

level at or above its banks for two consecutive days on only two occasions; and it never maintained that level for as much as three or more consecutive days. From the point of view of the Kantu', this is by far the most important characteristic of the Empanang in their territory: the river characteristically does not flood long enough to kill dry rice growing along its banks.

This ability to grow rice along the banks of a river is a critical factor - perhaps the most critical factor

- in the geographical distribution within Kalimantan of not only the Kantu', but all other swiddening peoples as well. The Kantu' cite the prolonged flooding along the banks of the Kapuas river as the main reason for their return to the Empanang valley, following their initial flight to the Kapuas in the late nineteenth century. They also cite this as their reason for a move earlier this century from a longhouse further downriver on the Empanang itself. The pattern of flooding that prevails in the vicinity of Tikul Batu does not prevail along the entire length of the Empanang. The rise and fall of flood waters becomes progressively quicker upriver from Tikul Batu and progressively slower downriver. The first of these two patterns is by no means inimical to swidden cultivation, and indeed most of the valley's human population (composed solely of swidden farmers) is located on stretches of the Empanang or its tributaries upriver from Tikul Batu. The second pattern of flooding, encountered progressively further downriver from Tikul Batu, is inimical to swidden cultivation and this too is reflectied in the distribution of the valley's population. From Tikul Batu to the mouth of the Empanang is a distance of approximately thirty-five statute kilometers. In the first eight kilometers downriver, which is still far enough upriver to be relatively safe for swidden cultivation, there are five other Kantu' and Iban longhouse. In the remaining twenty-seven kilometers, which is below this safe zone, there are no large, permanent settlements of swidden cultivators at all.

Edaphic Factors

There are several edaphic distinctions within the territory of Tikul Batu that are important to, and recognized by, the Kantu'. The distinctions based upon drainage (viz., dryland versus swampland) and upon elevation (viz., floodzone land versus highlands versus mountain land) have been mentioned earlier. A third distinction of great importance to the Kantu' is the character of the tanah 'earth' or 'soil'. They say that the soil has two components. The first or highest component is the kulit tanah 'skin of the earth', which consists of rotting leaves and tree roots. When the forest is cleared and burned, this humus layer is burned too, and its ashes contribute to the sustenance of the rice crop. While Kantu' do not believe that this layer increases continuously in depth as the forest ages, they do say that it tends to be very thin in very young forest and that this is one of the reasons for the failure of the rice crop in swiddens cut from such forest. Below this 'skin' lies the tanah 'soil' proper. The Kantu' recognize seven different features of the soil as critical to the success of their swiddens.

The first relevant feature is the structure of the soil. The Kantu' distinguish between ordinary tanah, which is loose in structure, and tanah batu-batu 'rocky soil', tanah kelilit 'clayey soil', and tanah pasir 'sandy soil'. Only the first of these soil structures is thought to be suitable to swidden cultivation. The second cannot be dibbled, the third is said to be somewhat infertile, and the fourth is said to be very infertile. A second critical feature of the soil is its friability. The Kantu' say that tanah lemah 'soft soil' is good, while tanah kering 'hard soil' is bad. They judge this feature by the ability to pierce the soil with a dibble stick on a hot day, when the soil is dry. Bound up with the friability of the soil is a third feature, its moisture content. Here, the Kantu' distinguish between soil that is or is not rangkai 'dried out'. In general, they believe that their crops will grow better in wetter soil than in drier soil. A fourth and also related feature is the temperature of the soil. The Kantu' believe that tanah celap 'cool soil' is good for their crops, while tanah angat 'hot soil' is bad. They judge a given area of soil to be 'hot' or 'cool' depending upon its feel beneath the soles of their feet on a hot day. However, if the rice crop of a given swidden fails, they then may call the soil there 'hot', regardless of how it feels. Indeed, both angat and celap have, in addition to their denotations of physical temperature, broader connotations of infertile and fertile, of bad and good. A fifth critical feature of the soil is its color. The Kantu' make a particular distinction between tanah celom 'black soil', with average agricultural potential, and tanah kuning 'yellow-orangish soil', in which rice usually grows poorly. A sixth feature is the taste of the soil. The Kantu' distinguish tanah masam 'sour soil' from all other soils. The Kantu' judge this (a highly acidic type of swamp soil) to be totally unsuitable for swidden cultivation. They sometimes speak as well of tanah masing 'salty/brackish soils', which also are unsuitable for cultivation. The seventh and final soil feature of importance to the Kantu' is its smell. They say that good soil bebau 'smells', the smell being characterized as either bau rom 'burnt smell' or bau lang 'rotten smell'. The Kantu' describe bad soil as Nadai idup agi' 'No longer alive', and Nadai bebau agi' 'No longer smells'. Within the territory

Tikul Batu, there are only a few patches of soil of this latter sort, having neither life nor smell; and they occur in gardens where pepper was raised continuously for five or six years.

Each of these soil features merits attention in deciding where to locate a swidden. The Kantu' usually observe them through indirect means, foremost among which is the character of the vegetation growing atop the soil. For example, if the land is covered with madang, a densely growing fern, the Kantu' know that the soil underneath is rangkai 'dry', angat 'hot', and likely to be composed of tanah kelilit 'clays'; and they consequently will be reluctant to locate a swidden there. If swampland is covered with purun, a sedge, the Kantu' know that the soil underneath is masam 'sour', and they avoid it on that account. Two other signs are the heath forest types kerapa and kerangas[55] . They reflect the presence of sour and sandy soils, respectively, both of which are unsuited to cultivation.

The distribution of good and bad soils within the territory of Tikul Batu roughly corresponds to the division of the territory into the zone that is susceptible to riverine flooding and the zone that is not. Soils in the latter zone tend to be hard, hot and dry, and generally less than ideal for cultivation; while soils in the former zone tend to be soft, cool and less dry, and generally more ideal for cultivation. Even within the floodzone, however, there is a great deal of poor soil. The Kantu' say that along the Empanang, and perhaps along the Kantu' as well, good soil is found for only two to three belah 'swidden-sized sections' to the left and right of the river. Beyond that distance, but still within the floodzone, they say that there is much of the kerapa heath forest and much of the poor soil that is associated with it[56].

Biotic Factors

The Kantu' divide the flora within their territory into opposing categories of utai ti ditanam 'things that are planted' and utai ti idup kediri' 'things that live on their own'. The cultigens are concentrated within the swiddens and the noncultigens are concentrated outside of them, where they are referred to collectively as babas 'woods' or 'forest'. In any given year, this forest covers almost 90 percent of the territory of Tikul Batu (excluding the current year's swiddens and the rubber groves). Within the forest the Kantu' make a basic distinction, critical to their system of property rights as well as to their system of swidden strategies, between kampung and memudai.

As earlier noted, kampung - which I am translating as 'primary forest' - refers to forest that has not been felled within the memory of the current Kantu' settlers. Just outside the territory of Tikul Batu, on Bo' mountain, there are stands of true primeval forest, meaning forest that probably has never been cleared by man. There is no such forest within the territory itself, however. The kampung that lies within the territory probably is not primeval. This is suggested by the presence of ceramic plates, bowls, and other human detritus in the earth under scattered sections of this forest. Some Kantu' view this as proof that this forest was cleared during the previous Kantu' occupation of the Empanang valley, which terminated in the 1882 flight to the Kapuas [57]. If true, all of this forest would have been a minimum of ninety-two years old by the time of this study (1974-1976) [58]. Individual trees that I measured in this forest averaged 24.5 centimeters in diameter (at chest height); but trees with a diameter of 75-100 centimeters were not uncommon. After the 1976 swidden season, twenty-four swidden-sized sections of this forest remained within the territory of Tikul Batu (figure 7).

After the 1976 swidden season there also were 257 swidden-sized sections of memudai in the territory of Tikul Batu. Memudai, which I am translating as 'secondary forest', refers to forest that has grown up after the clearing, within memory, of the primary forest or prior secondary forest on that spot. The Kantu' distinguish between these two cases. The former they call memudai bali' kampung 'secondary forest changed from primary forest', and the latter they call memudai bali' memudai 'secondary forest changed from secondary forest'. In the latter case, they also may state how many times regrowth has grown up since the initial clearing of the primary forest on the land. The Kantu' distinguish among these successive types of regrowth according to the species of trees and other plants that they know to characterize them.

The Kantu' also make distinctions within secondary forest based on tree size. The trees that I measured in secondary forest (excluding purely grassland successions) averaged 12.2 centimeters in diameter; but there was considerable variation from this mean, due not least to the fact that the secondary forest within the territory of Tikul Batu ranges in age from less than one year to as much as twenty to thirty years. To cope with this variation, the Kantu' distinguish between memudai

FIGURE 7
Primary Forest In The Territory Of Tikul Batu

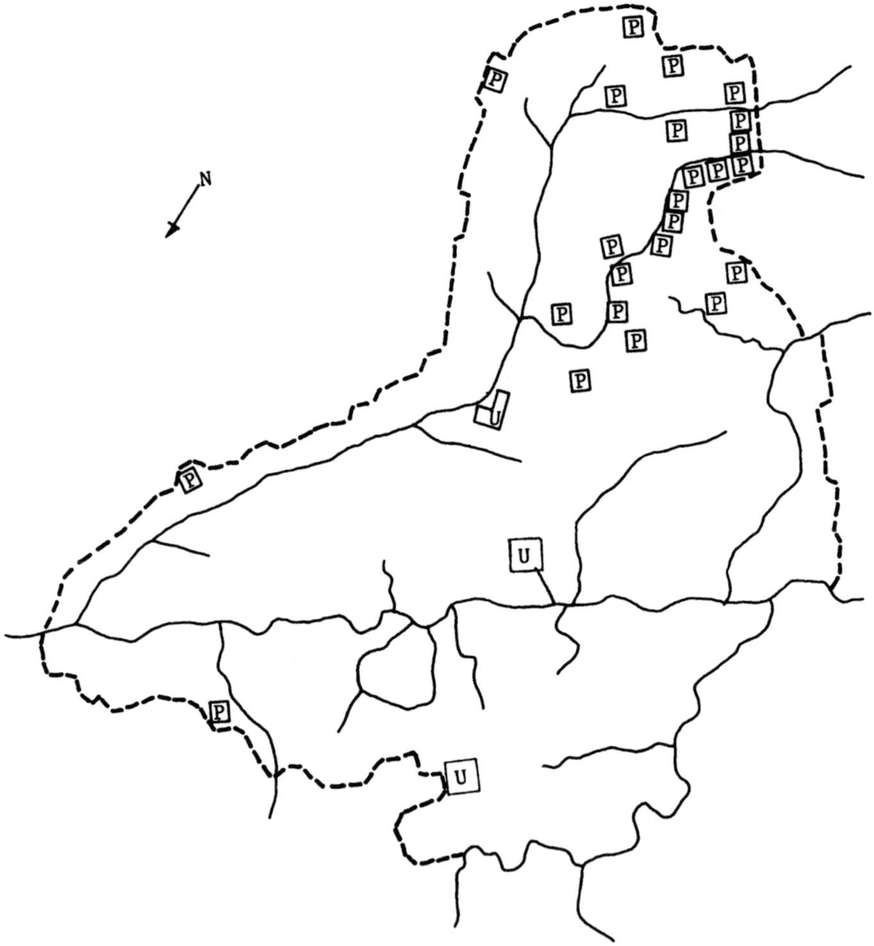

SCALE: 1:18,000

KEY: P = One swidden-sized section of primary forest (not drawn to scale).

U = Large area of uncultivable primary forest (not drawn to scale).

tuai 'old secondary forest' and memudai biak 'young secondary forest'. They make finer distinctions by comparing the largest trees in a given stand or forest to parts of the human body. In these comparisons, the critical dimension of both tree and body part is diameter. The body parts are ranked by all Kantu' in a standard order (here in decreasing order of size): tuboh 'trunk', pala' 'head', pah 'thigh', petih 'calf', lengan 'forearm', dekuan 'wrist', and tunyo' jari' finger'.

The smallest trees, like the largest, are called kayu. Herbaceous matter, especially grasses and reeds, is called rumput. Even the smallest tree usually is called not rumput, but sapi' kayu 'tree seedling'. Only when such seedlings are encountered in the swidden during weeding might the Kantu' call them rumput. The Kantu' include both rumput 'grasses' and kayu 'trees' in the broader category memudai 'secondary forest'. Thus, even a patch of swamp grass devoid of ligneous growth may be called memudai (which suggests that the translation 'secondary forest', although I will continue to use it for the sake of simplicity, actually is less accurate than 'secondary growth' or 'regrowth'). This conceptual lumping of grassland with forest probably is due to the predominance of the latter and rarity of the former in the territory of the Kantu'. Grassland successions occur only in swampland that is cropped intensively (viz., once in every two or three years). Even there the succession is not permanent, but gives way to ligneous growth if the fallow period is extended from one or two years to five to ten years. No grasslands have as yet been created on dryland, within the territory of Tikul Batu, as the result of swidden activities. Pepper cultivation alone results in dryland, grassland succession. The amount of land that is planted in pepper and thus exposed to this succession is very limited, however, and in any case the Kantu' have been able to reclaim all such land by planting rubber trees in the middle of the grass[59].

There is a rich assortment of fauna in the territory of Tikul Batu, and much of it plays an important role in the lives of the Kantu'. As noted earlier, fishing is a daily occupation, drawing on an astonishing variety of fish, amphibians, terrapins, crustaceans and mollusks in the local waters. A number of animals of water, land and air are important as game in hunting and trapping. These include (1) several hoofed animals, the rusa' 'sambar deer', kijang 'barking deer', pelandok 'mouse-deer', and jani' 'bearded pig'; (2) several primates, mpeliau' 'gibbon', nyumoh 'pig-tailed macaque', kera' 'long-tailed macaque', mayas 'orang-utan', and occasional rasong 'proboscis monkey'; (3) several carnivores, munsang 'civet', menturun 'bear cat', the rare beruang 'sun bear', and one or two species of jungle cat or leopard; (4) several rodents, including tupai 'squirrel', cit 'field rat' and landak 'porcupine'; (5) certain reptiles, notably sawa' 'python' and bwaya' 'monitor'; (6) certain insects, notably the mwanyi' 'honey bee', whose honey and larvae are both eaten; (7) a variety of game birds; and finally (8) the ntamba' 'flying fox' and the tingiling 'pangolin', whose flesh is eaten and whose scales are traded in the Sarawak bazaars. As mentioned earlier, domesticated animals include jani' laman 'yard pigs' (which interbreed with the feral stock), manok 'chickens', ukoi 'dogs' and mayau 'cats'.

Not all of the local animal life serves the Kantu', however. Civets and pythons take many of their chickens, as does the Brahminy kite. Some animals threaten the Kantu' themselves. When clearing secondary forest, the Kantu' sometimes are troubled with nests of the kesa' 'weaver ant', an aggressive red ant almost 1.5 centimeters in length. Even more troublesome are nests of keruai or pantak, wasps whose sting can cause a person to lose consciousness. The greatest threat to health from insects is posed by the nyamok 'mosquito'. They bite all year around in all parts of the territory, one result of which is that malaria is endemic among the Kantu'. Some threat to health also is posed by the melipan 'centipede' and kala 'scorpion'. Mortal threats are posed by the mpelawa' mpeliau 'gibbon spider' (the notorious bird-eating spider) and by two snakes, the tedung 'common black cobra' or 'king cobra' and the kenawang 'red-headed krait'. Among the larger animals, only the sun bear is feared by the Kantu'. Even armed, adult male Kantu' flee an encounter with this animal. The Kantu' also profess to know and fear the remaung 'tiger'. However, there is some doubt as to whether this animal now occurs, or ever did occur, in Borneo[60]. Perhaps the greatest threat posed by local animal life is not to the Kantu' but to their crops. These are preyed upon by a number of the hoofed animals, primates, rodents, birds, and insects (each of which will be discussed at some length in chapter seven)[61].

Summary

In summary, the environment of the Kantu' is characterized by, first, great diversity. Within the scant ten square kilometers occupied by the Kantu', there is variation from one place to another in (1) the drainage pattern of the land (viz., dryland versus swampland),

(2) the elevation of the land (viz., in or above the river's floodzone), (3) the type of soil present (viz., cultivable or not), (4) the type of forest on the land (viz., primary vesus secondary), and (5) the distribution of the native fauna. In addition to its diversity, this environment is characterized by similarly great uncertainty. From one day to the next there is great uncertainty as to (1) whether the winds will blow and from what direction, (2) whether the day will be hot or overcast, (3) whether there will be drought or rain, and (4) whether the river will flood its banks or fall far below them. The existence of such uncertainty and diversity in the environment has played an important role in the development of Kantu' society, especially in the development of their swidden technology. As will be seen in the chapters to come, the Kantu' swidden technology shields them from this uncertainty insofar as it is inimical to their welfare, and it exploits this diversity insofar as it can contribute to the richness and stability of their livelihood.

Chapter One
Selecting

The Kantu' call the first, distinct phase of the swidden cycle beburong 'using the birds', which refers to the initial seeking and receiving of omens from birds at the intended site of a swidden. This taking of omens actually constitutes but a small part of the complex of activities that go into the selection of the swidden sites. Selection ultimately encompasses decisions regarding four critical aspects of the swidden cycle, namely the location of each swidden, the size of each swidden, the total number of swiddens per household, and the shape of each swidden.

SWIDDEN LOCATION

The location of each swidden itself is based upon an analysis of three variables: the ownership of the site, its relative location, and its character or condition[1].

Ownership of Site

Considerations of ownership vary greatly according to whether a potential site lies in primary forest or secondary forest. In the case of secondary forest, there is typically one household or group of households (related through partition) that holds clear rights to each site. These households can make a swidden on that site at will. If two or more households hold equal rights to the same spot, they must alternate in their use of it. Any other household desirous of farming that site can do so only by (1) minyau 'borrowing' that section for one swidden cycle, (2) beganti 'temporarily exchanging' a section of their own forest for the desired section, (3) besilih 'permanently exchanging' a section of their own forest for the desired section[2], (4) nyewa 'renting' use of the desired section for one swidden cycle (which is a rare practice), or (5) meli 'buying' that section for its own permanent use. Rarely, but increasing in frequency as population/land pressure increases, two households or groups of households both may claim exclusive rights to a single section of secondary forest. This typically occurs in the case of land that long has lain fallow, since this allows for loss of memory and confusion as to ownership rights. Such cases are brought to the longhouse headman (or other person of rank) for resolution, before further clearing can be carried out[3].

When selecting a site in primary forest, the first consideration is territorial. No household has a right to clear primary forest that lies outside the hak 'territory' (or, more accurately, 'territorial rights') of its own longhouse and inside that of another. A household can do this only if the latter longhouse grants its permission. In the event that two or more households from two different longhouses attempt to clear (or otherwise lay claim to) the same section of primary forest, supralonghouse authorities are called upon to adjudicate the case and, if necessary, clarify the boundary between the two territories. While it is not permissible for a household to clear primary forest in the territory of another longhouse, however, it is permissible for a household to hold and exercise rights to secondary forest in another longhouse's territory. Such a circumstance may come about, for example, if the members of a given household move from one longhouse to another one nearby and retain their forest rights in the territory of the former. It also is possible for a household in one longhouse to be sold or given rights to secondary forest by a household in another longhouse. Thus, it is possible for rights to secondary forest in one longhouse's territory to devolve to a household in another longhouse; but it is not possible (without special permission) for a household to create such rights by opening up primary forest in another longhouse's territory.

Within the territory of its own longhouse, a household is allowed to clear any primary forest that has not been slashed previously by another household. A household that slashes any part of a swidden-sized section of primary forest holds exclusive rights to that section (to finish clearing it and make a swidden) for the succeeding three years. During this time, the partially slashed section of forest is called a tasau. After three years, if that household has neither made a swidden there nor renewed its slashing, then any other household has the right to clear that forest section. In practice, it is rare for a household to let its rights of slashing lapse, given that these rights can be renewed easily with a minimal amount of slashing. Because of this ease, slashing is an attractive means for the individual household to establish forest rights. Thus, households sometimes fail to complete slashing (and farming) in the first year due not to any unforeseen inability, but rather to a premeditated strategy. The household may know in advance that it will not be able (or willing) to clear and farm a particular section of primary forest in a given year, but by merely slashing a little of that forest it can establish rights to it, effectively neutralizing competition for that section from any other household.

If a section of primary forest has not been slashed previously by any household, but it is adjoined by a section of secondary forest or a current swidden belonging to a given household, then that section also is proscribed to clearing by all other households. The adjoining household is said to have rights because, the Kantu' say, _Memudai sida' bepala' ke kampung nya'_ 'Their secondary forest heads onto the primary forest'. These rights usually take precedence over rights of slashing. Thus, a household possesses rights of adjacency to primary forest adjoining its secondary forest regardless of whether or not it has done any slashing in that primary forest, and also regardless of whether or not any other household has done any slashing there. Rights of slashing in primary forest take precedence over rights of adjacency only if the primary forest in question was slashed before it was adjoined.

Rights of adjacency to secondary forest extend for only _sebelah_ 'one swidden-sized section' of primary forest. That is, they apply only to the section of primary forest immediately adjoining the section of secondary forest, and not to any other primary forest sections that adjoin the former but not the latter. An exception to this occurs when two households both claim a single section of primary forest that is not adjoined directly by secondary forest belonging to either of them. In that event, the household that owns secondary forest closest to the disputed section, even if it does not adjoin it, is recognized as having the greatest right to that section. All rights of adjacency perdure until the household decides to utilize them (viz., by clearing the primary forest in question), and they require no periodic renewal, unlike the rights of slashing. Further, rights of adjacency can be exercised only by the household that owns the secondary forest in question. According to what the Kantu' say is their oldest customary law pertaining to land use and land tenure, a household that borrows a section of secondary forest belonging to another household, for one year's farming, cannot exercise rights of adjacency to primary forest that adjoins that section, neither during that year nor any succeeding year.

It sometimes happens that a given section of primary forest is adjoined by two or more sections of secondary forest (viz., on two or more of its sides) belonging to two or more different housholds. In that case, foremost rights to the section are held by the household owning secondary forest on its downhill side. That is, the strongest rights extend from the uphill edge or *pala'* 'head' and not the *rusok* 'side' of a section of secondary forest (or swidden). Thus, in figure 8, household A has a greater right than household B to clear the primary forest section 2a, because the latter is adjoined by the 'head' of household A's secondary forest, whereas it is adjoined only by the 'side' of household B's secondary forest. However, if household A should decide not to exercise its right to section 2a then the fact that the section is adjoined by the secondary forest of household B, even though on the side, would give this household a greater right to the section than, for example, household C.

FIGURE 8
Rights Of Adjacency To Primary Forest

3a Section of primary forest	3b Section of primary forest	3c Section of primary forest
2a Section of primary forest	2b Secondary forest of household B	2c Secondary forest of household C
1a Secondary forest of household A	1b Secondary forest of household B	1c Secondary forest of household C

↑ Uphill (Annual movement is in this direction.)

River

In its narrowest and most literal sense, therefore, rights of adjacency apply to primary forest adjoining the uphill edge of secondary forest or swiddens. This derives from the customary Kantu' pattern of land usage, wherein each household from a given longhouse clears one swidden along the banks of a given river or stream, and then clears the second and successive years' swiddens inland and hence uphill from the waterway (figure 8). This pattern derives, in turn, from the fact that land on or near waterways is most desired by the Kantu' because of its typically greater fertility, proximity to the longhouse, and ease of transportation (by canoe); while land that is further from the waterways is least desired for all of the opposite reasons. In order to distribute both more and less desirable land equally among its constituent households, therefore, the longhouse community does not allow a household to make successive years' swiddens along the banks of a waterway, progressing upriver (e.g.) instead of uphill.

It also may happen that a given section of primary forest is adjoined by the uphill edges of two different sections of secondary forest (or swiddens), each belonging to different households (figure 9). In this case,

FIGURE 9
Rights Of Slashing To Primary Forest

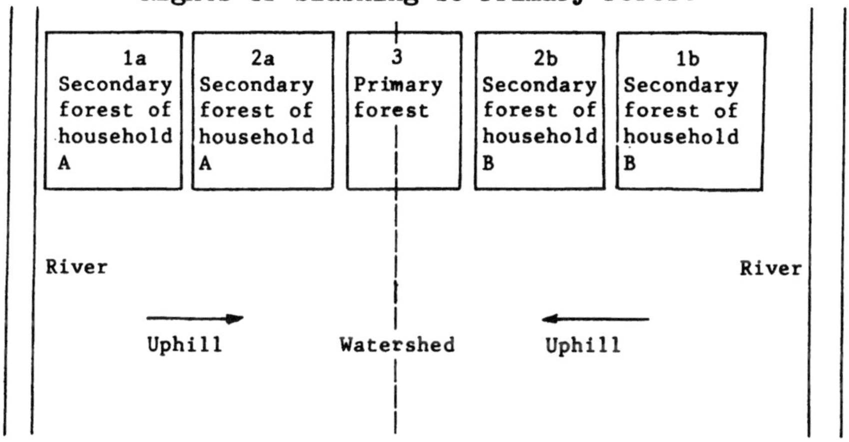

whichever household first cleared one of the forest sections immediately adjoining the disputed primary forest section has the right to farm the disputed section at any later date. If, in figure 9, the two households cleared the adjacent sections 2a and 2b at the same time (viz., in the same swidden year), then rights to the central section of primary forest will fall to whichever household first begins to clear it. If both households begin to clear that section at the same time (viz., in the same swidden year), then more subtle factors enter into consideration. For example, if household B did not farm section 1b itself but was aole to farm section 2b only by asking for permission from whichever household did clear 1b, then it has less of a right to farm section 3 than does household A, assuming that the rights of A to section 3 derive from its own clearing and ownership of sections 1a as well as 2a.

In anticipation of such a conflict, an especially acquisitive household may attempt to strengthen its position by employing a timeworn strategem. If households A and B clear forest sections 2a and 2b, respectively, in the same year, both households thereby would establish equal rights to clear section 3. Knowing this, household A, for example, might just clear a small portion of section 2a and then, on the pretext of an ill omen, <u>ngelangkah</u> 'pass on' to section 3, clearing and farming it instead that year. By so doing, household A establishes firm rights to section 3, thus eliminating any possibility of a future conflict with household B. Of equal importance, household A also thereby establishes firm rights to section 2a. This clearing, even though not completed, confers the previously mentioned rights of slashing on household A. In a single year, therefore, the household can establish rights to two sections of primary forest, although it farms only one of them.

If a household cannot establish rights to clear primary forest by any of the means discussed above, it can try to acquire these rights by <u>minta'</u> 'request'. This consists in asking permission to clear and farm a section from the household (or households) that possess the rights to it (whether rights of slashing or adjacency). Such permission is granted infrequently. When given, the grantee, after it has cleared and farmed the primary forest section, then possesses the rights to the secondary forest that succeds on that section as well as the rights of adjacency to any proximate primary forest. (However, as noted above, these latter rights are weaker than rights of adjacency from a forest section that was originally cleared itself by the right of adjacency as opposed to by request.) In no case can the rights to succeeding secondary forest or to any adjoining primary forest return to, or remain with, the household from which permission to clear that section was asked and obtained.

Unlike secondary forest, that is, primary forest cannot be borrowed by one household from another. The Kantu' justify this rule on the basis of the relative amount of labor involved in clearing primary forest. They say that no household would want to expend that labor only to see the succeeding secondary forest -

which can be cleared with much less effort - return to another household. For the same reason, one household cannot ask another to permanently cede its rights to a section of secondary forest, as can be done with primary forest. Rights to secondary forest usually are ceded only by sale; in contrast to which rights to primary forest cannot be sold. The fact that a household can ask for but cannot borrow or buy another household's rights to primary forest, in contrast to secondary forest, reflects the fact that rights to the former are weaker or more tentative than rights to the latter. In turn, this reflects the fact that it is only the clearing of primary forest, and the investment of labor in this clearing, that establishes rights that are sufficiently strong to be borrowed or bought (but too strong to be requested).

The preceding discussion of the creation, devolution and use of rights to primary forest applies, specifically, to swidden-sized sections of forest. The picture is slightly different as regards smaller, isolated stands or copses of primary forest, called lulong kampung. Lulong refers to the fact that such stands are dilulong 'surrounded' by secondary forest. Such copses are created, for example, when a household is unable to completely clear and farm a given section of primary forest in one year. If the remaining, uncleared portion of that section is adjoined by an additional section of primary forest to which the same household holds rights, then it may clear and farm this fractional section along with the latter whole section in some succeeding year. If this is not the case, if the fractional section is adjoined only by secondary forest or by primary forest belonging to another household, and if the fractional section is sufficiently large - one hectare or more in area - then the household may clear and farm this forest by itself in a later year. If the fractional section is below this critical size it is too small to be farmed efficiently by itself, because of decreasing returns to scale. In this case, the household will not farm this forest until it returns to farm the regrowth on the former primary forest swidden in which this fractional section of forest originally was to have been included. During the intervening period, after the isolation of this stand of primary forest and before its eventual clearing and farming, it is referred to as a lulong kampung.

One such copse was created in 1976 when household #8 failed to complete the felling of a section of primary forest. (This failure was due to the partitioning of the household, coupled with the mental illness of one of its members.) The section originally meaured approximately 2.5 hectares. Two-thirds of this, or 1.6 square hectares, was completely cleared and made into a swidden (#52); while the forest was left standing on the remaining 0.9 hectare. The senior male of household #8, Tigang, said that he would not clear this copse until he returned to the area to clear and farm the regrowth on the 1.6 hectares. That is, his intention was to eventually make a single swidden of the 0.9 hectare of primary forest plus the future regrowth on the 1.6 hectares. He said that the copse was too small to be farmed by itself (nor was it adjoined by any other primary forest to which it could have been joined in the making of a larger swidden).

There also is a second set of circumstances in which lulong kampung may be created. By chance, terrain and land use patterns may be such that two households clear primary forest swiddens that are close to, but not quite adjacent to, one another. The intervening strip of forest then becomes a lulong kampung, not to be cleared and farmed until one or both of the adjoining sections are later recleared as secondary forest swiddens. It is in cases such as this that the question of rights to primary forest copses becomes pertinent, for such strips commonly are disputed by the two (or more) households involved. Indeed, primary forest copses are at issue in approximately one-half of all Kantu' adjudication concerning primary forest. Conflicting claims are evaluated in terms of the ownership of adjacent secondary forest (or swiddens), just as with normal-sized sections of primary forest. If a given copse is adjoined by secondary forest sections belonging to two separate households, as is often the case, Kantu' legal authorities usually hold that it should be divided between the two households. It sometimes happens that the secondary forest sections of two separate households will adjoin two separate copses. In such cases, the two households typically agree to clear one copse apiece when they eventually come to reclear their adjoining sections of secondary forest. Such danyi 'agreements' are frequently violated, however, and themselves become subject to adjudication.

Location of Site

Some locational factors already have been discussed. Thus, because of Kantu' adat 'law' on forest rights, a household usually is restricted to clearing primary forest that is located adjacent to, and uphill from, one of its sections of secondary forest. Also discussed previously is the Kantu' preference for swidden sites along the banks of the major rivers, based on their relative fertility and accessibility. Other factors that can influence the selection of swidden sites sites are proximity to rubber groves, to past swiddens, and to other current swiddens.

Kantu' households like to farm forest and tap rubber groves in the same vicinity. While the household is living in its swidden, it can tap a nearby grove (either between or even during periods of work in the swidden, depending upon the size of the household's labor force), without any additional expenditure of travel time. Instead of the household locating its swidden near a rubber grove that it wishes to tap, however, the household may tap a grove only when it is located near a forest section that it wishes to farm. That is, the independent and dependent variables often are, respectively, swidden location and rubber tapping, and not the reverse. This is in accordance with the fact that a rubber grove can be tapped any time, any year, at will, whereas the forest that adjoins that grove only can be cleared at fixed intervals, separated by the requisite fallow period (which is subject to some modification, but not much). There are far more constraints involved in the selection of a forest section for farming than in the selection of a rubber grove for (the resumption of) tapping.

While it rarely happens that a household decides to farm a given section of forest because of its location relative to its rubber groves, it sometimes happens that a household decides not to farm a given section of forest for just that reason. Household #8 made such a decision at the beginning of the 1977 swidden year. This household had planned to clear a 1977 swidden in a section of primary forest to which it held rights, which was distant from the longhouse by over 1½ hour's walk. (Indeed, the forest was located beyond the boundaries of the longhouse territory.) Household #8 had to abandon this plan after reaping a very poor harvest from its 1976 Kenua' swidden. This harvest amounted to only 14 percent of the household's requirements for the coming year. To make up the shortfall, the household planned to tap rubber (trading the latex for rice) as intensively as possible during the coming year, commencing right after the 1976 harvest. Given this need to tap rubber on a regular basis, it was no longer feasible for the household to farm the distant section of primary forest. The distance between this forest and the rubber groves was too great for the household to simultaneously (e.g., on alternate days) tap rubber and work in the swidden. Nor would it have been possible for the household to split its labor force between the swidden and the rubber groves, as is often done, because household #8 had been reduced to two persons by this time. As Tigang of household #8 said, Nadai isi' bekal ti ngau kitai bumai jauh 'We don't have the supplies [viz., rice stores] to farm far away'. That is, because the household had no rice stores to speak of, its members were compelled to remain near the rubber groves, which alone could enable them to meet their subsistence rice requirements. As a result, the household put off farming the section of primary forest until another year and instead began to clear a section of secondary forest located closer to its rubber groves and the longhouse.

Another factor that influences the selection of swidden sites is proximity to the site of one of the household's swiddens from the previous year. There are several marked advantages to such a location. One advantage has to do with ease of access to crops in the year-old swidden. A number of different nonrice crops are harvestable for up to sixteen months after the completion of the main rice harvest in the swidden. These crops are available to the household, therefore, during the entire swidden cycle of the following year (see chapter 10). They are especially important to the household during the early phases of the following cycle (viz., selecting, slashing, felling, burning, and planting), when the new swidden is not yet producing any crops of its own. The harvesting of the previous year's crops is greatly facilitated if the new year's swidden, in which the household members are spending much of their time, is located near the previous year's swidden in which these crops are growing. This minimization of distance is important because the poor storability of most of these cultigens (e.g., cassava) necessitates frequent, small harvests.

There is a second advantage to the proximity of successive years' swiddens, which concerns the storage of the rice crop. In most cases, after a swidden's rice crop has been harvested and processed, it is carried from the swidden back to the longhouse for storage in the longhouse loft. However, the rice crop may be left in storage in the swidden house if the household plans to farm a nearby section of forest during the succeeding year. In such cases the rice stores are apt to be more accessible to household members if kept in the old swidden than if stored in the longhouse. One advantage of making proximate swiddens, therefore, is in obviating the necessity of carrying the rice crop back to the longhouse, a necessity that otherwise entails heavy labor and feasting costs.

A final advantage of the proximity of successive years' swiddens concerns travel time between the longhouse and the swiddens. Household members usually sleep in the longhouse at night and walk to and from the swiddens each day, during the swidden phases of selecting, slashing, felling, burning and planting. Typically not until after planting will the household have an opportunity to build a dwelling in the new swidden, allowing them to sleep there instead of the longhouse (see chapter 11). However, if the household makes its new swidden close to its past one, and if it previously built a house in this past swidden, then the household members may sleep there instead of the longhouse during the selecting, slashing, felling, burning and planting of the new, nearby swidden. The household thereby can achieve significant savings of travel time that would otherwise be expended during these first five phases of the swidden cycle. When preparing a secondary forest swidden of average size, the savings in time can equal as much as twenty-one workdays[4]. There will be additional savings if the household continues to use the old house through the remainder of the new swidden cycle (a decision that depends on the precise distance from the old house to the new swidden). This saves the household an additional eighteen workdays (viz., the amount of time devoted to the construction of a typical swidden house, less the time needed to refurbish an old house).

The advantages of making proximate, successive years' swiddens is reflected in actual practice. In 1976 twelve of thirty-five longhouse swiddens were located adjacent to 1975 swiddens belonging to the same household (figure 10). In addition, at least six of the remaining twenty-three swiddens were made sufficiently close to the households' previous year's swiddens so that some of the advantages detailed above also obtained. As a result of this adjacency or proximity of successive years' swiddens, the Kantu' were able to reuse seven old swidden houses in 1975 and ten old swidden houses in 1976 (figure 11, 12)[5]. The advantages of locating successive years' swiddens close to one another also are reflected in the interhousehold exchange of forest sections. Assume that two different households each have just harvested a swidden in different parts of the longhouse territory, neither of which adjoins a section of forest belonging to the same household in which the following year's swidden can be made, and both of which adjoin (or are close to) a forest section owned by the other household. Under these circumstance, it is common

FIGURE 10
Adjacency Of Same Household's 1975 And 1976 Swiddens

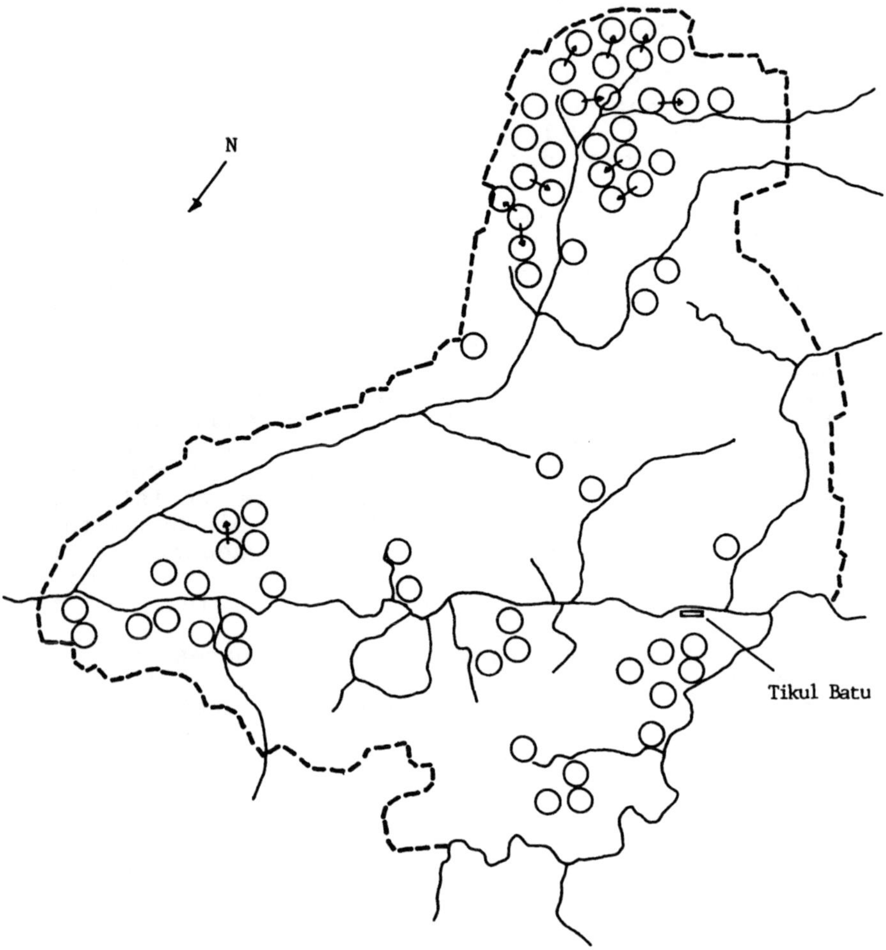

SCALE: 1:18,000 (swiddens not drawn to scale).
KEY: ○→○ signifies that the 1975 swidden at the base of the arrow adjoins the 1976 swidden - belonging to the same household - at the tip of the arrow. One case of adjacency is off the map.

FIGURE 11
The Use Of New Versus Old Swidden Houses In 1975

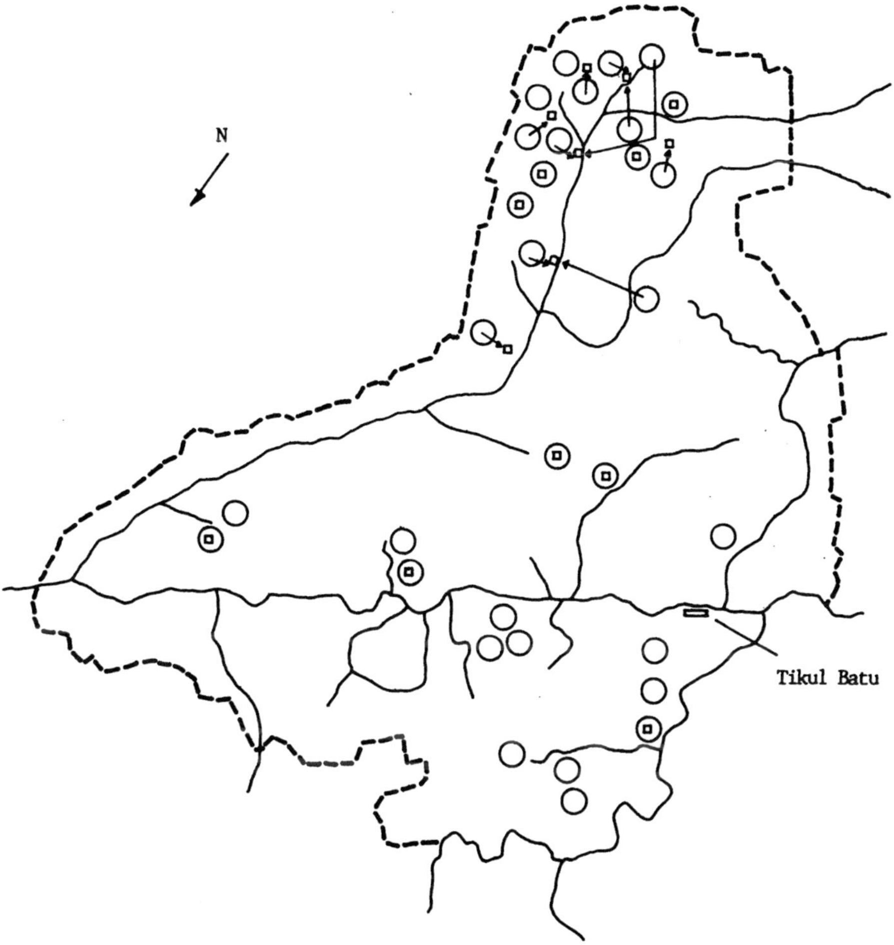

SCALE: 1:18,000 (swiddens not drawn to scale).
KEY: (◻) signifies that a new house was made in this 1975 swidden.
◯→◻ signifies that the old house at the tip of the arrow was used for the 1975 swidden at the base of the arrow.

FIGURE 12
The Use Of New Versus Old Swidden Houses In 1976

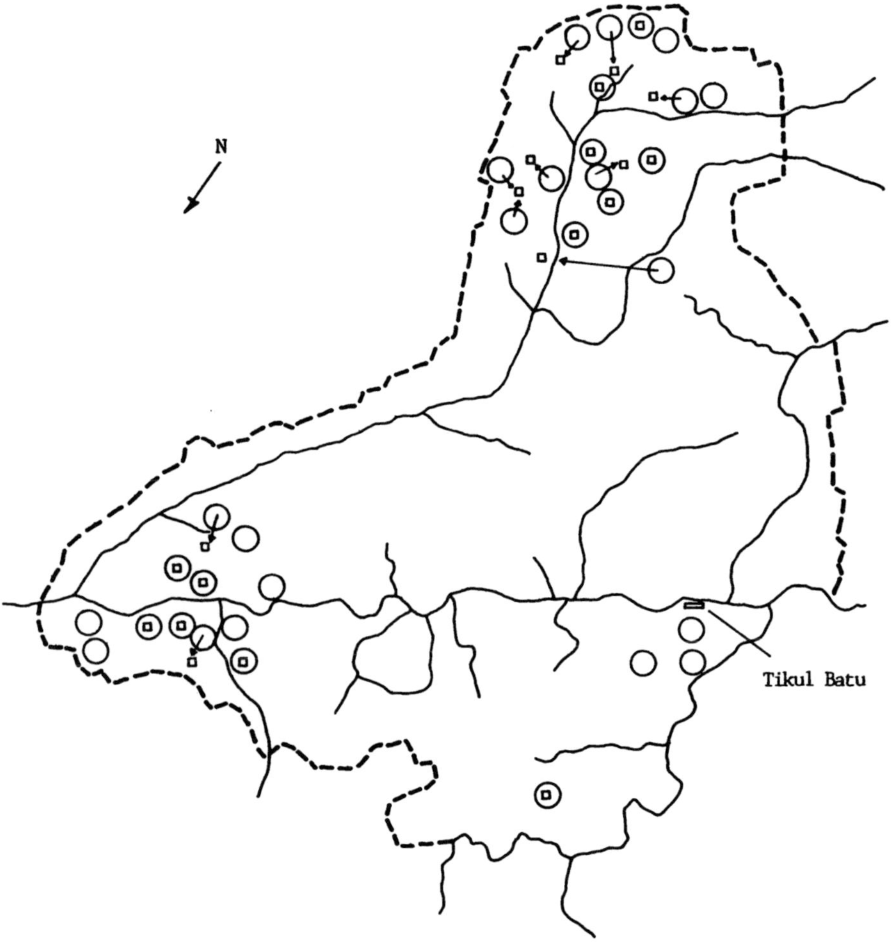

SCALE: 1:18,000 (swiddens not drawn to scale).
KEY: (◻) signifies that a new house was made in this 1976 swidden.
◯→◻ signifies that the old house at the tip of the arrow was used for the 1976 swidden at the base of the arrow. One case of reuse is off the map.

practice for the two households to temporarily besilih 'exchange' use-rights to the adjoining sections of forest, so that each can make the coming year's swidden adjacent to their past year's swidden. This exchange obviates the need, otherwise, to move to another part of the longhouse territory after just one year.

While each household tries to locate its new swiddens near its previous year's swiddens, it does not try to locate two or more of its new swiddens near one another. The proximity of current swiddens belonging to a single household is regulated by a formal prohibition that stipulates Umai nadai tau' dipeda' 'A swidden must not be visible [from another swidden belonging to the same household]'. In practice, the Kantu' interpret this proscription as prescribing the separation of any two swiddens belonging to a single household by at least three intervening sections of forest (or three swiddens) or two sections plus a river or stream. Some Kantu' maintain that this proscription is no longer in force. Whether this is true or not, it is still very rare for a household to make two or more swiddens in close proximity in the same year. Among the fourteen households that participated in the swidden cycle at Tikul Batu in 1975, and the sixteen that participated in 1976, there were twenty-six cases in which a single household made two or more swiddens in a single year. In only three of these cases were two swiddens located within the proscribed distance (viz., swiddens #47-48, 64-65 and 67-68).

This marked tendency for each household to widely separate its swiddens reflects the importance to the Kantu' of maximizing swidden diversity. To the extent that the swiddens of one household are located near one another, they are likely to share the same microenvironment (viz., the same pest populations, the same forest type, the same drainage pattern, and so on). If the success of the household's swiddens are dependent upon a single microenvironment and conditions in that environment happen to be good for swidden cultivation in a given year, then the household's economy will prosper exceedingly. On the other hand, if conditions in that microenvironment happen to be poor that year, then the household's economy will suffer exceedingly. The instability that attends reliance upon a single microenvironment is illustrated in the cases of those households that made but a single swidden in one year, namely household #9 in 1975, and households #8, 15 and 17 in 1976. The total harvests of these households (measured as a percentage of each household's annual requirements) ranged from the first, second and third worst (out of the thirty 1975 and 1976 harvests) to the second best. Thus, extreme high and low points in the economic fortunes of the household are associated with dependence upon a single microenvironment. Such extreme fluctuations are minimized by dependence upon two or more different microenvironments. This diversification is part of a broader household strategy of maximizing not the total economic returns to the household in a given year, but rather the likelihood of some returns every year. The separation of the household's swiddens is one of the critical components in this strategy[6].

This principle, wherein the swiddens of a single household are widely separated, does not apply to the swiddens of different households. In the latter case precisely the opposite principle prevails. Each household usually tries to make each of its swiddens close to the swiddens of one or more other household (figure 13). The contrary practice wherein each household locates its swiddens far from those of other households is derided as Nebok kiya', nebok kiya' 'To bore a hole [e.g., in the forest] here, to bore a hole there'. The practice of making swiddens belonging to different households proximate, as opposed to distant, does not mean that the just discussed principle of diversification is not important at the interhousehold level. On the contrary, just as it is important for the individual household to diversify the microenvironments upon which it depends, so is it important for the longhouse as a whole to do this. However, while the diversity of the Kantu' environment is sufficiently great to allow for diversity among the 2-3 swiddens of a single household, it is not sufficiently great to allow for total diversity among the 14-16 households of the longhouse. Diversity among households is based less on differences in the microenvironments farmed, than on differences in how and when they are farmed. Thus, the proximity of the swiddens of different households is not inimical to the principle of diversification at the longhouse level. Moreover, there are distinct advantages to this proximity.

One advantage of the proximity of swiddens belonging to different households is that this allows for social interaction among the members of these households when they are away from the longhouse and living in their swidden houses, in what otherwise would be comparatively lonely circumstances. Of more importance, the proximity of the swiddens of different households facilitates the participation of these households in reciprocal work groups by minimizing the amount of travel time involved. Table 13 presents data from the two swidden

FIGURE 13
Adjacency Of Different Households' Swiddens

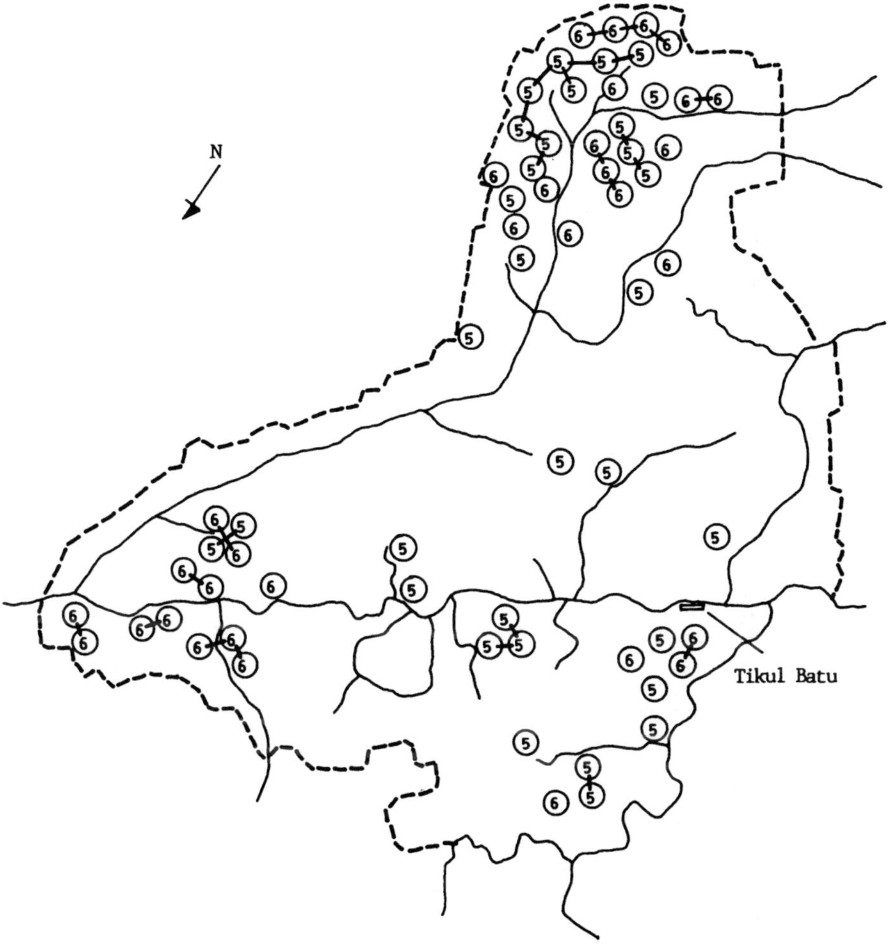

SCALE: 1:18,000 (swiddens not drawn to scale).
KEY: ⑤ signifies a 1975 swidden.
⑥ signifies a 1976 swidden.
○—○ signifies that the two swiddens were adjacent.

TABLE 13
Swidden Proximity And Participation In Reciprocal Work Groups

		At least one swidden of one household adjoined at least one swidden of the other household in 1975 or 1976	
		No	Yes
Two households participated in reciprocal work groups in 1975 or 1976	No	147	14
	Yes	40	10

N = 211 possible household pairs.
$X_c^2 = 3.78$
$P < .10$

years, 1975 and 1976. In each year, the proximity of the swiddens of two different households was measured by whether or not at least one swidden in one household adjoined at least one swidden in the other household. This is a conservative measure of proximity and its effects: if two households make swiddens that are close to although not adjoining one another, this also is likely to influence their participation in joint work groups. Participation in reciprocal work groups was measured each year by whether or not two households joined in reciprocal work groups during any one or more of four swidden phases, namely slashing, felling, weeding or harvesting. The use of reciprocal work groups during the other swidden phases is either so restricted (viz., during selecting, burning and guarding) or else so pervasive (viz., during planting and carrying) as to make variation in this use difficult to study. During the 1975 swidden year, with fourteen households making swiddens, there were a total of 91 possible relationships between two different households in the longhouse; and in 1976, with sixteen households involved, there were a total of 120 possible relationships, yielding a total of 211 over the two years[7]. According to the analysis of these 211 possible relationships in table 13, there is an association between participation in reciprocal work groups and the proximity of the swiddens of the households involved (although the value of X^2 is not quite significant at the [.05] level).

There is another ever more important advantage to the proximity of different households' swiddens, involving the problem of swidden predation. Many pest populations are relatively sedentary or territorial. To the extent that this is true (and to the extent that the pest population also is relatively finite in size), to that same extent the severity of predation will vary according to the number of swiddens simultaneously located in a given territory. The greater the number of swiddens in a given area, the fewer (cet.par.) the number of pests that will feed on any single swidden; and the reverse also holds true. Thus, simply by locating its swiddens near the swiddens of other households, the individual household can take a significant step towards reducing its pest problem. The Kantu' say that this is especially needed in the case of primary forest swiddens, where predation (from wild pigs and macaques) can be especially severe. Indeed, for this reason the Kantu' say Umai kampung nadai allah kediri' 'A primary forest swidden is impossible [to cultivate) in isolation'[8].

The strength of these several reasons for different households to make proximate swiddens is examined in table 14, by looking at actual practices in 1975 and 1976 (cf.fig. 13)[9]. This table shows that during both years the incidence of actual adjacency exceeded the incidence of adjacency expected by chance alone. The data also reflect the fact that the advantages of swidden proximity are greater in primary forest swiddens than in secondary forest swiddens (table 15). While the extent

TABLE 14
Swidden Adjacency By Chance Versus Design

	1975	1976
Possibility for any given swidden to be adjoined by at least one other swidden	43%	42%
Percentage of swiddens that actually were adjoined by at least one other swidden	53%	69%

of adjacency among secondary forest swiddens is either equal to or slightly greater than the extent expected by chance alone, the extent of adjacency among primary forest swiddens is 60-70 percent greater than expected by chance.

TABLE 15
Swidden Adjacency In Primary Versus Secondary Forest

	1975	1976
Percentage of primary forest swidden that were adjoined by at least one other swidden	69%	71%
Percentage of secondary forest swiddens that were adjoined by at least one other swidden	42%	55%

The Kantu' attach such importance to the proximity of primary forest swiddens that they will postpone an attempt to farm a section of primary forest, even if they have started to slash that section, if other households fail to join them in making swiddens in that area. The Kantu' commenced twenty primary forest swiddens in 1976 and 1977, farmed thirteen of them to completion, and aborted the remaining seven. In at least three of these latter seven instances, one of the reasons given for the abortion was the lack of other nearby swiddens (table 16)[10].

TABLE 16
Abortion Of Primary Forest Swiddens

Year	Household	Work Completed in Swidden at Time of Abortion	Stated Reason for Abortion
1976	# 12	Omen taking and slashing.	Death in Household in May, 1976.
	13	Omen taking and part of slashing.	No other swiddens nearby.
	14	Omen taking and part of slashing.	Illness in household.
1977	# 6	Omen taking and part of slashing.	Ill omen (dream of bird), and late start.
	8	Omen taking and part of slashing.	Lack of rice stores, and late start.
	13	Omen taking and part of slashing.	No other swiddens nearby, and early harvest festival.
	14	Omen taking and part of slashing.	No other swiddens nearby.

The importance of proximity to other households' swiddens may force not only the abortion, but also the relocation of planned swiddens. Thus, in 1976 household #13 aborted its plans to make a primary forest swidden in an area that was adjacent to the regrowth from its 1975 Kenua' swidden but near which no other household planned to make a swidden, and shifted to another site near to which household #4 also planned to make a primary forest swidden. (This latter swidden was the one eventually aborted by household #14 in 1976, much to the chagrin of household 13.) The need for proximity also may force the making of swiddens that otherwise were not planned. Thus, the Kantu' say that they sometimes farm a section of primary forest (to which they hold rights) before they desire to, because other households are proceeding with the farming of some of the surrounding sections. Failure to farm its own section at that time would cost the household all the benefits of swidden proximity in primary forest.

Character of Site

The third major consideration in the selection of a swidden site is the physical character of the site. Several different characteristics may be weighed in the decision to locate a swidden at a given site or not. One characteristic is the size and type of the local communities of swidden pests. It occasionally happens that the concentration of pests in a given area is so bad as to make it impossible for the Kantu' to successfully farm there. For example, the Kantu' maintain that they are never able to make swiddens near the mouth of the Selupai stream (figure 6), because the area is so heavily infested with sambar deer. A second characteristic of some importance in site selection is soil type. As discussed in the introduction, the Kantu' recognize a number of soil features that are inimical to the growing of rice. Whenever possible, the Kantu' avoid soils that possess any of these features when selecting their swidden sites. The most important site characteristics, however, relate to flooding and drainage, forest cover, and ritual benefits and constraints.

Flooding and Drainage Characteristics

In selecting a swidden site, some attention always is paid to the elevation of the land relative to the floodzone of the Empanang river. Whenever a single household makes two or more separate swiddens in the same year, there is a tendency - again part of the strategy of diversification - for them to be divided between the floodzone and the highlands or mountains. This is evident in table 17, which analyzes the thirteen households that made two or more swiddens apiece in 1976[11].

TABLE 17
Multiple Swiddens And Diversification Of Flooding Characteristics

		In 1976 the Household Made	
		2 swiddens	3 swiddens
The household made at least one swidden within the floodzone and at least one swidden above it in 1976	Expected by chance	12% of cases	24% of cases
	Actual incidence	43% of cases	100% of cases

This strategy is not blindly pursued year after year, however. As I previously noted, the Empanang river usually does not flood so severely in the Kantu' territory as to jeopardize rice growing in the floodzone. At intervals of some years, however, floods of this severity do occur. In 1976 the eldest male in household #7, Tikin, observed that a flood of these proportions had not occurred for several years. (I ascertained, minimally, that no such flood occurred during the swidden years 1976, 1975 or 1974.) Given this, Tikin believed that the likelihood of such a flood occurring in the coming year was particularly great, and so he determined to locate all of his 1977 swiddens on sites above the floodzone.

Each household also attempts to farm both dryland and swampland every year. In part this is achieved by the making, each year, of separate (predominantly) dry swiddens and separate (predominantly) swamp swiddens. Table 18 shows that among the twenty-six cases in 1975 and 1976 in which one household made two or more swiddens during the same year, there was a marked tendency to divide them between dryland and swampland .

In addition to mixing dryland and swampland among separate swiddens, the Kantu' also mix dryland and swampland within individual swiddens: thirty-five of the sixty-nine 1975 and 1976 swiddens contained both dryland and swampland (in varying proportions). The fewer the number

TABLE 18
Multiple Swiddens And Diversification Of Drainage Characteristics In 1975 And 1976

		In 1975 or 1976 the Household Made		
		2 swiddens	3 swiddens	4 swiddens
The household made at least one swidden on dryland and at least one swidden on swampland	Expected by chance	5% of cases	10% of cases	14% of cases
	Actual incidence	75% of cases	100% of cases	100% of cases

of swiddens made by a given household in one year, the greater the likelihood that the household will make each swidden on a mixture of dryland and swampland. The fewer the number of swiddens per household, the smaller the potential for interswidden diversity and the greater the need for intraswidden diversity.

There are some exceptions to this demonstrated desire on the part of the Kantu' to cultivate both dryland and swampland. For example, in 1974 the sole adult male in household #12 stayed away from the longhouse for months on a trading expedition, and in his absence the sole adult female in the household farmed swampland alone. All of the tasks involved in making a swampland swidden can be performed by a woman; and indeed often they all are performed by women by preference, even when there is a man in the household. On the other hand, some of the tasks involved in making a dryland swidden typically are performed by men alone, especially slashing and felling the forest, as well as some of the redressive burn operations. For this reason, and given the absence of its sole male worker, household #12 did not farm any dryland in 1974. The opposite case occurred in 1976 in household #17. This household made only one swidden that year, and this was made largely on dryland. It originally had planned to make, in addition, a swamp swidden; but this plan was aborted after the discovery early in the year of the pregnancy of the sole adult female in the household (she gave birth on 11 September 1975). Anticipating this birth and the constraints that it would impose (both before and after) upon the woman's contribution to swidden labor, the adult male of the household observed early in the year, _Nadai isi' ino' ti ngau ngapih utai d'umai paya_' 'There is no woman to look after things in a swamp swidden'. Given this temporary lack of female labor, and given the traditional prominence of such labor in the farming of swampland, especially as regards planting and transplanting, this household chose not to farm any swampland.

Character of Forest Cover

Each Kantu' household ideally makes at least one primary forest swidden and one secondary forest swidden each year, as part of the same diversifying strategy that causes them to mix swiddens on dryland and swampland, or in the floodzone and above it[13]. The actual tendency to mix the two forest types is shown in table 19, based on the same twenty-six cases analyzed in table 18[14].

TABLE 19
Multiple Swiddens And Diversification Of Forest Cover

		In 1975 or 1976 the Household Made		
		2 swiddens	3 swiddens	4 swiddens
The household made at least one swidden in primary forest and at least one swidden in secondary forest	Expected by chance	10% of cases	20% of cases	29% of cases
	Actual incidence	54% of cases	92% of cases	100% of cases

While the Kantu' ideal is to farm both primary and secondary forest in a given year, here too there are exceptions. Sometimes, circumstances cause a household to favor the clearing of primary forest at the expense of secondary forest. This may stem from anxiety on the part of a land-hungry household to establish rights - by the clearing of primary forest - to as much land as possible[15]. Other circumstances may cause a household to avoid clearing primary forest and clear secondary forest instead, as when it experiences an exceptionally ill omen. The Kantu' are reluctant to muka' 'open up' a section of primary forest - meaning to farm a section of natural forest for the first time - when this would be associated with some unfortuitous occurrence. During the ten years preceding my fieldwork, the entire longhouse twice aborted plans to farm primary forest, once upon the death of the senior male in household #1, and once upon the death of an adolescent in household #3. These two deaths had a special impact upon the longhouse because they were unexpected (viz., because they involved neither an infant nor an invalid elder) and, hence, at least superficially inexplicable. Following the death of a young child in household #12 in 1975, only the immediate household aborted a planned 1976 primary forest swidden. The abortion of a planned primary forest swidden in such circumstances is not always due entirely to the perceived inauspiciousness of the death. The Kantu' also note that in a year in which a death has occurred, and in which a major mortuary ceremony must be staged, there simply is little time to farm primary forest, given the extensive preparations as well as work prohibitions entailed by this ceremony.

Other decisions to avoid primary forest also involve the lack of ability to farm it. This may involve predicted problems with the burn. For example, when the Kantu' observe the _ngkabang_ 'candlenut' tree flowering at the beginning of the year, they infer from this that there will be a lengthy drought in which to burn their swiddens during the coming year. Thus, if they see that the trees are _not_ flowering, they may make only secondary forest swiddens, which do not need a lengthy drought in order to burn successfully. More important than a household's ability to burn primary forest, however, is its ability to clear it. This can be affected by a number of factors. If the household suffers some misfortune early in the year, requiring the presence of its members in the longhouse or otherwise disrupting its work schedule, any plan to farm primary forest is likely to be aborted. Previously discussed examples of this included the failed 1976 harvest of household #8, the death in household #12 in 1975, and the illness in household #14 also in 1975.

Special difficulties are created by the illness of any adult male. The felling of primary forest is the sole responsibility of the adult males in each household's labor force. Given an average of just one or two adult males per household, the illness or injury of one seriously affects the household's ability to clear primary forest. Proof of this can be seen in the farming history of household #8. Over the years, this household has established rights to a comparatively limited number of forest sections. Among the nine households that originally founded Tikul Batu (viz., households #1-9), the number of sections to which rights are held averaged thirty-one by the end of the 1976 swidden year, but the figure for household #8 was only nineteen. The members of household #8 attribute this to an illness or series of illnesses that plagued the founder of the household (Temunau) throughout his adulthood, compelling him to farm secondary forest as opposed to primary forest. The arduous and potentially dangerous task of clearing primary forest can be done only by someone in excellent physical condition, whereas it is possible for secondary forest to be cleared by an ill or injured male (or female).

If the household's work force for felling is of minimal size, meaning just one adult male, the household is in a weak position to clear primary forest even if this one is healthy. Because there are other tasks besides felling that also are the responsibility of adult males, it is difficult for a household with a single male to farm primary forest year after year. Thus Gayan - senior male of household #3 - claims that he kalah 'lost out' to Suat - senior male of household #2 - in the race to establish rights to forest (they respectively hold rights to 24 sections and 38 sections), because Suat begat sons 10-15 years before Gayan did, and these sons heightened Suat's ability to clear primary forest 10-15 years before Gayan's ability was similarly heightened.

This discussion has focused on the avoidance or abortion of swidden cultivation in primary forest. Secondary forest is far less subject to such pressures. Whereas seven primary forest swiddens were aborted in 1976 and 1977, there was only one instance during the two years in which a secondary forest swidden was aborted. (This instance, involving household #3, is discussed in a subsequent section on omen taking.) There are several reasons for this distinction. First, secondary forest is easier to clear following misfortune (e.g., poor harvest, illness or death). Second, it is better to clear following misfortune, because it does not carry the symbolic burden of being cut for the first time, as does primary forest. Third, the abortion of secondary forest carries a much higher cost than the abortion of primary forest, because of a faster rate of regrowth. Primary forest that is slashed one year can be farmed the following year after light slashing; but secondary forest that is slashed one year will be covered in impenetrable underbrush by the following year, and thus it cannot be farmed.

The question of whether to farm secondary forest often is less important, in the context of selecting forest for swidden sites, than the question of when to farm it. After a household has cleared and farmed the primary or secondary forest on a given stretch of land, they do not return there for further clearing and farming until the regrowth has attained a certain character. Attainment of the appropriate character is ultimately a function of the amount of time that has passed since the last farming of the land (viz., the fallow period). However, because the rate of tree growth varies within the longhouse territory according to variation in such factors as soil type and drainage, the requisite fallow period also varies from one area to another. For this reason, the Kantu' determine when regrowth can be farmed based not upon its age, but rather upon its character.

The character of regrowth comprises several distinct features, one of which is the type of vegetation. If most of the land is covered not with trees but with stands of dense brush, the Kantu' say that the land should not be farmed. Only when the vegetation has reached the stage known as jerong can the land be farmed. At this stage the brush has been succeeded by widely spaced trees, beneath which the forest floor is relatively clear. Once trees predominate, their size becomes a relevant feature. The Kantu' say that they should not farm forest until its trees have attained a girth equal to that of one's pala' 'head'. If the largest trees are equal in size only to one's lengan 'forearm' or petih 'calf', then the forest should not be farmed. (In practice, this rule is not infrequently violated, however.) In addition to measuring individual tree trunks, there is a second way in which the Kantu' determine appropriate tree size. This is the determination as to whether, once the forest has been slashed and felled, it will cover the ground beneath it with a rapit 'dense' layer of tree trunks and limbs. The density of this layer is directly related to how well the swidden will burn later and how much ash will be provided for the rice plants. The critical density is thus some (as yet unknown) ratio of the surface area of the timber to the surface area of the swidden (viz., the area of the land upon which the timber originally stood).

In all of these cases, the undesirable features or aspects usually are found only in very young forest. As the forest matures, it acquires desirable features. It is not possible for forest to be too old to farm, but it is possible for it to be too young. In the latter case, its farming is attended by several difficulties. First, it is difficult to clear forest that is young and thick with underbrush. Second, after being cleared, very young forest is initially and rapidly succeeded by herbaceous growth that will present stiff competition to any cultigens planted there. Third, and of greatest importance to the Kantu', if they farm forest having any of these features, the resultant rice crop will be matak 'unripe'. As the Kantu' describe it, Alai padi nyau bauh tang agi' tengah umur, ka' kuning 'When the rice is now tall but still half its age, it will yellow'. This is the most commonly reported symptom of crop failure. Its occurrence indicates that the nutrients provided by burning the forest were insufficient to sustain the rice crop to maturity[16].

This analysis of the type of regrowth that can be farmed, and the type that cannot yet be farmed, applies to the general case. In particular cases there is some variation (viz., there is variation in the minimum fallow period), first, according to whether the regrowth in question is succeeding primary or secondary forest. The Kantu' say that the fallow period can be shorter in the former than in the latter, for several different reasons. The felling of primary forest produces huge tree trunks that will neither burn nor decompose in time for their nutrients to be used by the swidden crops. However, after five years (e.g.), the decomposition of these trunks will have progressed and some of their nutrients will be available to any swidden (and its crops) that is cut from the regrowth on that land. This store of nutrients thus makes a shorter fallow possible. A shorter fallow also is facilitated by the greater speed with which primary as opposed to secondary forest swiddens are succeeded by ligneous as opposed to herbaceous growth. This is due to the initially greater scarcity of herbaceous growth in primary forest.

The requisite fallow period also varies according to the distinction between land within versus without the riverine floodzone. The Kantu' say that if in the same year they farm two forest sections, one in the floodzone and one above it, the latter will be the first to be recleared and farmed again - although reforestation will be quicker in the former section. The Kantu' state that the trees in a floodzone section must be much bigger before they can be cleared successfully for a swidden than is necessary in a section above the floodzone. In the latter case the forest can be farmed after the trees have attained the size of one's forearm or calf, they say, whereas in the former case farming must be postponed until the trees are as large as one's head. The Kantu' cite several reasons for this distinction, all bearing upon the need (or absence of need) to protect the growing rice plants from damage by floodwaters. Whenever a floodzone swidden is inundated by the river, there is a danger that some of the ash and earth at the base of the rice plants will be swept away, exposing the plants' roots (after the subsidence of the floodwaters) to the rays of the sun. Such exposure is injurious to the rice plant[17]. It is less likely to occur if the trees that are cleared for the floodzone swidden are of good size. Large tree trunks, lying in the swidden, break the flow of the floodwaters through the swidden, shelter the rice plants, and reduce the amount of ash and earth that is swept from their roots. Not only can small tree trunks not perform a similar function, but they will be swept to the edges of the swidden during flooding, which in itself poses a threat to the standing rice plants. There is a second way in which the farming of larger trees protects the rice crop from the effects of flooding. The Kantu' say that if they farm large trees as opposed to small trees the rice plants will grow taller, because more nutrients are made available to the plants from the clearing and burning of the former than the latter. The taller the plant, the smaller the likelihood that all or most of it will be submerged in any given flood and, hence, the smaller its chances of being killed. A final advantage to farming larger trees in the floodzone is that this ensures the presence of a normal amount of litter or humus. Some Kantu' maintain that while the reforestation of erstwhile swidden land is rapid within the floodzone, the restoration of its layer of humus is not, because some of this is swept away whenever the land is flooded. The loss of humus in floodzone forest and the effect of this on rice grown there is reflected in the Kantu' comment that, if a flood zone swidden is cleared from small trees (which might successfully be farmed above the floodzone), the rice crop will not ripen to maturity. This, again, is symptomatic of an insufficiency of nutrients for the crop.

Calculation of the requisite fallow period also varies according to whether the swidden is located in swampland as opposed to dryland. The Kantu' say that the norm is to allow swampland to lie fallow for just two or three years between croppings. This norm is reflected in the fact that seventeen of the twenty-six swampland swiddens made in 1975 and 1976 were made after such brief fallow periods that the vegetative cover was still predominantly herbaceous (viz., swamp grasses) rather than ligneous. Even briefer fallow periods occasionally are attained, as when swampland is cropped for two years in a row. The first year's cropping ends with the harvest in February and March. The swampland then is allowed to succeed to grassland. Six to seven months later, the grasses are cleared and the swampland is planted in a second rice crop. The fallow period between these two crops is, therefore, only six to seven months. The cropping of swampland for two successive years is not uncommon. Among the twelve predominantly swampland swiddens made at Tikul Batu in 1976, at least three were made on land that also had been cropped in 1975. This practice is not associated with any dramatic drop off in yields during the second cropping. The Kantu' say that yields from the second cropping generally are acceptable, providing only that they plant the second crop wholly using seedlings and not seed, hence minimizing nutrient demands.

Dryland swiddens, in contrast, never are cropped for two successive years[18]. When making a new swidden, the Kantu' occasionally overlap a small portion of the previous year's swidden, calling this betangkir ncipit 'adjoining a little'. The overlapped area thus is cropped during two successive years. The Kantu' do not make a practice of this, however, because they have observed that the rice planted in the overlapped area does not grow well. Nor do they ever attempt this over an entire swidden. The Kantu' do have a term for this latter practice, however: ngakah. Kakah, the nominative term, refers to last year's swidden; and hence ngakah (or also, bumai kakah 'farm kakah') refers to the farming of last year's swidden, as in the case of swampland that is cropped for two successive years. However, the term ngakah implies the cropping of dryland and is not applicable to the swampland case. The Kantu' refer to the farming of swampland two years in a row simply as Bumai dua taun terit 'Farming two years straight'. This latter expression also can be used in reference to the farming of dryland two years in a row, but ngakah is the preferred usage (although ngulang 'repeat' is heard as well).

The Kantu' say that some of the groups living along the Kapuas river practice second-year cropping. This, improbably, is the source of the Kantu' knowledge of this practice. Freeman observed the cropping of dryland (both primary and secondary forest) for two and even three successive years among Sarawak Iban[19]; but the Kantu', who are surrounded by Iban migrants from Sarawak, claim to have no knowledge of the Iban doing this. When told of it, the Kantu' in fact refused to believe it, maintaining that the reports probably referred to Iban farming of not dryland, but swampland.

Although the Kantu' apparently have little knowledge of second-year cropping being practiced elsewhere, and although they do not currently practice it themselves, there is some indication that they formerly did. Kantu' elders say that formerly, if a given household encountered an important omen and then made the requisite offerings (called pudas) when clearing primary forest, it acquired the right to crop that forest a second time, during the succeeding year. (There is some ambiguity on this point, an alternative interpretation being that the household in question thereby acquired the right to crop that land a second time during any succeeding year.) In the absence of the omen and the offerings, the household in question did not have a greater right than any other household to farm that land during the succeeding year or any other year. At that stage in Kantu' history, the principle wherein the clearing of primary forest established rights to the regrowth on the land had not yet been developed.

The recognition of limited rights on the basis of a ritual offering probably represented a transitional stage between the absence of any household rights at all to secondary forest and the system of rights, based on the clearing of primary forest, that exists today. Part of the impetus for this development of land rights came from the belief that the household that cleared the primary forest should be awarded the relatively easier task of clearing the resulting regrowth; but this development clearly also was stimulated by the Kantu' contact with the Iban and their already developed system of land rights. It therefore is significant that the initial development of household rights to land involved the right to crop a given piece of land two years in a row, given the reported Iban predilection for just this practice. It may be, indeed, that the Kantu' adopted the practice of cropping two years in a row because if they did not crop during the succeeding year, they feared that the Iban (if not other Kantu') would, thereby sponging off the labor that they had devoted to clearing the primary forest. This fear was no longer necessary once the Kantu' had further developed their own system of land rights and extended the use-rights of the household that cleared primary forest from the first succeeding year to all succeeding years. This development may explain why the Kantu', if they once did practice second-year cropping, no longer do so[20].

This discussion of cropping in successive years has focused on the cropping of rice. The fact that nonrice cultigens also are planted with the rice, and may continue to produce into the first or even second succeeding year, was noted earlier. However, there is an additional, distinct pattern of second-year cropping of nonrice cultigens, in particular of cassava and sweet potatoes. In this case, the Kantu' plant the tubers not while the rice crop is still in the ground, as is normal, but after the rice has been harvested. First they slash, dry and burn the rice stubble and weeds, then they plant their tubers. This clearly represents a second cropping of the swidden. They do not call this _ngakah_, however, as in the second-year cropping of a swidden in rice. Rather they call it _nempalai_, after _tempalai_, which is the name (not umai) for a tuber patch or garden. The success of these second-year tuber gardens is made possible by the fact that the tubers make relatively limited demands on the earth/ash layer, which will have been drained of many of its nutrients by the preceding rice crop. The nutrient demands of the tubers aside, the Kantu' insure that the maximum possible amount of nutrients will be available by making such gardens only in newly fallowed swiddens that originally were cut from primary forest or older secondary forest. In any case, these tuber gardens are not a major enterprise for the Kantu'. When they make such a garden in a just harvested swidden, it typically covers just 1 or 2 percent of the former swidden's area.

Ritual Characteristics

A final characteristic of the prospective swidden site that must be examined before preparations to clear and farm it can get underway is the attitude taken by the antu 'spirits' towards this undertaking. The spirits express their attitude primarily through the medium of seven, distinct species of forest birds, known to the Kantu' as burong antu 'spirit birds'. They are (1) Ketupong 'Rufous Piculet', (2) Nenak 'White-rumped Shama', (3) Beragai 'Scarlet-rumped Trogon', (4) Papau 'Diard's Trogon', (5) Memuas 'Banded Kingfisher', (6) Bejampong 'Crested Jay', and (7) Kutok 'Maroon Woodpecker'. The Kantu' believe that these birds are the sons-in-law of Singalang Burong, one of their principal deities . They further believe that they can divine the will of this and other deities through interpretation of the flight and song of the omen birds. This process is called beburong 'to use [or to go with] the birds'. Such divination must be carried out before the Kantu' can begin to clear forest that has been selected for a swidden site. This is especially true if the swidden being made is a 'stem swidden', in which the ritually preeminent 'stem rice' is planted. The Kantu' are apt to perform this initial divination less thoroughly when making a swidden in which no stem rice is planted. They are not likely to perform it at all when making swamp swiddens. The Kantu' justify the absence of omen taking in these latter cases by the absence of dry rice. Dry rice is always the focus of ritual activity in the swiddens.

Divination at an intended swidden site begins with an activity called ningah amo' 'listening to the claimed [site]'. (At this stage of the swidden cycle, the area that is to be farmed cannot yet be called umai 'swidden', but only amo' 'claim'.) This activity is the responsibility of the eldest male in each household, and he usually carries it out alone. Ideally, it is carried out early in the morning on a rainless day. If the sun is high in the sky, or if it is raining, omen birds are loathe to sing or fly about. This 'listening' consists in traversing the intended site and selecting a spot at which one is likely to hear birds. At that spot, the augurer then makes a pangal, a rough seat made of sticks laid parallel on the ground. The sticks must be dead wood gathered from the forest floor, because the Kantu' cannot cut living wood at the swidden site until after securing the preliminary omens. The augurer then sits on his seat and attempts to call to the omen birds regarded as most auspicious. If the augurer wishes to hear the Rufous Piculet, he coughs aloud and/or snaps his fingers against the wooden scabbard of his brush sword. If he wishes to hear the White-rumped Shama, he whistles. The augurer repeats these calls before and after sitting on the pangal, and in other parts of the swidden as well. If the omen birds reply to the calls, which imitate their own, their reply is received as an omen regarding the acceptability or nonacceptability of that site for a swidden. In the event that the omen is good, at the instant it is received the augurer uproots, by hand, a varying number of seedlings from the forest floor. The uprooted seedlings are called kayu burong 'omen sticks'. They are believed to embody the omen that occurred at the time of their uprooting. After being uprooted, the omen sticks are tied in a bundle at the base of one of the largest trees in that forest section. The augurer then slashes the forest a little (which he is now free to do), and replaces the dead wood of the pangal with newly cut wood. Then he leaves the swidden as quickly as possible for fear of hearing an ill omen, which would nullify his morning's efforts.

The first day's activities are collectively known as mangol 'commence'. The site henceforth may be called pangol 'commencement', at least until the clearing is well under way, at which time the site can be called tebas 'slash' or even umai 'swidden'. For the remainder of this first day, and for the two following days, entry into the swidden site is proscribed. On the third following day, the augurer returns to the site and slashes a little in the vicinity of the pangal 'seat'. This completes the sequence of events involved in securing the primary auspicious omen for the swidden, which is called the pun burong 'stem omen'. This may be followed by a second, similar sequence of events directed towards securing a secondary omen, called the pun nyuror 'stem that gives off roots'. Not until one or both omens have been secured can the household begin to slash the forest in earnest. There is no further seeking out of omens during slashing nor during any of the other succeeding swidden phases[21]. However, if the Kantu' encounter omens while working in or traveling to their swiddens during these later phases, they usually are required to cease work for one day[22].

The period prior to clearing, in which the initial omen taking is performed, may be longer or shorter according to whether the household is more or less thorough in its performance of this task, and also according to how quickly they are able to find the desired omen birds. During omen taking for the 1976 and 1977 swiddens, the length of this period ranged from as little as one day to as long as several weeks. The average was one week (during which a total of perhaps two work days were expended on omen taking). The time allowed for omen taking ultimately is limited by the chronology of the slashing phase, which cannot begin until after omen taking has been completed. During 1976 and 1977, slashing commenced, at one extreme, nearly a month after omen taking was completed and, at another extreme (when the time pressure on slashing was greater), on the same day that omen taking was completed. The Kantu' never entirely forego omen taking, even if hard pressed for time, because they view this as an integral phase of the swidden cycle. They also view it as a critical phase of the cycle. The Kantu' commonly ascribe crop failures or successes to, respectively, the burong jai' 'bad omens' or burong badas 'good omens' obtained at the start of the cycle. More generally, the Kantu' say Asa kami boleih burong badas, urang ti betangkir ngau kami pedih ngalah kami 'If we obtain a good omen, those who [whose farms] adjoin us will with difficulty outdo us [re. the relative success of their rice crops]'.

What is or is not considered a good omen varies somewhat from one person to another. The Kantu' say Beburong ngau ti rasi, ngau bidik kitai 'Omen taking is a matter of good fortune, dependent upon our luck'. For the most part, however, there is agreement among the Kantu' in interpreting the behavior of the omen birds. Thus, of the seven omen birds, only three - the Rufous Piculet, White-rumped Shama and Scarlet-rumped Trogon - are acceptable for either the primary or secondary omen in a swidden[23]. The Rufous Piculet provides, in general, the most desirable omens and is acceptable under any circumstances. The acceptability or desirablity of the other two varies according to their individual character and the particular needs of the swidden in question. The Scarlet-rumped Trogon, for example, is perceived as a burong angat 'hot bird', which often forecasts unwelcome, symbolically 'hot' phenomena. For example, it is associated with swidden pests, and so if either the primary or secondary omen is a Scarlet-rumped Trogon, a serious pest problem is expected in the swidden. On the other hand, the 'heat' of this bird may take on positive connotations in the floodzone, which is perceived as celap 'cool' and is associated with problems of flooding. The Kantu' actually may seek the Scarlet-rumped Trogon as the primary or secondary omen in a floodzone swidden, in the belief that it will make the swidden 'hot' and so reduce the threat of flooding there. The White-rumped Shama, in contrast, is perceived as a burong celap 'cool bird'. The qualities of this omen are most appreciated in highland swiddens, where a major problem is the heat and desiccation of the land. An omen from the White-rumped Shama is greatly desired in such swiddens, because it is believed to cool off the land and reduce the threat of drought as well as other pemedih 'troubles', which usually are thought of as 'hot'. In contrast, this omen is less desired in the floodzone where, the Kantu' say, it may make the swiddens terlalu celap 'too cool', exacerbating the flood problem.

The auspiciousness of an omen for a given swidden site varies not only according to the type of omen bird involved, but also according to the manner in which it is observed. There is a crucial distinction between the observer's right and left. If the observer hears or sees a bird on the left, it signifies that the land will be lemah 'soft', whereas a bird on the right signifies that the land will be rangkai 'hard' or 'dry'. For the reasons just discussed, the Kantu' most desire right-hand omens when intending to make a swidden in the floodzone and left-hand omens when intending to make a swidden above it. The auspiciousness of an omen from a given bird also may vary according to a number of additional factors, including the type of its call and the number of calls.

If the Kantu' do not encounter an omen that is auspicious for the farming of a given section of forest, they may proceed to clear it anyway. This is called beburong besi 'taking an omen from the iron' (the 'iron' referring to the commencement of slashing with the brush sword). If, however, they encounter an omen that is decidedly inauspicious, they may balang 'abort' their plans to farm that forest that year. Such an action would follow omens from either the Diard's Trogon, Banded Kingfisher, Crested Jay, or Maroon Woodpecker. The Rufous

Piculet and Scarlet-rumped Trogon, whose normal calls are positive, have variant calls that have very negative interpretations and also can lead to the abortion of a planned swidden. Equally ill omens also can be received from a variety of other fauna.

The Kantu' say that they must reject a forest section if one such ill omen is encountered during the first three days of omen taking. This rule may be honored in its breaching. For example, when Guyak of household #1 was seeking omens for his 1977 swidden, he heard a Diard's Trogon. At that time he had made his seat of sticks at the intended site and was in the midst of augury. Because of the Trogon, he was obliged to move and to begin again. However, he merely moved to another part of the same forest section. That is, he rejected not the whole section, but only the small area in the vicinity of the augural seat (intending to leave it out, apparently, of the planned swidden). This is not always (or even typically) the case, however. During the same year, when Gayan of household #3 was in his first day of omen taking at an intended swidden site on the banks of the Empanang, he too heard a Diard's Trogon. Gayan then rejected that site and moved elsewhere, upon the advice of two Kantu' elders whom he consulted on the matter. If this Trogon or other ill omen were encountered after the first three days of omen taking, however, neither the entire site nor any part of it would be rejected. Instead, the Kantu' would simply observe it by leaving the site and returning to the longhouse for the rest of the day. The exception to this is an especially ill omen, such as the death of someone in the longhouse, which might lead the Kantu' to abort planned swiddens (especially in primary forest) even after they have begun slashing, as noted earlier.

The role of the ill omen in the rejection of a swidden site is an extremely interesting and problematic one. Freeman has written the best analysis to date of the augural system of an Ibanic people[24]. He maintained that the function of the augural system is psychological: it mitigates anxiety. Indeed, the Kantu' frequently seem to use omens either to ratify a prior inclination by heeding a present omen, or to explain a present misfortune by remembering a prior omen that was not heeded. This suggestion that omens have a projective value or function also is supported, perhaps, by the fact that at least five of the seven omen birds are noted for the arresting and unusual character of their calls. Thus, the Rufous Piculet and Maroon Woodpecker both make the distinct noises characteristic of woodpeckers, while the Crested Jay and Banded Kingfisher are noted for the abundance and volume of their calls and the White-rumped Shama for the melody of its call[25]. In certain other respects, however, the data do not appear to support Freeman's interpretation. Of most importance, it often appears that omens exert an independent and autonomous influence on Kantu' behavior, a conclusion that also is supported by my reading of the literature[26]. Based on my belief in the autonomy of at least some omens, I suggest that they play a functional role within the swidden ecology and economy of the Kantu'. In the case of rejection of a planned swidden site, one of the most dramatic examples of the apparent autonomy of omens, my hypothesis (as yet untested) is that the omens actually encode critical information pertaining to the unacceptability of the site, information that otherwise is not accessible to the Kantu'[27].

SWIDDEN SIZE

A second important variable, which also is determined during site selection, is the size of the intended swidden. I construe swidden size (except where otherwise noted) as the areal measure of the land upon which forest is cleared for a swidden. This measure is most relevant to my analysis of agricultural strategies, since it reflects the swidden size that is desired or intended by the farmers. This intent is not reflected as clearly in other measures of swidden size, such as that fraction of the cleared land that burns when the swidden is fired and is planted in rice, or that fraction upon which rice survives to be harvested[28].

During 1975 and 1976, the swiddens at Tikul Batu ranged in size from 0.03 hectare to 7.95 hectares. A number of different factors account for this variation. One such factor involves an historic change in the capacity or desire of the Kantu' to make larger swiddens. When the Kantu' today clear primary forest, in which they are relatively free to make their swiddens as large or as small as they want, they make larger swiddens than they did when clearing primary forest earlier this century. The Kantu' themselves are cognizant of this distinction. They often speak of how small their primary forest swiddens used to be compared to the ones made today.

The magnitude of this historic change can be examined by comparing the size of contemporary primary forest swiddens to the size of contemporary secondary forest swiddens, the latter of which reflects the size of the primary forest swiddens originally cleared on that land. The boundaries of each belah 'forest section' are established when a given area of forest is first farmed as a single, separate, primary forest swidden. Successive farms, cut from successive stands of secondary forest on that land, typically follow the boundaries and hence dimensions of the initial, primary forest swidden. The measurements of contemporary swiddens produced the following figures: the mean size of 21½ 1975 and 1976 secondary forest swiddens was 2.3 square hectares, and the mean size of 21½ 1975 and 1976 primary forest swiddens was 3.4 square hectares[29]. According to these figures, there has been an historic increase in the size of swiddens on the order of almost 50 percent. This increase has occurred since 1888, when the Kantu' began their current occupation of the Empanang valley and began to make the primary forest swiddens that set the parameters for today's secondary forest swiddens.

There are several possible explanations for this increase in swidden size over the last eighty years. First, the gradual diminution of chronic warfare has enabled increasing amounts of labor to be devoted to farming[30]. The efflorescence of smallholder rubber cultivation, beginning in the 1930s, has had the same effect in that it has freed the Kantu' from the time-consuming trading and gutta percha expeditions that previously were necessary for the acquisition of consumer goods. Finally, it is possible that the current, near exhaustion of primary forest is stimulating households to clear larger primary forest swiddens - so as to establish rights to larger areas - than they would have done formerly.

In addition to the historic shift towards larger swiddens, there are other factors that determine the sizes of contemporary swiddens, involving the efficiencies of scale. As swidden size increases there are several different, corresponding increases in the efficiency of cultivation, the most important of which have to do with the inconstant association between swidden size and swidden circumference. As swidden area increases, there is a continual decrease in the ratio of circumference to area (viz., this ratio decreases by a constant 29 percent whenever the area is doubled). The magnitude of the ratio of circumference to area becomes an important determinant of farming success at several stages in the swidden cycle, as will be seen in later chapters. This pertains to the shading of the swidden's edge by adjoining forest and resultant problems with both the burn and the growth of the rice plants. It also involves the vulnerability of the swidden's edge to predation and the relative cost of fencing. The

Kantu' are cognizant of these benefits of scale, as is reflected in the tendency for households that clear more land to make bigger swiddens (table 20).

TABLE 20
Number Of Hectares Farmed Per Household And Individual Swidden Size

		Total hectares cleared per household	
		< 4.6 ha	> 4.6 ha
Size of each swidden	< Mean	24	19½
	> Mean	4	21½

N = 69 1975 and 1976 swiddens (mean size calculated separately for primary forest, secondary forest, and swamp swiddens).
4.6 = mean total number of hectares cleared by each household.
X^2 = 10.4
$P < .01$

As I noted during the discussion of land ownership, the Kantu' will not farm a small copse of primary forest until such time as they can include it within a larger swidden. Accordingly, the sizes of these unfarmable copses give some insight into the critical lower limits of swidden size. During 1975 and 1976, the Kantu' made seven predominantly secondary forest swiddens that contained primary forest copses. The copses in these swiddens averaged 0.7 hectare in area (while the swiddens averaged 4.0 hectares overall in area). In contrast, the seven smallest swiddens made during 1975 and 1976 (excluding predominantly swamp swiddens) averaged 1.1 hectares in area. The critical lower limit to swidden size thus can be estimated to equal one hectare. Among these seven smallest swiddens, only one (swidden #7) measured less than one hectare in area; and it is noteworthy that the owners said that they made this swidden not so much for the rice crop (although they did plant rice there) as for a source of nonrice relishes close to the longhouse. Among the seven primary forest copses measured above, only one measured over one hectare in area. This copse (in swidden #41) measured 1.1 hectares. A section of primary forest of approximately this size may be farmed by itself or it may be left until it can be included in another swidden. Thus, household #1 made a primary forest swidden (swidden #35) that covered only 1.0 hectare, while household #8 said that it would not farm a copse of 0.9 hectare (produced by its failure to finish felling swidden #52) until it could incorporate it into a larger swidden.

The critical minimum swidden size for primary forest is not the same as for secondary forest, nor are the limits on dryland the same as the limits on swampland. The limits vary according to variation in productivity. The more productive the swidden type, the smaller the swidden can be before severe inefficiencies of scale are incurred. This inverse correlation between average swidden productivity and average swidden size is seen clearly in table 21.

TABLE 21
Swidden Type, Size And Productivity

Swidden Type	Mean Size	Mean Productivity
Primary forest	3.4 ha	457 kg
Secondary forest	2.5 ha	752 kg
Swampland	0.4 ha	1322 kg

NOTE: These figures are based on the 69 1975 and 1976 swiddens. The yield figures refer to threshed, winnowed, and unhusked rice.

The total amount of land cleared per household per year (ranging from 1.1 to 10.4 hectares) is determined by several different factors. Perhaps the most important determinant is the consumption requirements of the household. The average household requires 2,820 kilograms of threshed, unhusked rice for consumption each year. As the requirements of individual households vary about this average, there is corresponding and consequential variation in the amount of land cleared (table 22).

A second important determinant of the amount of land cleared is the varying ability of the household to clear. This ability depends upon the availability of both land and labor, and of the two the latter appears to be most important among the Kantu'. There is a strong association between the size of the household work force and the amount of land cleared per year (table 23). These data suggest that labor is a scare good or lim-

TABLE 22
Household Consumption Requirements And Number Of Hectares Farmed

		Household consumption requirements	
		< 2820 kg	> 2820 kg
Number of hectares farmed per household	< 4.6 ha	13	3
	> 4.6 ha	2	12

N = 14 households that farmed in 1975 and 16 that farmed in 1976.
2820 kg = mean yearly rice requirement per household.
4.6 = mean number of hectares cleared per year per household.
$X_c^2 = 10.8$
$P < .001$

TABLE 23
Size Of Household Work Force And Number Of Hectares Farmed

		Number of workers per household	
		< 4.4	> 4.4
Number of hectares farmed per household	< 4.6 ha	12	3
	> 4.6 ha	3	12

N = 14 households that farmed in 1975 and 16 that farmed in 1976.
4.4 = mean number of workers per household capable of participating in wage labor or reciprocal labor in one or more stages of the swidden cycle.
4.6 = mean number of hectares cleared per household.
$X_c^2 = 8.5$
$P < .005$

iting factor in the Kantu' swidden system. This finding is in accord with most other studies of swidden agriculture, which emphasize that labor and not land is the crucial constraint on the system [31]. The association between the amount of labor and the amount of land cleared is not a simple, direct function, however. The amount of land cleared per unit of labor varies considerably among households, as well as varying within the same household from one year to the next. This is due in large part to variation in the household's consumption requirements as measured against its productive capability. When consumer/producer ratios are calculated, they correlate strongly with the intensity of labor of the producers[32].

There is one final factor that affects the total amount of land cleared by the household: the relative productivity of the land. Variation in productivity is associated with variation not only in the sizes of the individual swiddens, as discussed previously (table 21), but also with variation in the overall amount of land farmed. Table 24 shows that the higher the percentage of swampland that is cleared by a household, for example, the fewer the total number of hectares cleared.

TABLE 24
Amount Of Swampland And Number Of Hectares Farmed Per Household

		Swampland as percentage of total amount of land farmed per household	
		< 14%	> 14%
Number of hectares farmed per household	< 4.6 ha	8	7
	> 4.6 ha	13	2

N = 14 households that farmed in 1975 and 16 that farmed in 1976.
14% = Mean amount of swampland cleared as a percentage of the household total.
4.6 = Mean number of hectares cleared per household per year.
P = .0542

There is a similar, although weaker ($X_c^2 = 3.35$, $P < .10$) association between the percentage of land that is cleared from secondary forest and the total amount of land cleared. Thus, when the Kantu' clear relatively more productive land, namely swampland as opposed to dryland, or secondary forest as opposed to primary forest, they tend to clear less of it.

SWIDDEN NUMBER

The norm is for each household to make multiple swiddens each year. During 1975 and 1976, the households of Tikul Batu made an average of 2.3 swiddens apiece. The making of multiple swiddens is a key element in the Kantu' strategy of swidden diversification, enabling each household to farm two or more different microenvironments in a single year. This interpretation is supported by the rarity with which a household makes two or more of its swiddens in the same microenvironment in a given year. This occurred in only four instances out of thirty in 1975 and 1976[33]. Given this interest in diversification, any increase in the productive capacity of the household is used in part to make a greater number or swiddens (table 25). In addition, part of any such

TABLE 25
Number Of Hectares And Number Of Swiddens Farmed Per Household

		Total number of hectares cleared per household	
		< 4.6 ha	> 4.6 ha
Total number of swiddens made per household	< 2.3	13	5
	> 2.3	2	10

N = 14 households that farmed in 1975 and 16 that farmed in 1976.
2.3 = mean number of swiddens per household.
4.6 = mean number of hectares cleared per household.
$X_c^2 = 6.8$
$P < .01$

increase is used to make larger swiddens, as noted earlier (table 20). These two principles, of increasing diversification and increasing scale, are in basic opposition to one another. Nevertheless the Kantu' usually can strike a balance between them, so that they neither increase the number of swiddens (per household) at the expense of individual swidden size, nor increase the size of individual swiddens at the expense of the overall number of swiddens (table 26).

TABLE 26
Number Of Swiddens Per Household And Individual Swidden Size

		Number of swiddens per household	
		1 - 2	3 - 4
Size of each swidden	< Mean	19	23
	> Mean	13	14

N = 69 1975 and 1976 swiddens (mean size calculated separately for primary forest, secondary forest, and swamp swiddens).
X^2 = .06
$P < .85$

A somewhat less important determinant of the number of swiddens made is the social structure of the household. During 1975 and 1976 one-half of the households at Tikul Batu contained just one married couple, while the other half contained two (or, in a few cases, three) couples. The two-couple households typically consist of an older couple, their children, and the spouse and children of one of these latter. A two-couple household potentially is more productive (per unit of labor) than a one-couple household, due to the high threshold of labor inputs required for participation in the swidden system. (This threshold is related, in turn, to the need to achieve some minimal diversification of the household's swidden land, as well as some minimal economies of scale within each swidden.) Despite the advantages of multiple couple households, tensions between couples often make it difficult to keep such households together. These difficulties are mitigated if the household makes multiple swiddens, because this allows each couple to assume responsibility for, and live in, a separate swidden during the farming year. For example, it was a rise in intrahousehold tension that caused household #11 to make two major dryland swiddens (#55 and 56) in 1976, whereas they had made only one (#23) in 1975.

SWIDDEN SHAPE

The final aspect of swidden selection is the shape of the swidden. Most Kantu' swiddens have a characteristic shape, this being square or slightly rectangular with rounded off corners. This shape is most conducive to the minimization of boundary disputes, to the practice of chain reaction felling, and to the utilization of wind movement during the burn. These several benefits are lost if the swidden takes on an irregular shape, with bulges in its perimeter. This eventuality is guarded against by a ritual proscription that states Umai nadai tau' betelok 'Swiddens may not have bays'. There are two exceptions to this proscription. It is permissible to make a swidden with an irregular perimeter if that perimeter follows the banks of a watercourse. It also is permissible to make a swidden with an irregular perimeter, even if not alongside a watercourse, if the irregularities do not break up the pattern of planting. That is, it is permissible to make a swidden with a finger of uncleared forest extending into it if this land also is sown (even though there is no chance that this seed will produce plants) during planting (figure 14)[34].

FIGURE 14
Prescribed Planting Pattern In A Swidden With An Irregular Perimeter

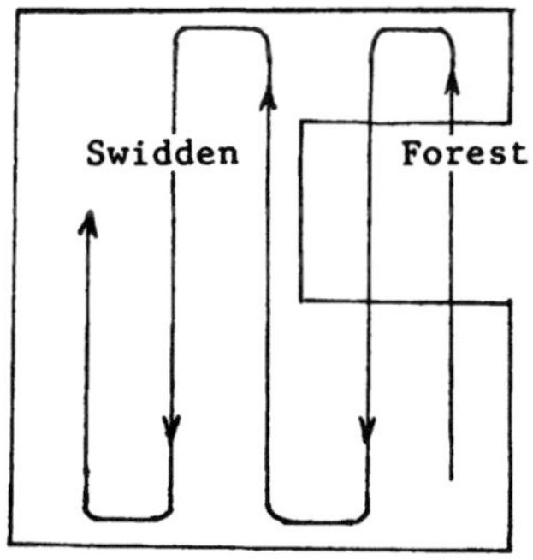

Chapter Two
Slashing

Nebas 'to slash' constitutes the second phase of the swidden cycle and is the first step in actually clearing the forest. The Kantu' say that there are two principal objectives in this operation. One objective is simply to kill the vegetation so that it can dry out and then burn easily when it comes time to fire the swidden. This is especially important because the slashed vegetation serves the function, during the burn, of igniting the large trees that are felled during the second stage of clearing and are more difficult to dry out and burn. The second objective of slashing is to provide a clear space, free from undergrowth, in which the Kantu' can work when later felling these large trees. If the forest floor is not cleared of undergrowth, it is more difficult for the workers to flee to a safe distance when a tree on which they have been working begins to fall.

TECHNOLOGY

The principal tool used in slashing is the *duko' penebas* 'slashing sword'. This is a heavy, slightly convex, single-edged sword, hafted to a wooden hilt, and carried on the left hip in a wooden scabbard[1]. Both hilt and scabbard are finely lashed with rattan. An additional tool is the *pengait* 'hook', which is cut from the fork of any young tree. When the Kantu' slash very thick underbrush or grasses, they use this to gather the vegetation together so that it can be more easily slashed with the brush sword.

The brush sword always is held in the right hand, and if a brush hook is used it is held in the left. When the Kantu' slash sizable trees they stand erect, though bent slightly at the waist, and wield the sword with both forehand and backhand strokes. The latter is preferable, because it is both stronger and safer. In either case, the Kantu' always make (and they speak of this) a diagonal stroke, preferably entering the tree trunk at an angle of 45 degrees to the vertical. By taking advantage of the vertical grain of the tree, this maximizes the horizontal bite into the trunk. After the initial stroke on a trunk, successive strokes are angled slightly off the 45-degree angle so as to remove chips from the cut, until the trunk finally is cut through. Whether slashing seedlings, saplings or sizable trees, care is taken to sever the trunk completely, thus insuring the death of the tree.

If the trunk is large the Kantu' make cuts on two sides, but if it is small the trunk is cut through from one side only. The height at which the cut is made varies according to the age of the forest. In older forest the workers make their cuts at waist level, because this is easiest for a man standing erect. However, in younger secondary forest they make their cuts below knee level. If they made their cuts any higher, they say, that would leave tall stumps that later would pose a problem during the weeding of the swidden, because the weeders have to squat on the ground while they work.

When the Kantu' slash young saplings and underbrush instead of trees, they use a different technique. Bending sharply at the waist, they sever the vegetation near the ground with single strokes of the brush sword. When they slash stands of grass (found only in swampland), they also use a different technique. In this case, they work while squatting on their heels. This posture is necessary because they have to cut the grasses through slightly below ground level in order to kill not only the grass but also its root system, which otherwise would quickly sprout a new stand of grass. The workers wield their swords with flat, horizontal (not diagonal) strokes. The absence of ligneous matter in the stands of grass permits the use of this stroke. The slashing of swamp grasses also is characterized by some postslashing treatment of the vegetation, to prepare it for drying. The workers remove the cut grasses from all watery depressions in the swidden and place them on higher bits of land, where they spread them evenly about. Both actions are designed to facilitate the drying and later burning of the slashed vegetation. Nothing similar is done in the dryland swiddens.

The pattern of movement of the workers, through the area that has been designated for the new swidden, is governed by two factors. First, because slashed trees tend to fall downhill, the workers always try to proceed in an uphill direction. If they proceeded downhill, the trees that they slashed would fall against and obstruct the vegetation that was not yet slashed. Second, the workers also attempt to slash all of the shorter vegetation in their immediate vicinity, before slashing any of the saplings and larger trees. If the trees were slashed first, they would fall atop the undergrowth, obstructing it and making it impossible to be slashed.

Although the workers cover the entire swidden during this stage of the cycle, they usually do not slash all of the vegetation therein. Two types of vegetation are exempted: the largest trees and the smallest plants. In the latter case, the decision to slash or not depends upon whether the swidden is located in or out of the riverine floodzone. If it lies within this zone, the Kantu' say that they must slash all vegetation no matter how small (even including seedlings on the forest floor), or else the swidden will not burn well when they fire it. They expect a weak burn in a floodzone swidden, because periodic inundations by the river do not allow the swidden to thoroughly dry out. What is thus at best a weak burn will be made even worse if unslashed and hence still living and green seedlings are present in the swidden. In swiddens located above the floodzone, a strong burn is less problematic, and so the Kantu' do not slash small seedlings.

With regard to larger trees, there are several factors that determine whether they are cut with the brush sword during this slashing stage or left to be cut with the adze during the following stage of felling. One factor is tree size or diameter. My measurements indicate that most trees under eight centimeters in diameter are cleared during slashing, while most trees with a diameter greater than this are not[2]. This dividing point is determined by whether a tree of given size can easily be cut through with a brush sword, or not. The dividing point of eight centimeters is not invariable, however[3]. The Kantu' recognize that different species of trees differ in terms of the hardness and friability of their wood. Given a large tree with soft and friable wood, and a smaller tree with hard and nonfriable wood, the Kantu' might slash the former with the brush sword and leave the latter to be felled with the adze. The angle of lean of a tree also can be important. Less cutting is required to topple a tree that leans considerably away from the perpendicular. Therefore, large trees that lean in this manner may be cut with the brush sword, whereas if they were perpendicular they might be left to be cut with the adze.

In the absence of any large trees, in very young secondary forest, the Kantu' slash all standing vegetation, thus eliminating any need for the next stage in clearing, namely felling. The Kantu' call this practice bai' sekali' 'take at one time'. The Kantu' characterize this, first, by the failure to reserve any trees from slashing for later felling and, second, by the absence of any use of the adze. Bai' sekali' is distinguished thereby from the ordinary slashing and felling sequence. (They speak of it as a type of slashing, nonetheless.)

On an average day, each worker spends 322 minutes at work slashing the household's swiddens. I arrived at this figure after subtracting time spent in bathing, eating, resting, and smoking or chewing betel, as well as the time that is spent traveling to and fro between the swidden and the longhouse. During this stage of the swidden cycle, most households live in the longhouse rather than in separate swidden houses. In the exceptional cases in which the household members do live in a swidden house, they work a slightly longer day, averaging 350 minutes, because of the time that they save by not having to travel back and forth between the longhouse and the swidden.

When slashing is carried out by several households working together, under either reciprocal or wage labor arrangements, the work day averages 495 minutes in length. The much greater length of this work day is due to the desire of each household to appear industrious in the eyes of the other participating households. A poor performance by the members of a given household in such circumstances opens them to the exceedingly embarassing (to a Kantu') charge of being lazy and of taking advantage of this labor arrangement. Because each household strives to perform well in these circumstances,and because this has measurable effects on the work pattern (viz., not only in terms of lengthening the workday, but in increasing the work rate as well), the Kantu' like to work in multihousehold groups whenever possible. They say themselves that swidden work is completed quicker if performed by a multihousehold group as opposed to the members of a single household[4].

During either the 350-minute or 495-minute workday, slashing in the swidden constitutes the workers' major activity during the daylight hours. Other activities, some of an economic nature, are pursued during the several hours in the evening, after nightfall and before sleeping. Some minor activities are pursued by the worker during the day as well, when the household members work on their own, during a 350-minute day in their own swidden. These activities, carried out in the morning before work begins, in the late afternoon after the work is finished, and during the midday meal break, include checking fish traps and nets, gathering nonrice cultigens from the previous year's swiddens, cooking, and plaiting baskets, mats and other gear. The pursuit of these activities partially explains the difference between the 350-minute and 495-minute workdays.

It sometimes happens that the workers are prevented from working a full day in the swidden. If they see or hear an omen bird while they are slashing the forest, they usually are obliged to cease working for the rest of that day. The Kantu' believe that if they ignore these omens, the swidden involved will fail. On average, at least one out of every seven work days is cut short in this manner[5]. The experiences in a five-swidden sample are summarized in table 27.

TABLE 27
Ritually Foreshortened Workdays When Slashing The 1976 Swiddens

Swidden #	Number of Days on Which One or More Workers Slashed in the Swidden	Number of These Days That Were Shortened by an Omen
41	33	3
43	15	3
44	12	5
52	15	2
67	19	5

The function of omen taking at this stage and during subsequent stages of the swidden cycle appears to be largely calendrical[6]. That is, this observation and reaction to omens is part of the system whereby the Kantu' apportion their hours and days among different types of work, as well as between work and nonwork. For example, a worker whose slashing is interrupted by an ill omen might then spend the rest of the day hunting, fishing, or engaged in craft work in the swidden house or longhouse. If the interruption comes in the latter part of the day, the worker might spend the remainder of the day resting. This is arguably an important function of omens in a society in which no time is formally set aside for relaxation per se. An important aspect of this time allocation is that it has ritual sanctity. When the Kantu' do or do not engage in a given activity on the basis of a given omen, they can be at ease about the propriety of their decision. This calendrical interpretation of omen functions is supported by the Kantu's comparison of their belief system to that of Christianized Dayak. The Kantu' say, Sida' Kristen diau k'ari Minggu, kami diau ke burong 'Those Christians observe work proscriptions on Sundays, we observe work proscriptions according to the birds'.

LABOR

On average, eight work days and seven work days, respectively, are devoted to the slashing of one hectare of primary forest and one hectare of secondary forest. These totals assume a workday of 350 minutes, and they assume that the workers are all adult males. Most scholars of this subject have maintained that it takes much less time to slash primary forest than secondary forest, because of the supposed lack of dense undergrowth in the former. The absence of undergrowth in primary forest is open to question, however. In the Kantu' territory, for example, there is a phenomenon called rampa'. The Kantu' use this term to refer to patches of dense undergrowth that spring up in primary forest following the natural felling of a large tree. Such trees, when felled in storms, for example, also pull down numerous surrounding trees and can open up quite a large space in the forest canopy. This lets sunlight into the forest floor, which in turn allows brush and saplings to grow in. The evidence from the Kantu' territory suggests that rampa' is not an uncommon phenomenon in primary forest, especially on steeply sloped land. Its occurrence partially equalizes the burdens of slashing primary forest and secondary forest. Even in the absence of rampa', primary forest is not easy to slash. This is due to the fact that primary forest is composed largely of different species than is secondary forest, and these species tend to have denser, comparatively harder wood, and accordingly are more difficult to slash than trees of the same size in secondary forest.

There are only two sets of circumstances in which the slashing rate in secondary forest climbs well above the rate in primary forest. First, if the secondary forest consists of grass covered swampland, the work rate climbs to approximately 11½ days per hectare. These dense stands of grass take much more time to slash than underbrush and saplings. Second, the work rate also will rise to approximately 11½ days per hectare in secondary forest if the workers practice bai' sekali', in which they slash every tree present in the swidden. This is done in approximately one-half of all secondary forest swiddens, so the average labor input in these swiddens is 9.25 days per hectare[7].

Adult males perform most of the slashing. The adult females of the household usually play a minimal role in this stage of the swidden cycle (and adolescents of either gender usually play no role at all). In 1976, females in eight out of sixteen households participated in the slashing of fourteen out of thirty-five swiddens, and in eleven of these fourteen cases they performed only a minor part of the slashing. The limited use of female labor at this time is probably due to two factors. First, because the body mass of the average adult female is less than that of the average male, she puts less force (cet.par.) into each stroke of the brush sword. This distinction is reflected in the fact that in those instances in which females do participate in slashing, it is almost always secondary forest (which is easier) that they slash and not primary forest (which is harder). In 1976, the females of Tikul Batu participated in the slashing of only one out of eight primary forest swiddens, while they participated in eleven out of twenty-seven secondary forest swiddens. The lesser ability of female slashers also is reflected in the fact that they never participate in reciprocal or wage labor groups in the swiddens of other households. Acceptance into such groups is reserved for those workers who are fully competent at a given task. This typically excludes the very old and the very young - and, during stages such as this one, the females.

The distinction between males and females in terms of ability to slash is enhanced by a second factor, the absence of any real pressure at this time upon the household work force. In the absence of such pressure, the household need only send to its swiddens those workers who are best suited to slashing (viz., its adult males), while workers who are less suited to slashing (viz., its adult females) are freed for other tasks. In particular, the females are free to engage in rubber tapping, a task that is of considerable importance to the household economy. Those females who do participate in slashing usually do so only on rainy days when they are unable to tap rubber. The absence of any need for female slashers is reflected in the absence of any relation between the participation of a given household's females in slashing and the slashing burden (viz., the number of hectares of forest to be slashed) on each adult, male slasher in that household ($P = .50$).

The household may not rely totally on its own labor force when slashing the swiddens. Labor is recruited from other households under both bedurok 'reciprocal' and bekuli 'wage labor' arrangements. In 1975 and 1976, reciprocal work groups were used in the slashing of nineteen of sixty-nine swiddens, and wage laborers were used in the slashing of fourteen of sixty-nine swiddens. In the case of wage labor groups, participation is usually determined by extant trading debts. Of the total amount of wage labor hired to slash the 1975 and 1976 swiddens, 62 percent was hired by just four households (#2, 4, 5, and 9), each of which engages extensively in trading. In the case of reciprocal work groups, the mutual participation of different households is determined both by the proximity of their kinship relationships and by the proximity of their swiddens. The most proximate kinship relationships are direct consanguineal or affinal ties (figure 4). Table 28 shows that there is a statistically significant association between the existence of such kinship ties and participation in reciprocal labor arrangements during the slashing phase of the swidden cycle. The second factor in

TABLE 28
Kinship Ties And Participation In Reciprocal Slashing Arrangements

		Paired households have direct consanguineal or affinal tie(s)	
		No	Yes
Paired households joined in reciprocal labor arrangements to slash their swiddens in 1975 or 1976	No	159	25
	Yes	18	9

N = 211 possible, different pairs of households.
$X^2 = 4.5$
$P < 0.05$

such labor arrangements, swidden proximity, was discussed in general terms in chapter 1. I noted that there is a strong association over the swidden cycle as a whole between the adjacency of the swiddens of two different households and their participation in reciprocal work groups. This association is especially strong during the slashing stage ($X^2 = 20$, $P < .001$)[8].

Reciprocal labor accounted for 42 percent of the

total labor input in those swiddens in which it was used in 1975 and 1976, and wage labor accounted for 24 percent. The attraction of reciprocal labor lies in speeding up the work, while the attraction of hiring wage labor lies in having trade debts paid off. Given these attractions, the fact that greater use is not made of such labor arrangements is due to the relative absence of time constraints during this swidden phase. This is reflected, for example, in the absence of any association between the use of either reciprocal labor or wage labor, on the one hand, and the severity of the slashing burden on the household's labor force, on the other (P = .71 and P = .47, respectively)[9].

The extent to which reciprocal and wage labor is used also is constrained by the omen system. Most omen related work proscriptions occur early in the slashing phase (as in all other swidden phases), because the Kantu' are required to proscribe only one day during each swidden phase for each type of omen (with occasional exceptions). Once a household has proscribed/aborted a work day for a given omen while slashing a given swidden, it usually can ignore that omen during all remaining slashing in that same swidden. For this reason, the Kantu' prefer to postpone use of reciprocal or wage labor until the later stages of slashing. By that time they will have observed a large proportion of the possible omens, and the likelihood of a work day being forcibly interrupted by an omen-based proscription will be relatively low. If a household is using reciprocal or wage labor in a swidden and an omen is encountered for which they have not yet observed the work proscription, everyone, including the reciprocal and wage laborers, is required to leave that swidden for the rest of the day. In such cases, however, the debt of the swidden owners to the guest workers still is reckoned at a full day. The Kantu' explicitly state that the avoidance of such losses is one of the reasons why they usually will not employ such labor during the first stages of slashing. The other way to avoid such losses is simply to ignore omens encountered when a multihousehold work group is in the swidden. This course of action is not infrequently chosen, but it makes the Kantu' anxious and they clearly prefer to just postpone the use of extrahousehold laborers until such time as this dilemma is unlikely to arise.

In addition to reciprocal and wage labor, there is a third, though less important source of extrahousehold labor. This is betolong 'gift labor'. During 1976, some such labor was involved in the slashing of three of the thirty-five swiddens made at Tikul Batu. This very slight incidence is due to the fact that one household typically gives such labor to another only if the two are very intimate and only if the recipient is in dire need of such assistance. Kantu' households rarely experience such a need during the slashing phase.

CHRONOLOGY

There are two principal determinants of the overall chronology of slashing. First, the slashing stage must precede the felling stage. If the larger trees are felled beforehand, it is no longer possible to slash the underlying and now obstructed smaller trees and undergrowth. In addition, the prior slashing of the smaller trees and undergrowth facilitates the later felling of the larger trees, as noted earlier. Second, slashing must precede felling by a sufficiently large interval to allow the slashed vegetation time in which to dry out before felling begins. If the vegetation has not dried by the time felling begins, it never will dry. If trees are felled atop it, the slashed vegetation will begin to rot and will not burn when the swidden is fired. Just as slashing must precede felling, so felling must precede burning and planting. For a number of reasons (to be discussed in detail in chapters 4 and 5), burning and planting must be carried out during the months of August and September, respectively. It is this timing of burning and planting that determines the timing of felling and thus slashing, and not the other way around[10].

Within the broad range determined by these two constraints, there is considerable variation in the chronology of slashing among the different swidden types. Thus, in 1975 the average primary forest, secondary forest, and swampland swiddens were slashed during the respective periods of 21 February to 11 April, 9 June to 30 July, and 5 August to 22 September. The early slashing of primary forest may be prompted, in part, by anxiety to establish claim to a given section of forest before a rival household can do so. This was certainly a factor in the very early slashing dates of households #2, 4 and 6 in their 1976 primary forest swiddens (viz., swiddens #38, 43 and 47). For the most part, however, the differences in timing among the swidden types are due to the differing importance of two problems: drying out the vegetation, and minimizing regrowth before the swidden is burned.

Thorough drying of slashed vegetation is crucial to a good burn. It is most difficult to attain in primary forest swiddens, due to both the density of the wood in primary forest trees as well as the minimal amount of sunlight that reaches slashed vegetation on the forest floor. Accordingly, the Kantu' try to slash their primary forest swiddens as far in advance of felling and burning as possible. The reverse holds true in the secondary forest swiddens, because of the problem of rapid regrowth. If the Kantu' allow too much time to pass after the swidden has been slashed but before it has been felled and burned, new growth will spring up and make it more difficult to burn the swidden. This problem is most severe in secondary forest swiddens, because shade intolerant, quick growing pioneer species initially are more abundant in secondary forest than primary forest. Accordingly, the Kantu' try to slash their secondary forest swiddens as soon as possible before felling and burning. This difference between primary and secondary forest swiddens is demonstrated in tables 29 and 30[11].

TABLE 29
Swidden Type And The Slashing-Felling Interval

		Swidden made from predominantly	
		Primary forest	Secondary forest
Length of slashing-felling interval	< 41 days	2	12
	> 41 days	6	0

N = 20 1976 dryland swiddens.
41 days = mean slashing-felling interval for this sample.
P = .0007

TABLE 30
Swidden Type And The Slashing-Burning Interval

		Swidden made from predominantly	
		Primary forest	Secondary forest
Length of slashing-burning interval	< 96 days	1	12
	> 96 days	7	1

N = 21 1976 dryland swiddens.
96 days = mean slashing-burning interval for this sample.
P = .005

The problems of drying and regrowth have a more equivocal effect on the timing of slashing in the swamp swiddens. For example, the drying of slashed vegetation certainly is difficult in swampland, but because the success of the burn is so much less important in such swiddens (the nutritive importance of ash being replaced by that of water), there is no tendency to lengthen the slashing-burning interval in response. Regrowth also is a potential threat in swampland, given that the original vegetative cover often is composed entirely of quick growing grasses and brush. However, the rapidity of regrowth is inhibited considerably by the presence of standing water in these swiddens, as a result of which there is little concern to minimize the slashing-burning interval. Overall, the impact of these problems causes the swamp swiddens to be slashed somewhat later in the year than either of the other swidden types.

Manipulation of the slashing chronology usually is a quite successful means of dealing with the problem of drying in primary forest swiddens and the problem of regrowth in secondary forest swiddens, nor is one problem solved at the expense of the other. The late slashing of secondary forest usually does not result in a drying problem, nor does the early slashing of primary forest typically result in regrowth problems. There are exceptions, however. Some Kantu' maintain that if the slashing-burning interval in primary forest exceeds two months, the slashed stumps of brush, saplings and trees will begin to produce new shoots and thereby threaten the coming burn. The Kantu' deal with such threats by slashing the new growth before burning the swidden. In practice, this operation is never performed unless the slashing-burning interval far exceeds two months. Thus, despite a mean slashing-burning interval of 154 days among the eight 1976 primary forest swiddens, reslashing was performed in only one swidden. This was the upstream Kenua' swidden of household #6, in which the slashing-burning interval - at 228 days - was the longest in the longhouse. This household reslashed the swidden on and about 4 August, 1975, just one week before it was fired.

Regardless of when it is carried out, slashing tends to be carried out rather slowly. As a result, the slashing phase tends to be quite lengthy, if compared with other swidden phases. For example, the average household at Tikul Batu devoted a total of thirty-seven workdays to slashing in its 1976 swiddens. Given an average household work force for slashing of 1.8 persons, this represented a work load of twenty-one days per worker. Despite this low work load, the average household carried its slashing out over a period of 103 days, meaning that the workers slashed on only one day out of every five (approximately) during the slashing phase. This is a very low ratio, compared with other phases of the swidden cycle.

There are several reasons for the apparently leisurely pace at which slashing is carried out. First, it often happens that the household members have to interrupt their slashing to work at something else. For example, the slashing of new, primary forest swiddens often coincides with the harvesting of the previous year's swiddens. In such cases, slashing must be interrupted until the harvesting is finished. The slashing of secondary forest swiddens may coincide with the previous year's post-harvest festival, occasioning an interruption in the work of perhaps two weeks. A second factor that affects the duration of the slashing phase is the mixture of vegetation types in many of the swiddens. If a swidden is cut from both primary and secondary forest, for example, the former will have to be slashed earlier in the year and the latter later in the year, so as to address the respective problems of drying and regrowth. One result is an elongation of the overall slashing period. Among four 1976 swiddens consisting of both primary and secondary forest, the slashing period averaged eighty-six days, whereas in sixteen homogenous swiddens it averaged just forty-two days. A third factor affecting the slashing timing is ritual. As noted earlier, on an average of one out of every seven workdays spent in slashing the swiddens, the workers must quit early because of bird omens. In addition, on an average of one out of every fifteen days, omens prevent the workers from even reaching their swiddens. Such omens are encountered either in the morning, on the trail to the swidden, or else during the previous night, in dreams. Table 31 presents the incidence of such omens in 1976, in a five-swidden sample[12]. Slashing is similarly proscribed on certain days in connection with the phases of the moon (as discussed in the Introductory chapter).

The most important determinant of the duration of the slashing phase, however, is the relative lack of pressure to complete the work in a shorter amount of time. Compared with most other phases of the swidden cycle, there are relatively few time constraints as to how quickly slashing must be carried out. The absence of any constraints on slashing is reflected in the 1976 swiddens in the absence of any association (P = .79) between the duration of the phase in each household and the onerousness of the work[13].

TABLE 31
Ritually Proscribed Workdays When Slashing
The 1976 Swiddens

Swidden #	Total Number of Days on Which One or More Workers Slashed in the Swidden	Total Number of Additional Days That Were Proscribed by Omens
41	33	2
43	15	0
44	12	7
52	15	3
67	19	0

Chapter Three
Felling

During the felling phase of the swidden cycle, the Kantu' cut all of the trees in the swidden that were not cut during the preceding slashing phase. The purpose of this is twofold. First, it is necessary to kill the trees so that they can be dried, then burned and transformed into ash. Second, it is necessary to let the sun into the surface of the swidden, to sustain the crops that will be planted there.

TECHNOLOGY

Two distinct technologies are employed during this phase: one during felling proper, and one during redressive and alternative operations.

Felling Proper

The principal tool used in the felling stage is the beliong 'adze'. Like the brush sword, the adze is hand forged by the Kantu'. It is notable for its convex cutting edge and the oblique angle at which the blade is hafted (with rattan) to the wooden shaft (figure 15).

FIGURE 15
Beliong 'Adze'

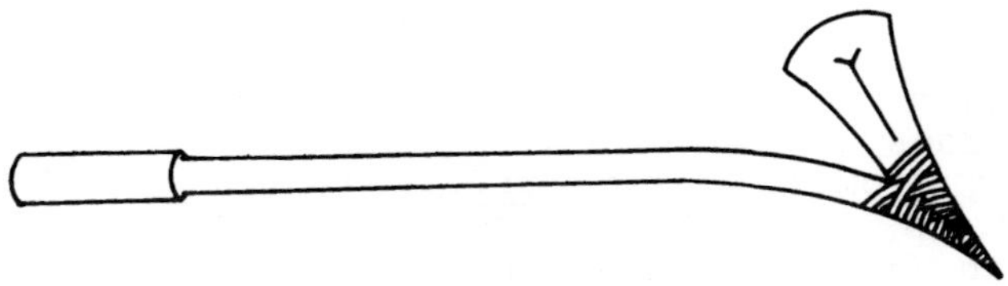

The iron blade weighs approximately $1\frac{1}{2}$ kilograms. The Kantu' also use their brush swords during this stage, to clear vines or saplings away from the larger trees in the cases where they would otherwise obstruct the felling. The sword also may be used to cut any smaller trees that were overlooked during the slashing stage. The third type of tool or instrument used during this phase is the felling stand, of which there are several varieties. The first is the tanga' (figure 16a). This

FIGURE 16
Types Of Felling Stands

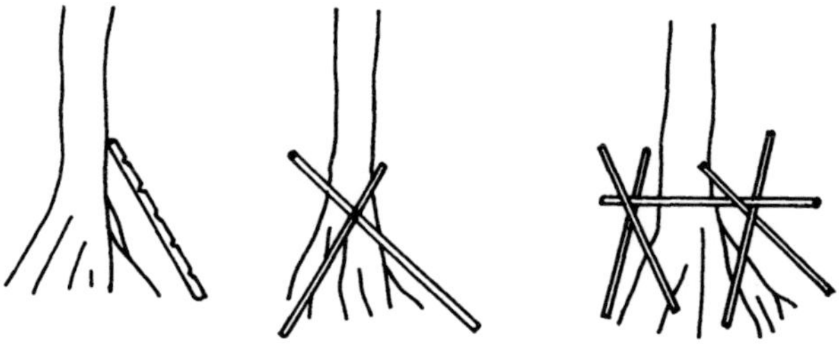

a. Tanga' 'Felling Ladder' b. Tunai 'Felling Platform' c. Tunai 'Felling Platform'

is a log, three meters in length, with steps cut all along one side. It is placed diagonally against the tree trunk, and the feller stands on the top step while he works. When that tree has been felled, he carries the tanga' to the next tree, and so on. The second type of stand is the tunai (figure 16b). This consists in two long poles lashed together in the form of 'X', and then lashed diagonally against the tree trunk. The worker stands in the crotch at the center of the 'X' while he works. Alternatively the worker may lash four poles in pairs against a single trunk, and then stand on a horizontal bar run across the two crotches (figure 16c). By moving along this crossbar, the worker can chop on both sides of the trunk, which is difficult to do when working from a single tunai. The tunai differ from the tanga' in that they cannot be moved easily from one tree to another and so are used only once. Their labor cost consequently is higher than that of the *tanga'*. On the other hand, most Kantu' find it more difficult to work from the latter. For both of these reasons, the Kantu' usually make *tunai* when felling the largest trees in the forest, while using a *tanga'* for all other trees.

The use of felling stands is a function of the necessary height of the cut. If possible, the Kantu' fell trees while standing on the forest floor, cutting the tree trunk at the level of their chest or, although this is slightly less desirable, at the level of their head. However, when felling primary forest, they often are required to cut higher up on the tree trunk, using a felling stand. There are several reasons for the higher cut in primary forest. The first reason involves *adat* 'tradition' or 'customary law'. The Kantu' say that their *adat* requires the use of a high cut on at least one or two trees in every primary forest swidden, because this produces tall tree stumps. The rice that is planted there will 'see' these tall stumps and will try to match that height in their own growth. The second reason for the high cut involves the presence of buttresses on the lower part of the trunk of the larger trees in primary forest. If the feller cuts low on the trunk, he will have to cut through these buttresses; whereas if he can cut a little higher up, he can avoid them. The difference in the amount of wood that must be cut through can be considerable. I measured one buttressed tree in primary forest that had an area of 0.42 square meter in cross section at the normal cutting height of 1.2 meters above the forest floor, but had an area of only 0.25 square meter at the actual cutting height of 2.7 meters that was made possible by standing atop a felling platform. This represents a reduction of 39 percent in the area of wood needing to be cut. Even in the absence of buttresses the Kantu' may use a felling stand, saying that the wood is less hard higher up on the trunk than it is lower down to the ground. All of these factors - the hardness of the wood, the presence of buttresses, and customary law - necessitate the use of felling stands in primary forest only. Rarely are they employed in secondary forest. During the 1976 swidden year, felling stands were employed in at least twelve primary forest swiddens but only one secondary forest swidden.

Height aside, a second important characteristic of the felling cut is its angle. The ideal type of cut, in this regard, is one called *pesunan* 'alternation'. To make this cut, the worker stands in front of the tree and alternately makes a diagonal cut (with the lower end of the cut nearer to him and the higher end further away) on the side of the tree to his right, using a forehand swing, and a similar diagonal cut on the side of the tree to his left, using a backhand swing. The cuts are diagonal because the worker's adze swing is directed upwards, for the reasons of wood density and area just discussed. In addition, a diagonal cut (as opposed to a perfectly horizontal cut) enables the worker to take partial advantage of the vertical grain of the tree. By using a backhand as well as forehand swing, the worker can remain in the same spot while cutting on both sides of the trunk before him. In addition to saving time (especially when the worker has built a felling stand against one side of the tree), this produces two parallel cuts on either side of the trunk. Because these cuts are parallel, the Kantu' say the tree is certain to fall in one of only two directions, namely towards one cut or the other. The fewer the number of directions in which the tree may fall, the greater the personal safety of the worker. They further increase their safety by making the two cuts on the uphill and downhill sides of the trunk, and by making the latter slightly lower than the former (figure 17). This practically ensures that the tree will fall in one direction only, namely downhill.

FIGURE 17
Standard <u>Pesunan</u> Cut For Felling

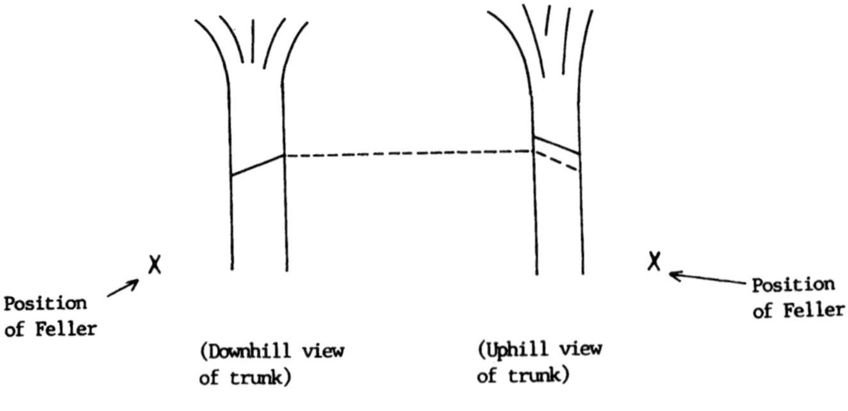

This ideal cut is not invariably employed. One variant, called <u>peling</u> 'go round', involves making a continuous cut all around the circumference of the trunk. A second variant, called <u>nebang kanan dua sepiak</u> 'fell right-handed on two sides', involves making a forehand cut on both sides of the trunk. This necessitates walking halfway around the trunk to make the second cut, and it results in two cuts whose angles cross one another as opposed to being parallel. The Kantu' say that either of these alternative cuts is dangerous, because the direction of tree fall cannot be predicted. These cuts are only employed by that (somewhat disparaged) minority of workers who have not mastered the backhand swing that is employed in the standard cut.

The third and final, important characteristic of the cut is its depth. This varies according to whether the worker intends to <u>nebang</u> 'fell' the tree directly or merely <u>nangkal</u> 'notch' it. The most common notched cut, called <u>nangkal dua sepiak</u> 'notching two sides', is identical to the standard <u>pesunan</u> cut earlier described except for being shallower - so shallow indeed, that it does not topple the tree. Notched trees are felled en masse and indirectly by the felling of a single, large, appropriately situated tree. This is called <u>nyerah</u>. If properly planned, the felling of this single tree will initiate a chain reaction of felling among the notched trees below it. This chain reaction typically spreads through a 'V'-shaped swath of forest, with the initial, felled tree located at the point of the 'V' (figure 18). Once the chain reaction has started, the workers urge it on with piercing war cries.

FIGURE 18
Chain Reaction Felling

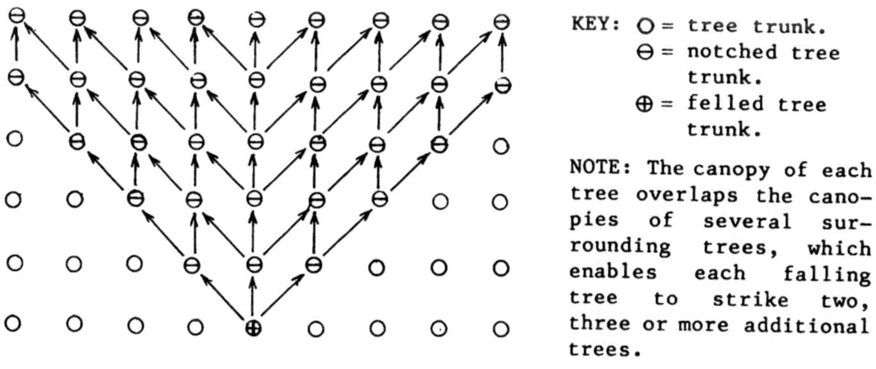

The advantage to notching a tree as opposed to felling it is that less time can be spent on the tree because less wood needs to be cut. When a tree is notched preparatory to chain reaction felling, it is cut through sufficient to cause it to fall when hit by other falling trees, but not sufficient to cause it to fall on the strength of this cutting alone. The difference between the two cases can be considerable. In one instance, I timed a worker spending 94 minutes to fell a tree, the cross section of which (at cutting level) measured 3537 square centimeters. This amounted to a felling rate of 38 cm^2/minute (based on the total cross section of the tree, not just that part of its wood that actually was cut through with adze). I then timed the same worker spending just 38 minutes to notch a tree with a cross section measuring 2516 square centimeters. This yielded a notching rate of 66 cm^2/minute, or more than 75 percent faster than the felling rate.

The advantage of notching and felling by chain reaction is reflected in the fact that during 1974-1976 the Kantu' notched an average of 77 percent of all trees in their swiddens, while felling just 23 percent. In individual cases the extent to which notching was performed varied greatly, according to the type of terrain in which the swidden lay and the type of forest from which it was cut. The Kantu' notched more often and more successfully in steeply sloped terrain, because trees fall more easily and with greater force down a slope than across level ground. The Kantu' ensure this advantage by cutting one notch high on the uphill side of each trunk and a second notch low on the downhill side (as in the pesunan cut described earlier). The Kantu' also notched more often in primary forest than in secondary forest. In the former, an average of 85 percent of all trees were notched, whereas the average was just 69 percent in the latter. This difference is due to the greater range of variation in tree size in primary forest, compared with secondary forest. It is easier in primary forest to notch the small and medium-sized trees and then find a single, very large tree to fell, to start the chain reaction. This is reflected in measures of tree size. Notched and felled trees in primary forest averaged sixteen and thirty-one centimeters in diameter, respectively. In contrast, notched and felled trees in secondary forest were almost identical in size, averaging just 12.3 and 12.6 centimeters in diameter, respectively.

Any time a tree fails to topple during a chain reaction felling, the worker must return and fell it directly and individually. The labor cost of doing this, added to the initial cost of notching the tree, is greater than if the worker had visited the tree just once and felled it directly. However, since this higher cost is incurred in fewer than 10 percent of the cases, on average, chain reaction felling as a whole is a successful labor-saving strategy.

The various types of cuts employed by the Kantu' are designed not merely to fell each tree with a minimum of effort, but also to ensure that each tree falls in a single, predetermined direction - because this minimizes the potential danger to themselves, as previously noted. This direction is almost always downhill; first, because less wood needs to be cut to topple a tree that already is leaning downhill and, second, because the added momentum of trees falling downhill is desirable in chain reaction felling. In this latter case, the workers also try to direct the fall of the initial tree so that as many notched trees as possible topple in the ensuing chain reaction. In all cases, the workers try not to fell a tree in the direction of any tree or trees that are still standing and have not been notched. If one tree is felled against another, the latter becomes more difficult and more dangerous to fell. Finally, the workers try to ensure that trees near the edges of the swidden-to-be do not fall into the surrounding forest. The Kantu' say they do this to ensure that the trees in question will burn - and therefore contribute their ash to the later sustenance of the rice crop - when the swidden as a whole is burned. Trees that fall outside of the swidden probably will not burn, but whether they do or do not their ash is lost to the swidden and its crops. This concern to fell trees within the swidden reflects the fact that the goal of the Kantu' is not merely to clear away the forest from a given plot of land, but rather to make the nutrients in that forest available to the crops that will be planted there.

It is just as important for the Kantu' to determine or predict the timing of tree-fall as its direction. While felling any large tree, the worker pauses occasionally to listen to the noises that the trunk makes as it sways, however minutely, in the gentle breezes that always blow over the forest canopy. Certain patterns of noise will alert him to the imminent falling of the tree, at which point he can flee to a safe distance. Sometimes, nonetheless, a tree will start to fall before it is expected to, whether due to a change in the direction or strength of the wind or to other factors. This poses a real danger to the worker still cutting at its base.

Equally important is determination of the timing of the eventual fall of notched trees. The Kantu' make their cuts on such trees with considerable precision, trying to make them deep enough so that the tree will fall when struck by other trees in a chain reaction, but not so deep that it will fall if struck by a strong gust of wind. A premature fall during the minutes, hours or even days after the workers make these cuts, but before they start the chain reaction, is problematic. This is true less because of any threat to the workers' safety than because of the threat of initiating an undesirable pattern of chain reaction felling among the surrounding trees. The chance falling of one notched tree will never initiate the large-scale, 'V'-shaped chain reaction that can be accomplished by the purposeful felling of a single tree, strategically located at the apex of the 'V', and selected on that basis. Rather, a chance fall is likely merely to set off a limited chain reaction within the 'V', making it impossible thereafter to fell the trees in that area through a single, massive chain reaction.

The length of the workday during this phase is similar to the length of the workday during slashing. It similarly varies according to whether the workers sleep near their swiddens or at the longhouse, and according to whether they work alone or in reciprocal or wage labor groups. Further, the felling workday can be cut short by ill omens just like the slashing workday.

Redressive and Alternative Techniques

The Kantu' have two major techniques for improving tree falls that are incomplete or otherwise unsatisfactory: secondary felling and trimming. Both use the brush sword as opposed to the adze. Secondary felling is performed, for example, if a tree does not fall completely to earth but comes to rest on the trunks or stumps of other, already felled trees. In that event the trunk is cut through a second time so that the leafy crown, at least, falls to earth. The Kantu' say that if this is not done, the crown may not burn when the swidden is fired. There also are cases, during chain reaction falls, in which trees pull out of the earth at their roots instead of breaking in two at the notches on their trunks. In such cases some part of the root system may continue to give the tree life, again making it difficult to burn when the swidden is eventually fired. To avoid this, the Kantu' return to such trees and deepen the original notches until the trunk is severed and the death of the tree is ensured.

After the Kantu' have felled a tree, in some cases they _ngeredak_ 'trim' the main branches away from the trunk. The Kantu' say that this hastens the death and drying out of the tree and, in addition, produces a denser mass of timber, closer to the ground. In both respects, the likelihood that the swidden will burn well thereby is increased. Even if the swidden does not burn well, the fact that trimming was carried out will make any secondary burning that much easier. Trimming is not performed equally in all swiddens. First, it is associated with swiddens located within the floodzone of the Empanang as opposed to above it (P = .000001 for 27 1976 swiddens). Because swiddens within the floodzone tend to burn poorly and tend to require secondary burning, the Kantu' say, the benefits of trimming usually justify the labor devoted to it within the floodzone but not above it. (The only factor that regularly constrains the practise of trimming in floodzone swiddens is the presence of _kesa_ 'weaver ants', whose trees can be felled, but not trimmed, with impunity.) Trimming also tends to be associated with swiddens cut from secondary as opposed to primary forest (P = .006 for 27 1976 swiddens). The Kantu' say that they do not trim in primary forest swiddens because the same effect is achieved by the mere fall of the trees. They say that when a tree falls whose trunk is the size of a _lanyi'_ 'carrying basket' (approximately thirty-five centimeters in diameter, a figure surpassed by many of the larger primary forest trees), its weight and momentum alone will break off many of its branches upon impact with the ground. The second function of trimming, that of facilitating redressive burning, is irrelevant in primary forest swiddens since, as the Kantu' say, the size of the tree trunks renders this latter operation too difficult to perform in any case.

Under certain circumstances, instead of felling a tree the workers climb it and _nyarau_ 'pollard' its branches with a brush sword. Two distinct techniques are employed. One, called _mengkah_, consists in lopping off the larger branches near their junctures with the main trunk. The second technique, _nutoh yepit_, consists in lopping off the tree's secondary or tertiary branches at their junctures with the larger, primary branches. This latter technique involves more climbing and the cutting of a greater, overall number of branches. This is reflected in their respective labor rates. I saw a tree fifteen centimeters in diameter lopped by the _mengkah_ technique in 5.7 minutes, while a second tree of identical size required 16.9 minutes to lop using the nutoh yepit technique. However, even the faster technique is much slower than felling the tree by cutting through (or notching) its main trunk. The two pollarding techniques achieve rates of 10 and 31 square centimeters of wood per minute, respectively (based on the area of a cross section of the main trunk at normal felling height), which compare with an average rate of 102 square centimeters per minute for felling a similar type of forest. Nor does the added labor devoted to pollarding result in a more thorough attainment of the normal goals of felling, namely to cut down the forest so that it no longer shades the ground beneath it and will burn when fired. This suggests that the return on labor is not the main, motivating factor in pollarding.

Some Kantu' say that the purpose of pollarding instead of felling some trees is to provide vertical supports for the various climbing legumes to be planted later in the swidden. However, this explanation is questionable in light of the small number of trees actually lopped, averaging just two per swidden in 1975 and 1976. A more likely reason for pollarding, and one more often cited by the Kantu', is adat 'custom'. I earlier noted that the Kantu' use felling platforms on at least a few trees in every primary forest swidden, so that there will be a few tall tree stumps to inspire the growth of the rice plants. Many Kantu' say that, for the same reasons, they must pollard at least a few trees in every secondary forest swidden. Some Kantu' add that, more generally, they must pollard in such swiddens so as to mark each one with the klai' 'symbol' of a swidden, namely a tree shorn of its branches.

This symbolic importance of pollarding suggests that it was once more central to the swidden technology than it is today[1]. This can be explained in terms of probable changes in both swidden technology and in the type of forest exploited by this technology. It is probable that the ancestors of the Kantu' developed or acquired the ability to make and use the brush sword before the ability to make and use the adze. Even today, when compared with the sword, the adze must be made from a larger piece of iron, of better quality, and its forging requires more skill. It also is probable that the ancestors of the Kantu' formerly farmed older primary forest, containing larger trees, than do their descendants today. Either or both circumstances make it likely that the Kantu' possessed the ability to pollard trees before they possessed the ability to fell them. As they developed the latter ability the practice of pollarding arguably declined, due to greater danger and higher labor costs, to the point where its practice today is limited and symbolic. As part of this decline, pollarding arguably became restricted to secondary forest because here its symbolic benefits can be reaped without incurring its risks. Indeed, the Kantu' themselves say that primary forest trees are too tall to climb and lop, an appreciation of risk that has in fact been formalized in a ritual proscription against pollarding in primary forest.

It is not uncommon for a few trees in each swidden to be neither felled nor pollarded but simply left standing, either because they have some unusual value or because they present special problems or costs in felling. The latter category includes any tree containing a nest of the greatly feared wasp, pantak, as well as certain primary forest trees with the largest boles and the densest and least friable wood (e.g., Melaban). Such trees are least likely to be felled if they have relatively small crowns. The smaller a tree's crown, the smaller the return on the labor devoted to felling it, both in terms of the amount of additional sunlight thereby let into the swidden, and in terms of the amount of ash later produced during its burn - recalling that most ash is produced by the burning of the foliage and small branches, the larger branches and trunk never being consumed. (This calculation of not only the cost of felling labor, but its very purpose, also is reflected in the fact that the Kantu' exempt from felling all pungo', the erect, branchless trunks of dead trees.) The exemption of a tree from felling does not mean that it will live, however. Such trees usually die during the swidden burn and, while they may not contribute much ash to the swidden, at the least they lose their foliage and so do not block off the sun.

Only when the trees involved have some special value are the Kantu' likely to spare large numbers of them from felling. This value may be either secular or sacred. Examples of the former include fruit and nut trees (including stands of rubber and candlenut trees), bee trees, old boundary trees, and trees with exceptionally valuable timber (e.g., Penyaut, used in making canoes, and Tebelian 'Iron wood', used in making roof shingles and house posts). Also included here is the Keira'[2]. The Kantu' say that they will not fell this tree because rice planted near it thrives especially well, which they attribute to the fact that its root system maintains an unusually high concentration of water in the topsoil. Trees with sacred value include those growing in ground in which human remains have been interred, any trees surrounding the sepan 'watering holes' of sambar or barking deer, and any trees in the vicinity of which a domestic cat has died. In all of these cases, the Kantu' do not merely spare the trees from felling, they actively try to preserve them from the ensuing burn. They accomplish this by also sparing all of the trees immediately surrounding the tree or trees in question. This technique, called ngelulong 'surround', establishes a firebreak around the valued trees. The Kantu' use this technique in both primary and secondary forest, but - as the Kantu' say and as my observations confirm - with much more success in the latter. When a primary forest swidden burns well, the intensity of the fire is such that any tree still living in the swidden will perish, whether surrounded by firebreaks or not.

LABOR

Spot measurements indicated that the Kantu' fell at a mean rate of 65 square centimeters of wood per minute in primary forest and 102 square centimeters per minute in secondary forest (again based on the total square area in cross section of the felled trees, not just on the area actually cut by the adze). Overall felling rates were, respectively, twelve days and eight days per hectare (assuming adult male workers, and a workday of 350 minutes)[3]. A number of different factors account for this primary/secondary forest distinction, one of which is the character of the timber involved. Wood that is relatively hard and nonfriable (qualities that the Kantu' explicitly associate with primary forest trees) takes longer to cut than wood with the opposite qualities. A second important determinant of the felling rate is the total amount of wood that must be cut. I calculated this by taking the circumference of the tree at its cutting height, calculating the area of that circle, repeating this for all of the trees in a sample plot, and then extrapolating to one hectare. This summed area ranged up to eight times as high in primary forest as in secondary forest. In addition to the amount of wood to be cut, also important is its distribution, meaning the number and size of the trees in the forest. This varied markedly between primary and secondary forest. Trees in the former averaged 473 square centimeters in cross section while those in the latter averaged just 118 square centimeters[4]. The larger the trees, the fewer of them usually found in a given area; whereas the smaller the trees, the more of them to be found. The felling time is not the same in both cases, because it takes the Kantu' less time to fell several small trees than one large tree of equivalent area. For example, in one stretch of primary forest I observed a worker notch one tree, with a cutting area of 0.25 square meter, in thirty-eight minutes, and then notch thirteen smaller trees, whose combined cutting areas equalled 0.25 square meter, in twenty-eight minutes.

The participation of the household work force in felling resembles its participation in slashing, with one difference. Whereas adult females and adolescent males often participate in slashing, albeit to a limited extent, they almost never participate in felling. For example, in 1976 the adult females of Tikul Batu participated in the slashing of fourteen out of thirty-five swiddens, but not one of them participated in the felling of even a single swidden[5]. This difference is noteworthy, given that time pressures are twice as great during felling as they are during slashing (as will be discussed shortly). The Kantu' explain the nonparticipation of females and adolescents in terms of the difficulty of the work. Many Kantu' say that felling is the hardest work that they ever have to perform. The difficulty of this work is reflected in the fact that the standard wage for felling labor is 50 percent higher (viz., three kolak of unhusked rice) than that paid during any other phase of the swidden cycle except carrying.

Like slashing, the Kantu' view felling as a difficult task, the performance of which can be eased through the use of reciprocal and wage laborers. Overall, however, extrahousehold labor is used somewhat more intensively during felling than during slashing, which is in accordance with the fact that felling is both more difficult than slashing and must be carried out under greater time constraints. As in slashing, the mutual participation of different households in reciprocal work groups is strongly influenced by the proximity of their kinship ties ($X^2 = 15.8$, $P < .001$ for 211 possible paired households in 1975 and 1976) as well as the proximity of their swiddens ($X^2 = 18.3$, $P < .001$ for 211 possible paired households in 1975 and 1976).

CHRONOLOGY

The most important determinant of the timing of the felling phase is not the length of the prior slashing-felling interval, but the length of the subsequent felling-burning interval. The length of the latter is determined by concerns with drying and regrowth, identical to those of the slashing-felling interval. As discussed in the chapter on slashing, the problem with drying is greater in primary forest swiddens while the problem with regrowth is greater in secondary forest swiddens. For this reason, the felling-burning interval tends to be longer in primary forest swiddens and shorter in secondary forest swiddens (P = .0007 for twenty-three 1976 swiddens). During the 1976 swidden year at Tikul Batu, this interval averaged eighty-seven days in primary forest swiddens and thirty-nine days in secondary forest swiddens. This corresponds very well with what are regarded as ideal intervals. The Kantu' say that they try to allow for a felling-burning interval of 2½-3 months in primary forest swiddens, while trying to allow for an interval of just one month in secondary forest swiddens. An interval of one month would be totally ineffectual in a primary forest swidden. The Kantu' say that the leaves alone of a felled, primary forest tree require more than one month to dry out. This is reflected in the failure of household #8 to complete felling of the forest section selected for its 1976 Kenua' swidden. They ceased felling a full forty-two days before the date on which the burn eventually took place, despite the fact that one-third of the swidden site had not been felled (although it had been slashed). They said that it was pointless to continue since there would no longer be enough time to dry felled trees before the burn.

Given that the ideal felling-burning interval varies between primary and secondary forest swiddens, and given also that both types of swiddens must be burned in the same month, the Kantu' necessarily fell their primary forest swiddens early in the year and their secondary forest swiddens late in the year. The average primary forest swidden harvested in 1976 was felled during the period from 23 April to 18 June 1975; while felling was carried out in the average secondary forest swidden between 10 July and 28 July of the same year.

The duration of the felling phase is much briefer than the duration of the slashing phase. In 1976 the average household at Tikul Batu devoted a total of forty-four workdays to felling in all of its swiddens. Given an average household work force for felling of 1.7 persons, this represented a work load of twenty-six days per worker. The average household performed this work over a period of fifty-one days. Thus, the intensity of utilization of labor during felling was 51 percent (viz., 26/51). This can be contrasted to the rate of just 25 percent attained during the slashing phase. In short, felling is carried out twice as fast as slashing. Because felling is timed closer to the burn than is slashing, it is more strongly influenced by the concerns with drying and regrowth than is slashing. As felling progresses and the burn approaches, the Kantu' increasingly worry about drying out their primary forest swiddens and preventing regrowth in their secondary forest swiddens, as a consequence of which they try to complete the felling of the former quickly and early (viz., well before the burn) and the felling of the latter quickly and late (viz., closer to the burn).

The pace of felling could be even quicker than it is, but for the fact that it is subject to some of the same constraints that were noted in the discussion of slashing. Thus, felling is apt to be interrupted by other swidden activities, notably the harvesting, processing, transporting, and storing of the rice crop from the previous year's swiddens. In addition, just as slashing often is interrupted for ritual reasons, so too is felling. Out of every ten days during the average 1976 felling period, four days were proscribed for work as a result of ill omens or the phases of the moon. Assuming that felling was performed on five of the remaining six days (based on a labor intensity of 51 percent), ill omens would have forced the workers to quit early on two or three of those days as well.

One final aspect of timing concerns the postponement of felling. Under certain circumstances, the Kantu' slash all or part of a section of primary forest but do not go on to fell it that same year (which practice is called nasau, after tasau 'a slashed but not felled section of forest'). Such failures to carry out felling may be intentional or unintentional. In the first case, the household involved may slash the forest in question for just a day or two, solely to establish ownership rights, without any intention of farming it that year. The second case, of unintentional failure, occurs when a household's farming schedule is disrupted by death, illness, or an exceptionally ill omen. In such cases, the slashing of a section may be well along or even completed when the decision to postpone is made. For example, household #12 had just completed the slashing of a proposed 1976 primary forest swidden (at a cost of twenty-seven workdays) when the death of a child in the household caused them to postpone any further work, including felling, until the following year.

If the Kantu' slash all or part of a primary forest swidden one year and then return to fell and plant the swidden the following year, no reslashing is necessary. In the space of one year, only minimal regrowth will occur in slashed, primary forest vegetation; and any regrowth up to a height of one meter will, according to the Kantu', be crushed when the primary forest trees are felled. If they do not return until the second or third following year, however, the regrowth will have progressed to the point where it will require reslashing, meaning that the labor previously devoted to slashing there will have been wasted. For this reason, if the Kantu' slash but do not fell in given year, they always attempt to fell during the following year. In the example cited above, household #12 felled and farmed in 1977 the swidden that had been slashed in 1976. Only when the Kantu' have devoted little labor to slashing, will they allow more than a year to pass before returning to the site. The postponement of felling in any case is restricted to primary forest. If the Kantu' slash a section of secondary forest and then set is aside for just one year, that single year's regrowth will be so extensive as to necessitate complete reslashing of the forest before they can fell it. However, this initial regrowth also will be so dense as to make any reslashing impossible until an additional four to five years have passed, and the undergrowth has matured and thinned out.

Chapter Four
Burning

The burning of the swidden serves several different purposes. The most important of these is conversion to ash of the slashed and felled vegetation, as well as the layer of humus on the forest floor. It is this ash that largely will sustain the crops later planted in the swidden. A second purpose of burning the swidden is to eliminate obstructions to later work therein. If not burned or if not burned well, the slashed brush and felled timber will hamper the workers who later plant, weed and harvest the swidden. Indeed, the Kantu' cite this as one of the main reasons for not planting any part of a swidden that does not burn well. Unburned brush and timber also pose a threat to the rice plants if the swidden ever is flooded. The third and final purpose of the burn is to kill or retard the growth of any vegetation still living in the swidden, which would otherwise compete with the swidden crops for scarce sunlight, moisture and nutrients. For example, a good burn will kill trees that were deemed too difficult to fell. A good burn also will kill (or at least retard the growth of) the sprouting trunks of felled trees, which otherwise must be slashed during the weeding phase. Finally, a good burn will kill many of the germinating seeds whose growth has been stimulated by removal of the forest canopy. Following a poor burn, the Kantu' say, the growth of pioneering weeds - which must be weeded out before the rice harvest in many cases - is much more rapid. This can be seen within a single swidden: thirteen days after swidden #55 was burned, seedlings ten to thirty centimeters in height were present in unburned portions of the swidden, but no seedlings at all were present in the portions that burned well.

The several purposes served by the burn are of singular importance in the swidden cycle. Indeed, the Kantu' say that the success or failure of the rice crop is determined chiefly by the success or failure of the burn. When a household experiences a bad burn they expect the worst in consequence, and they plan accordingly. For example, household #13 said that they decided to make a third 1976 swidden, in swampland, only after experiencing a poor burn in one of their two dryland swiddens (viz., their Kenua' swidden). That is, the harvest from this additional swamp swidden was intended to offset the poor harvest expected from the inadequately burned Kenua' swidden. The Kantu' believe that a poor burn presages a poor harvest except under the most extraordinary circumstances. One of the most benign omens in the Kantu' world is the fat lempedu 'gallbladder' of a domestic pig. When this is encountered, during hieromancy, the Kantu' say that it signifies a good harvest will be reaped during the coming year 'even if the swidden does not burn well'.

The importance to the Kantu' of a successful burn is reflected in the influence of the burn on the chronology of the entire swidden cycle. In chapter 1, for example, I noted that an important determinant of the minimal length of the fallow period is the combustibility of the vegetation. And in chapters 2 and 3 I noted that the timing of slashing and felling is largely a function of the timing of burning and planting. The timing of planting itself may be varied, somewhat, according to the needs of the burn. Thus, in a year in which there is a long but slightly early drought, the Kantu' may take advantage of it by burning earlier than usual, even though this means planting earlier as well.

The priority accorded to the burn is further reflected in Kantu' customary law, which effectively frees anyone burning a swidden from all encumbrances. Thus, a burner is not required to honor any ritual proscriptions against work in the swidden. Nor must a burner be concerned with any threat that his/her blaze may pose to the property of others. As the Kantu' say, Asa nunu umai, kitai nadai tau salah 'When burning swiddens, we can do no wrong'. All that is required of the burner under customary law is token protection of any property (viz., a hardwood tree) within the swidden that belongs to someone else, plus prior warning to anyone owning property adjacent to this swidden. For example, when household #8 burned its Kenua' swidden in 1975, the flames inadvertently burned one-quarter of an adjoining and as yet unburned swidden belonging to household #12. The lack of direction and hence spottiness of the burn in this latter swidden rendered most of it unsuitable for subsequent burning and planting. Household #12 sought legal redress from household #8, but the latter was held - in a longhouse moot - to be not responsible.

In addition to being an important determinant in the swidden cycle, the burn also represents an irrevocable point of commitment within it. During or following the prior phases of selecting, slashing or felling, the completion of the swidden may be postponed for a year or more, or even aborted. This is never done following the burn. Once a swidden is burned, it is always planted, weeded and so on, to the end of the cycle. The extent to which the burn marks a household's commitment to a given swidden is reflected in Kantu' customary law governing land disputes. Following adjudication of a dispute, a swidden that has been selected, slashed and even felled by one household may yet be awarded to another household; but this is never done once a swidden has been burned.

TECHNOLOGY

The only tool used during the main burn is a torch. These measure approximately 1½ meters in length and are made from dead and dried bamboo (of the variety betong), or else a rolled section of tree bark from the siding of an old swidden house. In either case, one end of the torch is beaten with a wooden club until it is finely split, so as to light better initially. Then, the Kantu' place inside each torch a small bundle of objects wrapped in rice straw. This includes pieces of each of the following: adze handle, bark rice bin, drum body, rice pestle, rice mortar, bamboo (of the variety temiang), and porcupine quill. Each of these objects symbolizes 'dryness', 'hardness', and 'strength', all qualities that the Kantu' believe will assist them in burning their swiddens.

The swiddens are burned by small parties of workers, numbering from one to four persons, usually all from within the owning household. The burns typically are carried out at midday. Between sunrise and the time of the burn, the persons intending to participate in the burn are forbidden to either drink or bathe. To quench their thirst, they are permitted only to chew sugar cane. Any contact with water, it is believed, will jeopardize the forthcoming burn. The poorest burn among the 1976 swiddens, that in the Kenua' swidden of household #13, was attributed to a violation of the bathing proscription by two members of the household. A second proscription must be observed just before the burn is begun. Namely, the workers must remove all garments from the upper part of their bodies. The Kantu' believe that if they fail to remove this clothing, the fire similarly will fail to remove (viz., consume) the kulit tanah 'earth's skin' or humus on the surface of the swidden. When the workers finally are ready to start the burn, they light their torches and then walk or run along the edges of the swidden, putting the torch to the mass of dried leaves, brush and timber at intervals of several meters. When the torch is used, it is dipped towards the ground. Otherwise, its burning end is held aloft, so that it will burn slower and last longer. As the workers burn, they cry out to the winds, beseeching them to blow hard and thereby ensure the success of the burn.

The movement of the workers during the burn is determined by the prevailing wind direction. The Kantu' always begin a burn at the windward edge of the swidden. This entire edge is set on fire first, and then the workers fire the two perpendicular sides, in a technique called ngerentai (figure 19). In a variant technique, called

FIGURE 19
Ngerentai Technique Of Burning The Swidden

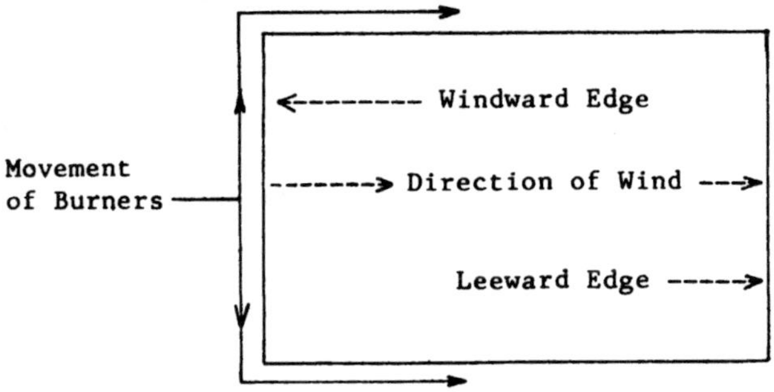

ngelayang 'skim across', one of the workers leaves his/her fire line on the perimeter and cuts through the center of the swidden, firing as he/she goes (figure 20). The purpose of this is to reduce the maximum distance that the fire must travel (viz., must be pushed by the wind) in order to reach all points in the swidden. Ngelayang also is practised in response to a change in wind direction, in an attempt to maintain a fire line upwind from the unburned portion of the swidden. At all times the objective of the burners is to maintain a fire line perpendicular to the direction of the wind, so that the direction in which they wish the fire to burn is the same as the direction in which the wind is blowing. This is important, because wind direction will determine burn direction regardless of other variables (e.g., the direction of the fire lines, the slope of the swidden, etc.). The Kantu' also say that it is

FIGURE 20
<u>Ngelayang</u> Technique Of Burning The Swidden

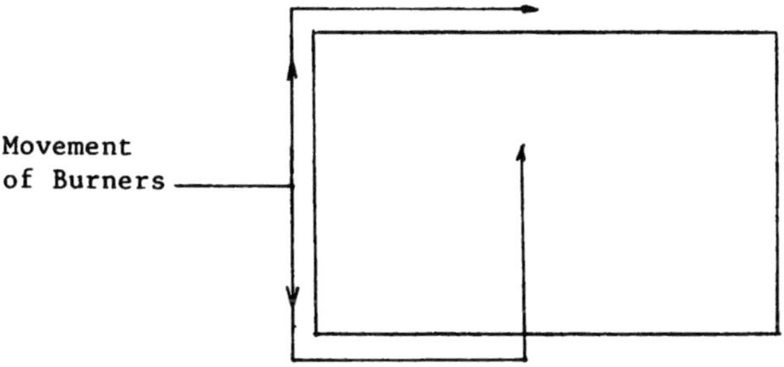

important to fire with the wind, because this precludes any possibility (unless the wind direction changes) of the workers being caught downwind from the fire and placed in danger from the smoke, heat and flames.

The Kantu' always burn their swiddens at midday. The observed range of variation is very limited, from 11:30 to 12:30, with each burn typically lasting one hour. There are two, principal advantages to burning at this time of day. By waiting until midday as opposed to burning earlier in the morning, the swidden receives five to six additional hours of sunshine just before it is burned. These final, few hours of sun improve the burn by evaporating any dew or light rainfall that accumulated in the swidden during the early morning or previous night. Timing the burn at midday also allows the Kantu' to take advantage of thermal updrafts, which are produced by the heat of the sun over any open, cleared land - such as a swidden waiting to be burned. These updrafts tend to be stronger at midday when the sun's heat is greater (air temperatures average 28.2° C), than early in the morning, when the sun's heat is weaker (air temperatures average just 23° C). Any air movement tends to fan the flames of a swidden burn, and the greater the movement the greater the benefit, so this is another reason for timing the burns at midday rather than in the morning.

It might appear that if a midday burn is advantageous, a late afternoon burn would be even more advantageous. If not burned until late afternoon, the swidden would receive a full day of sun and hence would be drier before being burned. In addition, the sun's heat is greatest in the afternoon (from 15:00 to 16:00) and hence thermal updrafts over the swidden also would be strongest at that time. The reason the Kantu' do not time their burns in the late afternoon, despite these advantages, involves the increasing likelihood of rainfall. At any given time, the greater the number of preceding consecutive hours of sunshine, unbroken by rainfall, the greater the likelihood that it will rain. On this basis it would be safest to burn early in the morning, after just an hour or two of sunshine. For example, on 9 September 1975, households #1 and 11 were preparing to fire their respective, secondary forest swiddens (#36 and 56) when a light shower commenced, at 11:55. Eight days passed before they had another opportunity to fire their swiddens, which represented a significant delay in the already late schedules of these two households. Timing the burn at midday represents a compromise, therefore, between the threat of rainfall - which is met by not burning too late in the day - and the benefits of additional sunshine - which are secured by not burning too early.

During the burn some effort may be devoted to protecting trees with either secular or sacred value that lie within the boundaries of the swidden. The method of protecting such trees from the fire is twofold. First, the immediately surrounding trees are not cut during slashing and felling, as noted earlier. Since these surrounding and still living trees will not burn very well, if at all, they ensure that no strong flames burn in the immediate vicinity of the valued trees. This is further ensured, shortly before the burn is to take place, by scraping the ground litter away from each copse of standing trees, to a distance of setengah depa' 'one-half fathom' (approximately eighty centimeters). The responsibility for these measures falls upon the household that owns and burns the swidden, even when the trees in question are owned by some other household[1]. While the former household is required to make this effort, it is not liable should it prove to be unsuccessful. In most cases, however, these protective measure are successful, with some exceptions in primary forest swiddens. The great mass of timber in such swiddens makes them difficult to dry out, and hence difficult to burn. However, when there is an opportunity to thoroughly dry out a primary forest swidden, the great mass of the timber has an opposite effect and ensures a fierce burn, against which all such protective measures are unavailing.

Some effort also may be devoted to protecting standing forest and old swidden houses that lie outside of but proximate to the swidden being burned. In the case of swidden houses, the protective measure consists in sprinkling some rice beer over the house. According to the

Kantu', rice beer is not only wet, it also is ritually 'cool'; and they believe that this property enables it to combat a threat from anything with the opposing qualities of hotness and dryness, such as a swidden fire. In the case of standing forest, protective measures consist in removing all fallen timber, brush, and leaves from a narrow strip of ground running along the common border of the swidden and the section of forest. These two protective measures rarely are carried out, however, because they rarely are needed. Among the sixty-nine separate swiddens made at Tikul Batu during 1975 and 1976, there was not a single instance in which a swidden fire burned into adjoining forest for a distance of more than four or five meters. Given the volume and regularity of rainfall in the Empanang valley, it is difficult for the forest to burn even after it has been slashed, felled and dried, and it is next to impossible for it to burn while it is still alive and green.

The greatest threat from swidden fires, and the one that stimulates the greatest preventive measures, is to adjoining, unburned swiddens. It may happen that an adjoining swidden is <u>matak</u> 'green' - not yet dry and hence not yet ready to be burned - as a result of being cleared later in the year than the swidden being burned. In such a case, if the former inadvertently catches fire during the burning of the latter, the Kantu' say that it is unlikely to burn well; but it may burn just enough, in scattered patches, to ruin its own chances of later being successfully burned. This is because burned patches will act as firebreaks during any later attempt to burn the whole swidden. Even if an adjoining swidden is ready to be burned, it is unlikely to burn well if the burn is started by accident. A burn cannot be successful unless it is carried out according to a calculated strategy, primarily based on wind direction. The consequences of an accidental burn can be seen in the previously mentioned case of the 1975 Kenua' swidden of household #12, which caught fire during the burning of an adjacent swidden belonging to household #8. Only 75 percent of this swidden burned, which compared with a mean figure of 96 percent for the other dryland swiddens burned that year.

The Kantu' employ a number of measures to avoid the accidental firing of adjoining swiddens. When two different households clear adjoining swiddens, they may agree beforehand to leave standing a narrow strip of brush and trees along the common side, called an <u>antara</u> 'division', to function as a firebreak in the event that one household burns its swidden before the other. This is likeliest to be done when different burn dates are predicted from different clearing schedules in the two swiddens. Whether a firebreak exists or not, a household preparing to burn a swidden must inform all households owning adjacent swiddens of its intended burn date. This prior warning allows the other households to prepare to burn on the same date, if their swiddens are ready to be burned. One or more of these households may request a postponement of a day or two if the date is inconvenient. For example, households #2, 4 and 14 initially intended to fire their proximate 1976 Kenua' swiddens on 6 August, 1975. When they communicated their intentions to several Iban households that had cleared adjacent swiddens just over the boundary to Tikul Batu's territory, the latter requested and received a one-day postponement. A request to postpone a firing date is not obligatory, however, and in no case, according to the Kantu', would a household agree to a postponement of more than one or two days. In the event that a household is informed of an intended burn, and their own, adjacent swidden is not ready to burn, prior notification enables them to undertake protective measures. If a firebreak of living trees does not exist, they will make the functional equivalent by removing all felled timber, brush, leaves and humus from a narrow strip along the common border of the two swiddens. On the day of the burn, the household that is not ready to burn will guard this firebreak, putting out any flames that cross it to their own swidden. The responsibility for making and guarding the firebreaks lies with the owners of the swidden adjoining the one being burned. The household that carries out the burn is required to share in this responsibility only if it burns unusually early in the year (e.g., in the fifth or sixth moons of the Kantu' calendar). As noted earlier, the household carrying out a burn has minimal responsibility to contain it, which reflects the fact that the Kantu' worry more about achieving adequate burns than about protecting themselves from overly fierce burns.

CHRONOLOGY

There is considerable interswidden and interhousehold variation in the timing of the burn. Among twenty-nine 1976 swidden for which data are available, the range in burn dates was eighty-eight days (from 1 August to 3 November 1975), with the mean date being 31 August and the standard deviation from this being twenty days.

The average burn dates for primary forest, secondary forest and swampland swiddens were 20 August, 1 September and 15 September, respectively. This variation is associated with three other variables: (1) the length of the interval between slashing/felling and the burn (viz., the 'drying interval'); (2) the length of the interval between the last rainfall and the burn (viz., the 'drought interval'); and (3) the length of the interval between the burn and planting (viz., the 'burning-planting interval')[2].

The Kantu' believe that a swidden will burn well only if the correct amount of time has passed since slashing and felling the forest. If this interval is too short, the vegetation may not be dry by the time of the burn and it will not burn well. If this interval is too long, the cleared swidden may be covered by regrowth before the burn can be carried out and, again, it will not burn well. Among the 1976 swiddens, the drying interval (measured here and throughout this discussion as the interval between the midpoint of the combined slashing and felling phases and the burn date) averaged 120 days in primary forest swiddens, 56 days in secondary forest swiddens, and 37 days in grass covered swampland swiddens. The relative brevity of the drying interval in secondary forest and grassland swiddens and the relative length of this interval in primary forest swiddens is due to the fact that regrowth poses the greatest problem in the former swiddens whereas drying poses the greatest problem in the latter. The relationship between tree size and the problem of drying can be explained, in part, in terms of the ratio of the surface area of the tree trunk - which is one measure of the ability of moisture to leave it - to the volume of the trunk - which is a measure of the amount of moisture initially contained in it. This ratio of surface to volume, which thus is a measure of the abililty or speed of a felled tree to dry out, decreases as the size of the trunk increases (by a constant 50 percent whenever trunk girth is doubled)[3].

The importance of a drying interval of proper length is reflected in the way the Kantu' select their burn dates. The ideal time in which to burn the 1976 primary forest swiddens was during the thirteen-day drought at the beginning of August 1975, since this was the longest dry spell by far of that burn season. Of nine primary forest swiddens, five were burned during this drought and the remaining four were burned later in the year. By the first day of this drought, the former had been drying out for an average of 132 days, while the latter had by then only dried for an average of

55 days. Because the drying interval was still insufficient in these latter cases, the advantages of this lengthy drought had to be passed up. These four swiddens were not burned until late August or early September, by which time their drying intervals were appreciably longer, averaging eighty-eight days at the time of their burns. The length of the drying interval similarly was involved in determining when the 1976 secondary forest swiddens were burned. The ideal burn date for these swiddens was seen by the Kantu' as 28 August 1975. This date was preceded by five days of little rain, and it also was close enough to the anticipated date of planting such that regrowth would not pose a problem. Among eleven as yet unburned dryland, secondary forest swiddens, eight were burned on this date and three were burned at a later date. The former had been drying for an average of fifty days and the latter for an average of only thirty-six days by August 28. When the latter three swiddens eventually were burned, their drying intervals had attained the more adequate length of, on average, fifty days.

There is a limit as to how late in the year the burn can be pushed, however, due to restrictions on the timing of planting. If planting is pushed back beyond its normal date by more than a week or two, a variety of problems arise (e.g., increased predation by swidden pests). For the most part, therefore, achievement of a proper drying interval involves varying not the timing of the burn or planting (by pushing them back in the year), but the timing of slashing and felling (by pushing them forward in the year). That is, the importance of a correct drying interval typically affects the timing of the preceding swidden phases more than it does the timing of the succeeding phases.

The overall drying interval is a less important determinant of the timing of the burn than the shorter interval of drought that occurs immediately before the burn. There is considerable variation in the length of droughts from year to year. During the traditional burn months of August and September, there was a thirteen-day drought in 1975, a drought of almost one month in 1974, and no drought at all in 1973. The principal means whereby the Kantu' attempt to predict this variation involves the observation of certain key indicators in their environment. One such indicator is the fruiting cycle of the candlenut tree. If these trees give signs that they are going to bear fruit (as they do approximately one out of every four years), the Kantu' take this as an indication that there will be a long drought that year. A second means of predicting drought length is based on probability logic. If a lengthy drought has occurred each year for two or more years, the Kantu' expect the likelihood of a drought in the coming year to be relatively low. However they make their predictions, and depending upon what these predictions are, the Kantu' then decide whether to make swiddens that require a lengthy drought (viz., those cut from primary forest and/or located in the floodzone) as opposed to swiddens that do not (viz., those cut from secondary forest and/or located above the floodzone).

Just as the length of the yearly droughts vary, so too does their timing. The Kantu' say that they used to be able to count on a drought occuring during the seventh and eighth moons of each lunar year (viz., the period from early August to early October). Today this expectation no longer holds. As the Kantu' say, Taun na' tentu nto' 'The year is not regular now'. Indeed, as noted in table 10, today there is little clear monthly or seasonal pattern of variation in the rainfall and droughts at Tikul Batu. The Kantu' have but limited means to deal with the uncertain timing of the latter. If a lengthy drought occurs early in the year, the Kantu' may try to burn earlier than usual. The Kantu' cannot move the burn forward by more than a few weeks, however, because any appreciable lengthening of the interval between burning and planting results in problems with regrowth. If they attempt to maintain a constant interval between burning and planting by moving the date of the latter forward as well, problems with pest predation will result. It is similarly difficult for the Kantu' to move the burn back in the year, by more than a week or two, in order to wait for a late drought. This would necessitate moving the planting date back and, again, this would result in pest problems. These constraints are exemplified by the last burn among the 1976 swiddens (excluding the swamp, grassland swiddens), namely that of swidden #45. This swidden was burned on 21 September 1975, even though two and one-half millimeters of rain had fallen during the previous twenty-four hours, and thirty-seven millimeters had fallen during the previous forty-eight hours. Burning at this time was an act of desperation, reflecting the fact that concern for the correct timing of planting had taken precedence over concern for the correct drought interval.

There is one way in which the Kantu' can respond to the ill timing of rainfall and drought without encountering these problems. If a household making a primary forest swidden fails to achieve a drought in good time, it can postpone the burn until the following year.

This practice, called ngerukoh (from kerukoh, the term for a swidden in which the burn has been postponed), is possible only in primary forest swiddens. Slashed and felled primary forest can be left to sit for one year and then, after just minimal slashing of resprouting trunks, it can be burned and planted. This one-year postponement even carries some advantages because, according to the Kantu', it ensures that the huge, partially burned branches and trunks of the primary forest trees will be more decomposed by the time that the swidden finally is planted. This, in turn, means that more of their nutrients will have been released into the environment - and potentially will be available to planted crops - than is the case when the swidden is burned and planted in the same year that the forest is cleared[4]. After a postponement of two or more years, however, the regrowth would have progressed too far to make successful burning of the land possible; so the Kantu' never postpone the burns of their primary forest swiddens for more than one year. Nor do they ever postpone the burn in secondary forest swiddens. After a postponement of just one year, the rapid regrowth in these swiddens would be so considerable as to make burning impossible without further clearing; and further clearing would be impossible without the passage of several years time and the thinning out of the undergrowth.

Postponement of the burn is a long-term response to the length and timing of droughts. More often the Kantu' are limited to short-term responses. These consist, first, in trying to time swidden burns after a drought as opposed to after a rainfall. If a small but significant amount of rain (viz., anything greater than one or two millimeters) falls during the twenty-four hours preceding the planned burning of a swidden, the Kantu' say that the burn should be postponed for one day in order for the swidden to dry out. If it is not postponed, they say, the burn will not be completely successful. Table 32 examines the occurrence of rainfall on each of the forty-six days, from 7 August to 21 September 1975, that comprised the 1976 burn season[5]. For each day, rainfall was measured for the twenty-four hour period previous to noon of that day (given that burns always are carried out at midday). The findings of the table support the thesis that the Kantu' try to avoid burning within twenty-four hours of a rainfall greater than 1.5 millimeters. The strength of this support is greater than indicated (viz., not significant at the .05 level), because the decision to burn on the seven rainless days actually was made not just seven times, but twenty-five times, since twenty-five separate swid-

TABLE 32
Rainfall During The Twenty-Four Hours Preceding The 1976 Swidden Burns

		Amount of rain during preceding twenty-four hours	
		< 1.5 mm	> 1.5 mm
Number of swiddens burned each day	≥ 1	7	2
	0	14	23

N = 46 days (7 August - 21 September 1975).
P = .04

dens were fired on these seven days; whereas the burning on the two rain days represented just two decisions in two separate swiddens.

While a light rainfall (viz., from 1½ to 15 millimeters) usually necessitates a postponement of the burn by at least twenty-four hours, a heavy rainfall (viz., greater than fifteen millimeters) usually necessitates a postponement of at least seventy-two hours, according to the Kantu'. They say that this is necessary for the swidden to completely dry out again before a burn is attempted. The extent to which the Kantu' follow this second rule is demonstrated in table 33. Again, the significance of the association between burning and rainfall is greater than shown, because the decision

TABLE 33
Rainfall During The Seventy-Two Hours Preceding The 1976 Swidden Burns

		Amount of rain during preceding seventy-two hours	
		< 15 mm	> 15 mm
Number of swiddens burned each day	≥ 1	6	3
	0	9	28

N = 46 days (7 August - 21 September 1975).
P = .02

to burn on the three days preceded by heavy rain represented burn decisions in only seven swiddens, whereas the decision to burn on the six days preceded by no heavy rain represented burn decisions in twenty separate swiddens.

The Kantu' try to pick dates for burning their swidden that not only are preceded by relatively much drought (and little rainfall), but that are followed by relatively little drought (and much rainfall). That is, when a drought occurs during the burning season, the households that are ready to burn their swiddens try to hold off burning until the last day or two of the drought, so that their swiddens receive its fullest benefit. However, if they misjudge the length of the drought and it is broken by rainfall before they decide to burn, then they lose not just part but all of its benefits. The considerable ability of the Kantu' to avoid such losses, by correctly predicting the ending of drought periods and timing their burns accordingly, is demonstrated in table 34 (based, again, on the forty-six day season in which twenty-seven 1976 swiddens were burned). Table 34 shows that whereas 46 percent of the

TABLE 34
Pattern Of Drought Following The 1976 Swidden Burns

	Followed by 0 Days of Drought	Followed by 1 Day of Drought	Followed by 2 or More Days of Drought	Total
Total number of days during burn season	54%	26%	20%	100% (46 days)
Number of swiddens burned on days	89%	7%	4%	100% (27 swiddens)

days during the burn season were followed by one, two or more days of drought, only 11 percent of the swiddens were burned on such days. In contrast, 54 percent of the days during this season were followed by no drought days at all (meaning that rain fell on the following day), but fully 89 percent of the burns were carried out on such days.

The Kantu' utilize a number of different climatic and biotic indicators to predict the end of a drought and the onset of rainfall. These indicators include, for example, the rian 'durian tree', the fruit of which is known to fall when a long period of drought is about to be broken by rainfall. In 1975, the durian fruit were first reported falling on 26 August; and indeed, while 56 millimeters of rain fell during the ten days preceding this date, more than three times as much (viz., 183 millimeters) fell during the succeeding ten days. The shortest-term indicators of rainfall, which forecast rain within a few hours, include the appearance of the daytime sky, the opacity of the night sky, wind noise in the upper tiers of the forest in distant parts of the valley, and the sudden, massed humming of cicadas.

If on the morning of a given day, the Kantu' forecast rainfall later in the afternoon on the basis of these indicators, they may decide to burn at midday. Rainfall that comes even an hour or two after a burn is of no consequence, and the close timing of burning and rainfall is common. The burning of twenty-four out of twenty-seven wooded (viz., nongrassland) 1976 swiddens was followed within twenty-four hours or less by rainfall. Errors in timing do occur, however. As noted earlier, on 9 September 1975 households #1 and 11 were within one hour of beginning to burn their respective downstream Kenua' swiddens, when a light rain began to fall (at 11:55), forcing them to cancel their plans. If they had anticipated this rain, they could have burned one hour earlier with satisfactory results. As it was, their mistiming forced an eight-day postponement of their burns.

In addition to predicting the ending of droughts, the Kantu' must be able to monitor their progress. Monitoring is a problem because the household members usually do not sleep in or near their swiddens at this stage of the swidden cycle. Given that the swiddens may be distant from the longhouse by three kilometers or more, it is possible for rainfall to wet a swidden without being observed by prospective burners at the longhouse. For this reason, the Kantu' must be adept at monitoring rainfall after the fact. For example, on the morning of 6 August 1975, three members of household #13 left the longhouse with the intention of burning their 1976 Kenua' swidden. No rain had fallen in the vicinity of the longhouse during the preceding night. However, upon arriving at their swidden shortly after 9:00, the three observed glistening leaves as well as standing water in the hollows of logs and in fallen, curled up leaves lying on the forest floor near the periphery of their swidden. On the basis of these signs, the party concluded that a light but still damaging rain had fallen there in the late afternoon of the previous day. To enable the swidden to dry out from this rainfall, they postponed the burn one day.

A final determinant of the timing of the burn - which probably has an equal effect on the timing of planting - is the length of the burning-planting interval. The desired interval varies according to the type of swidden involved. In the case of secondary forest swiddens, the Kantu' say that the burning-planting interval should not exceed one week. If the interval greatly exceeds one week, the growth of weeds will get too much of a headstart over the growth of the rice plants. Actual burning-planting intervals approach the cultural ideal: they averaged 9.5 days among secondary forest swiddens in 1976. In primary forest swiddens, postburn weed growth poses less of a problem because it does not commence as quickly. According to the Kantu', however, primary forest swiddens have their own problems involving toxic residues from the burn. These residues, which the Kantu' call bisa sepok 'smoke poison', are said to be both peculiar to the primary forest swiddens and injurious to rice plants. The danger from these toxins, according to the Kantu', is reduced with time and rainfall. They say that though rice will not grow well in primary forest swiddens if planted soon after the burn, it will grow all right if planted after an interval of one to two months. This stipulated interval is approximated in fact: actual burning-planting intervals averaged thirty-three days among the 1976 primary forest swiddens[6].

SUCCESS OF THE BURN

Because of a belief that boasting invites supernatural retribution, the Kantu' invariably describe their swidden burns in disparaging terms to anyone from another household. They will say of the swidden, Nadai ala' 'Not burned' or even Nadai isi' ti celom 'Nothing is blackened [by the fire]'. This public stance notwithstanding, there are several measures by which each household privately judges the success of its own burns. The Kantu' say that a poor burn is one in which even the humus layer is not consumed in the fire. Land that has burned this poorly - termed tanah matak 'raw earth' - cannot be planted unless redressive operations are first carried out on it. In a better burn, the humus layer is consumed and the yellow topsoil beneath it is exposed. All parts of a swidden that burn to this extent are planted. In the best possible burn not only is the humus layer consumed, but so too is the brush and light timber lying atop it. The sign of this is tree limbs that have been mutus 'broken in two' by the fire. The Kantu' say that after such a burn, the surface of the swidden is so clear as to be tawas bejalai 'open for walking' (in contrast to the preburned swidden, which is practically impassable). It also is possible for a swidden to burn too well. In an overly fierce blaze, the flames not only consume the humus layer, exposing the soil beneath, but they also scorch the soil itself, hardening it and making it less suitable for cultivation.

Causes of Success or Failure

During the 1975 and 1976 farm years, the percentage of the swidden that burned varied between 50 and 100 percent in dryland swiddens, and between zero and 100 pecent in swampland swiddens. The factors that account for this variation can be divided into two groups: those that determine the combustibility of the vegetation on the day of the burn, and those that determine the influence of the wind on the day of the burn.

The combustibility of a swidden varies, first, according to the type of forest from which it was cleared. As I suggested earlier, secondary forest is easier to dry out than primary forest and so it usually is more combustible (table 35). Among the primary forest

TABLE 35
Forest Type And The Success Of The Burn

		The type of forest from which the swidden was cut was predominantly	
		Primary forest	Secondary forest
Percentage of the swidden that burned	< 90%	4	0
	> 90%	4	7

N = 15 1976 nonfloodzone swiddens.
90% = mean percentage of the swiddens that burned in this sample.
P = .05

swiddens, combustibility and the success of the burn increase in direct relationship with the length of the drying period (table 36)[7]. The strength of the burn is

TABLE 36
Length Of The Drying Period And The Success Of The Burn In Primary Forest Swiddens

		Length of the drying period	
		< 120 days	> 120 days
Percentage of the swidden that burned	< 84%	4	0
	> 84%	1	3

N = 8 1976 primary forest swiddens.
84% = mean percentage of the swidden that burned in this sample.
P = .07

a function not only of the dryness of the felled timber but also its volume. Because this volume is greater in primary forest than in most sections of secondary forest, a long dried primary forest swidden will in fact burn better and more fiercely than the average secondary forest swidden. The ferocity of combustion that can be attained in a primary forest swidden under ideal conditions is exemplified by swidden #43. This primary forest swidden was burned on 7 August 1975, and the burn was said to be a good one by Kantu' standards. Fourteen days afterwards some of the largest tree trunks in the swidden were still smoldering, despite the fact that over 150 millimeters of rain had fallen in the interim.

The extent to which a swidden burns also is affected by the extent to which it is exposed to moisture before the burn. One source of moisture is the river and its adjoining swamps. There is a marked association between the relative success of the burn and location within, as opposed to without, the floodzone (here eliminating the swampland swiddens from consideration), as shown in table 37. On the other hand, the burns do not invariably fare poorly in floodland. The Kantu' can mitigate the effect of poor drainage on the burn both by slashing thoroughly and by trimming all felled trees. A second mitigating factor is the pattern of rainfall and flooding. After a period in which no heavy rains have fallen in or near the territory of Tikul Batu, even land in

TABLE 37
Floodzone Location And The Success Of The Burn

		The swidden was located partially or wholly in the floodzone	
		Yes	No
Percentage of the swidden that burned	< 83%	5	5
	> 83%	0	12

N = 22 wooded, dryland 1976 swiddens.
83% = mean percentage of the swidden that burned in this sample.
P = .01

the floodzone begins to dry out. If swiddens located there are fired at this time, they may burn as well as swiddens located in normally drier areas.

The pattern of rainfall also has a direct effect on the success of the burn, regardless of where the swidden is located. As I previously noted, even the lightest rainfall jeopardizes any burn that follows it by less than twenty-four hours; and a heavy, soaking rainfall jeopardizes any burn that follows it by less than seventy-two hours. Light rainfalls usually are not problematic, because it usually is easy for the Kantu' to obtain one dry day after one or more days of light rain. Heavy rains are more problematic, because it is not easy to obtain three successive dry days following any given heavy rain. During August and September of 1975, for example, there were only three periods of three or more consecutive days during which less than 2.5 millimeters of rain fell each day. As a result, the Kantu' often are forced to burn within seventy-two hours of a heavy rain, even though this precludes the possibility of a good burn. Preburn rainfall has a decisive impact on the success or failure of the burn and, as a result, the success or failure of the eventual harvest. The Kantu' attributed the disastrous 1974 harvest to the lack of any drought before the burns, while the very good 1975 harvest was attributed to the occurrence of a one-month drought before the burns, and the average 1976 harvest was attributed to the occurrence of a ten-day drought before (many of) the burns.

Another factor that affects the combustibility of a cleared swidden is the extent to which it is shaded by adjoining forest. This in turn varies according to the type of forest involved. Towering primary forest shades more of the swidden and has a greater impact on its burn than does shorter, secondary forest. The impact on the burn also varies according to which sides of the swidden are adjoined by forest. The surface of the swidden is not greatly shaded by forest on its north or south sides, but only by forest on its east and west sides (figure 21). Nor is the impact identical in the latter two cases, according to the Kantu'. Adjoining

FIGURE 21
Shading Of The Swidden By Adjoining Forest

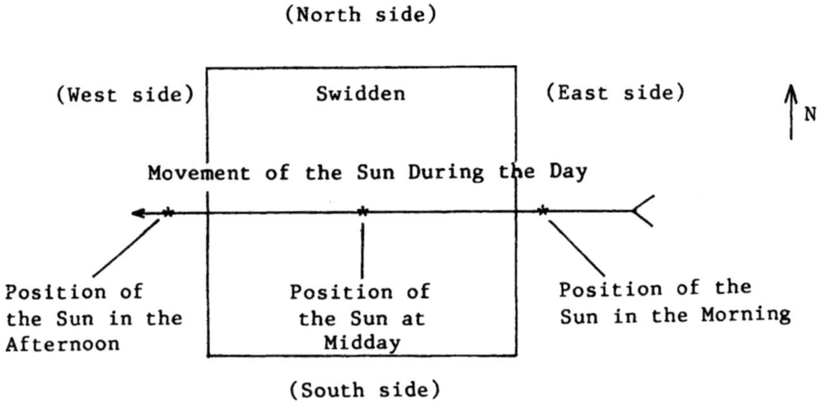

forest shades the western edge of the swidden from the rays of the afternoon sun, but exposes it to the morning sun; whereas the reverse is the case with forest adjoining the eastern edge. This difference has a varying impact on the burn, because the afternoon sun is much hotter than the morning sun. Measurements over a twelve-month period yielded an average afternoon temperature (at 15:00) of 32° C, compared with an average morning temperature (at 9:00) of only 25° C. During the month of August, which is usually the last month of drying before the swiddens are burned, an even more impressive difference of 9.5° C was measured. A final factor that affects the impact of shading from adjoining forest is the size of the swidden. As discussed in chapter 1, the ratio of swidden circumference to swidden area decreases by a constant as the area increases. Consequently, the percentage of the swidden vulnerable to shading from adjoining forest also decreases by a constant as swidden area increases. For example, a ten-meter strip shaded by adjoining forest along two sides of a swidden represents 14 percent of the area of a two-hectare swidden, but only 10 percent of the area of a four-hectare swidden.

The second, major determinant of burn success is the strength and direction of the wind on the day of the burn. The Kantu' are explicit in their belief that a strong wind is essential to a good burn, as evident in their practice of calling out to the wind during a burn and summoning it to their swidden. On the other hand, the Kantu' say that if the wind blows too strongly during a burn, it will blow the flames against the ground. This is undesirable because in such a case the flames may not burn tree trunks and branches lying slightly off the ground, but only the lower-lying leaves and humus. The earth may be scorched as well under these circumstances, rendering it less arable. The Kantu' cited these problems in the poor burns of at least three 1976 swiddens (viz., #39, 50 and 67).

The strength of the wind varies from one day to the next, and from one swidden to another. One factor in this variation is the altitude of the swidden. The higher the altitude, the greater the likelihood that it will be exposed to prevailing winds. The Kantu' say, in particular, that there is more wind (for burning) high up on the sides of the Empanang river valley than there is down in the valley basin. This difference may be reflected in the fact that the 1975 and 1976 burns in the former area averaged 88 percent (of the swidden surface burned), while those in the latter area averaged only 72 percent. A second factor that affects the strength of the wind is the extent to which it is blocked from the swidden by adjoining forest, in particular primary forest. Such blockage always is most severe along the windward edge of the swidden[8]. In figure 22, for example, assuming that the swidden is adjoined on all four sides by forest and given a wind blowing out of the east, the burn will be seriously obstructed only on side B of the swidden, its windward edge. Table 38 suggests

FIGURE 22
Blockage Of Wind From The Swidden By Adjoining Forest

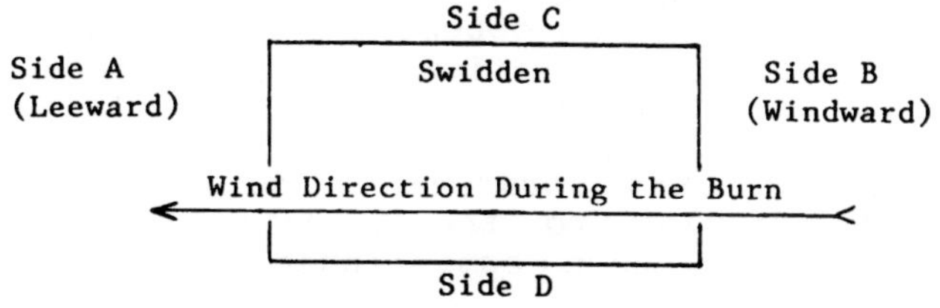

that such obstruction has a negative effect on the burn (although the association is not significant at the [.05] level).

TABLE 38
The Effect Of Adjoining Forest On The Burn Along The Windward Edge Of The Swidden

		Forest present along the windward edge of the swidden	
		No	Yes
Character of the burn along the windward edge of the swidden	Poor	1	4
	Good	3	0

N = 8 1975 & 1976 swiddens.
P = .0714

The direction of the wind is perhaps even more important than its strength. Already mentioned are the undesirable consequences of a wind that blows down against the surface of the swidden. Far more important is what the Kantu' term salah ribut 'incorrect wind'. This is a wind that blows from the direction in which lie the bases of the felled trees. The Kantu' say that felled trees will burn well only if the wind comes from the direction in which their tips lie. (They say the effect on the burn is uncertain when the wind blows from either side, called melintang 'cross'.) The basis for these beliefs lies, I believe, in the Kantu' practice of felling all trees in a downhill direction. Because of this the bases of felled trees always point uphill, while their tips point downhill. If the wind comes from the direction of the tree bases, therefore, it blows downhill; whereas if it comes from the direction of the tree tips, it blows uphill. In the former case the wind hampers the burn, because it blows contrary to the updrafts of heated air that otherwise would push the flames uphill. But in the latter case the wind, since it is blowing uphill, also pushes the flames uphill, thus enhancing the burn as opposed to hindering it.

Whatever the wind direction at the commencement of the burn, it is essential that it be maintained until the burn is finished. I earlier noted that the primary objective of the burn strategy is to maintain a line of fire perpendicular to the direction of the wind. Once this line has been established at the start of the burn, any subsequent change in wind direction necessarily alters this optimal relation between the line of fire and the direction of the wind. In the event of such a change, the Kantu' respond with a variation of ngelayang 'skim across' (figure 20). This involves one of the workers cutting across the swidden from one side to the other in order to establish a new fire line abreast of the new wind direction. Even given this redressive technique, the Kantu' regard a sudden shift in wind direction as one of the principal hazards of the burn. In the case of swidden #52, for example, the failure of one-quarter of the swidden to burn was attributed by onlookers to a change in wind direction midway through the burn (even though this was a change from blowing downhill to the otherwise desirable direction of uphill).

Neither the direction of the wind nor any other determinant of the burn is very susceptible to prediction and/or control by the Kantu'. As a result, there is a genuine element of serendipity in the success or failure of the burn. For example, the household with the least experience of swidden farming in Tikul Batu, #17, decided to burn its 1976 Sempinih swidden on 13 August 1975. This followed a total rainfall of just three millimeters during the previous fourteen days. The burn was acceptable. The household with the greatest knowledge of swidden farming in the longhouse, #5, also had a swidden ready for burning at this time, their 1976 Setungkus swidden. Instead of burning on 13 August, they decided to let their swidden dry out for another day or two. On 14 August, however, the drought was broken by seventy millimeters of rainfall. As a result, when household #5 finally burned, on 15 August, it was with far less success than that achieved by household #17. The Kantu' recognize the existence of this serendipitous factor in burning (as well as in other aspects of the swidden cycle), but they attribute it to the intervention of angry spirits. The anger of the spirits is attributed, in turn, to violations of either burn proscriptions (viz., drinking or bathing before carrying out a burn) or sexual proscriptions (viz., adultery, illegitimate birth, or incest).

Redressive Measures

As noted earlier, an overly weak burn does not produce sufficient ash, it does not sufficiently inhibit regrowth, and it leaves unburned timber to obstruct succeeding activities in the swidden. The Kantu' have several means for redressing the effects of such a burn, one of which is called <u>nunu seria'</u> 'burning again'. This usually is done on the day the swidden is planted (although it can be carried out any time after the main burn and before the germinating rice plants thrust their first shoots above the ground). As the planting party moves through the swidden and encounters the occasional clump of unburned but combustible vegetation, they fire it and then shortly thereafter sow it with rice. This is the extent of reburning. The Kantu' never attempt to reburn an entire swidden. They give two reasons for this. First, they say <u>'umai ngai alla'</u> 'a swidden will not burn' the second time around, even if large portions of it remain unburned from the first attempt. Indeed, any attempt to fire these unburned portions will be frustrated by the presence of the already burned portions, which will act as firebreaks. The second reason for not reburning entire swiddens is that this is ritually proscribed. The Kantu' justify this proscription by saying that after a swidden has been burned once, the vegetation in it is <u>bari'</u> 'musty'. This term usually refers to the bad smell of cooked food that has been stored too long. The Kantu' also honor a ritual proscription against reheating food after it has become <u>bari'</u>, which arguably safeguards their health. It is possible that the proscription against reburning the swiddens similarly safeguards the health of the rice plants, perhaps by preventing the scorching of the soil. This analogy also is suggested by the existence of a broader pattern of anthropomorphism, in which the Kantu' speak of the ash from the burn as the rice crop's <u>pemakai</u> 'food' and the ground in which the crop is planted as its <u>dapor</u> 'hearth'.

A more common redressive measure than reburning is stack burning, which takes two forms. If the burn was very bad, the Kantu' gather the unburned humus, leaves and branches into stacks and then burn them. This is called <u>manas</u>. If the burn was a little better, such that the layer of humus and leaves burned, the Kantu' only gather and burn the unburned branches, which is called <u>mandok</u>. In either case the stacked material can be burned after a day or two without rainfall, following which the ground covered by the stack is sown with rice. Stack burning always is carried out by the household's own labor force, largely by its adult males, although females and adolescents may assist in firing the stacks. The labor requirements of either type of stack burning are minimized, the Kantu' say, by any trimming that was carried out during the felling phase. In any case, stack burning can be quite time consuming and it rarely is possible to complete it during the immediate, postburn period, because of the need to devote most labor to planting. Most households work only sporadically at stack burning during the planting season, and do not complete it until after most planting in the longhouse is finished. Not atypical is the case of household #16, which burned its 1976 Lubuk Menserai swidden on 28 August 1975, planted it on 5 September, and did not finish stack burning (and additional planting in the vicinity of the burned stacks) until 4 October. The Kantu' maintain that stack burning can be carried out without any threat to the rice sown in the swidden during the main planting, so long as this rice has not yet sprouted above the ground.

The performance of stack burning varies greatly from one swidden to the next. It never is perfomed in a swidden that burns completely, and it almost always is performed in swiddens that burn incompletely (P = .002 for twenty-six 1976 swiddens). this association is so strong as to be used metaphorically. If the Kantu' dream of entering a burial ground, they will say Mesti sepuluh-duapuluh ari mandok 'Ten to twenty days of stack burning are certain'. That is, such a dream foretells the occurrence of a bad burn, requiring a lot of stack burning. The Kantu' are especially likely to perform stack burning in poorly burned swiddens located in the floodzone. The Kantu' say that the practice of manas is entirely restricted to the floodzone and that mandok is most common there as well. This is due to the fact that the burns tend to be worse within the floodzone, as also do the problems associated with poor burns (e.g., weed growth and the movement of unburned timber by flood waters). The Kantu' are least likely to carry out stack burning in a poorly burned swidden cut from primary forest, primarily because of the size of the timber involved.

Another technique used to redress the effects of a bad burn is ngesai 'clear away'. This is performed either as an alternative to stackburning in swiddens that are too wet for the latter to be successful, or as a supplement to stackburning in swiddens where the stacks burn poorly. In swamp swiddens, ngesai consists in gathering up the slashed but unburned grasses and carrying them out of the swidden. In dryland swiddens, ngesai consists in gathering and removing only the unburned timber. Ngesai is common within the floodzone - where a swidden that has not been picked clean of unburned timber is rarely encountered - and unheard-of above it. It is never performed in primary forest swiddens, whether located above or within the floodzone.

If a grassland swamp swidden burns very poorly or not at all, the grasses are cleared from it by ngesai; but if the burn is not that bad, the grasses are mulched into the soil instead, which is the final technique for redressing a bad burn. One mulching technique consists in slicing up the grasses (as they lie, already cut, atop the soil) and the soil beneath them, using a long brush sword. This is called nesap 'slice up'. After each small area has been worked over, the worker completes the operation by tramping up and down in place, mixing the cut grasses into the soil with his/her feet. If the household owns a hoe, it may perform a second operation called mepat 'chop up', as either an alternative or supplement to nesap. Mepat consists in turning over the topsoil and thereby turning under the cut grasses lying atop it. This also is followed by treading of the earth.

The Kantu' say that the main purpose of mulching (by either method) is to remove the layer of unburned grasses from the soil, thus allowing it be sown. A second purpose is to deter regrowth from the root system of these grasses, part of which will have survived the earlier slashing phase. The Kantu' say that the slicing, turning and treading of the mulching operation breaks up the root system and renders any later weeding unncessary. A third benefit of mulching, according to the Kantu', is that it renders arable swamp soils that otherwise are too acidic to be cultivated. The final benefit of mulching also involves improvement of the swampland soil, in particular the isolated patches of higher and hence drier ground. The Kantu' say that by mulching such patches, they are broken up, leveled off, and thereby transformed into swampland too. The value to the Kantu' of these additional benefits, over and above the primary purpose of clearing the layer of grasses from atop the soil, is reflected in the fact that the Kantu' mulch whenever possible. They do not mulch only when the layer of grasses is prohibitively thick, in which case they clear them out of the swidden by the ngesai technique. The fact that the Kantu' do not practise ngesai when the grass layer is thin as well, but instead prefer the more labor costly operation of mulching, reflects the value they place on these secondary benefits of mulching.

LABOR

One workday is devoted to burning the average swidden (including all immediate preparations), regardless of size. Since primary forest, secondary forest and swamp swiddens average 3.4, 2.5 and 0.4 hectare in area, respectively, this one workday represents an input of 0.3 workday, 0.4 workday and 2.5 workdays/hectare, respectively. Labor inputs into redressive burning operations average an additional 1.1 days/hectare in primary forest swiddens, 3.2 days/hectare in secondary forest swiddens and 7.9 days/hectare in swamp swiddens, which yield total labor inputs during burning of 1.4 days/hectare, 3.6 days/hectare and 10.4 days/hectare, respectively.

Chapter Five
Planting

SWIDDEN CULTIGENS

The Kantu' make a basic distinction between two types of swidden cultigens: padi 'rices', and ngkayu 'nonrices'. Of the two, the former is the more important, as is evident in the expression for reaping a good swidden harvest, Boleih padi 'Get rice'. Rice is the staple of the Kantu' diet. When the Kantu' want to know what someone else is eating, they ask Nama ngkayu nuan? 'What is your relish'?. They ask about the relish because the type of relish varies. They do not ask about the main dish because it is always rice, except during a famine[1].

Rice

The first ancestor of the Kantu' to cultivate and eat rice was Bui Nasi, one of the human offspring of the spirit couple Bintang Muga and Ruai Mana. These latter ate arang 'charcoal', not rice, but Bui Nasi refused to follow suit[2]. In order to satisfy Bui Nasi's desire for rice, Bintang Muga and Ruai Mana died, and from their grave sprang the first two stalks of rice. The Kantu' believe that this is the ultimate origin of all rice planted by themselves and related ethnic groups, with the possible exception of swamp rice[3]. They also believe that in the not so distant past a benign spirit left some grains of padi rampo' in the forest, as a gift to the immediate ancestors of the Kantu' at Tikul Batu. It is this variety of rice that is planted as the 'stem rice' of most households in Tikul Batu.

Diversity of Rice Stocks

During the 1975 and 1976 seasons, a minimum of forty-four different, named rice varieties were planted by the households of Tikul Batu, an average of seventeen different varieties being planted by each household each year. All of these varieties are subsumed under a single system of terminology and classification, which provides up to six different levels of contrast (figure 39). In level I, *padi* - here meaning all types of rice - is contrasted to *ngkayu* 'relish', when the question of staple versus nonstaple food is at issue. In level II, *padi* - here more narrowly defined as nonglutinous types - is contrasted to *puloi*, with glutination at issue. In level III, adaptation of the rice plant to *darat* 'dryland' versus *paya'* 'swampland' is at issue. In level IV, two different points are at issue, depending upon whether glutinous or nonglutinous varieties are involved. When the latter are involved, ritual priority is at issue. *Padi pun* 'stem rice', the variety of greatest ritual importance, is contrasted to the two *padi*

TABLE 39
Rice Classification System Of The Kantu'

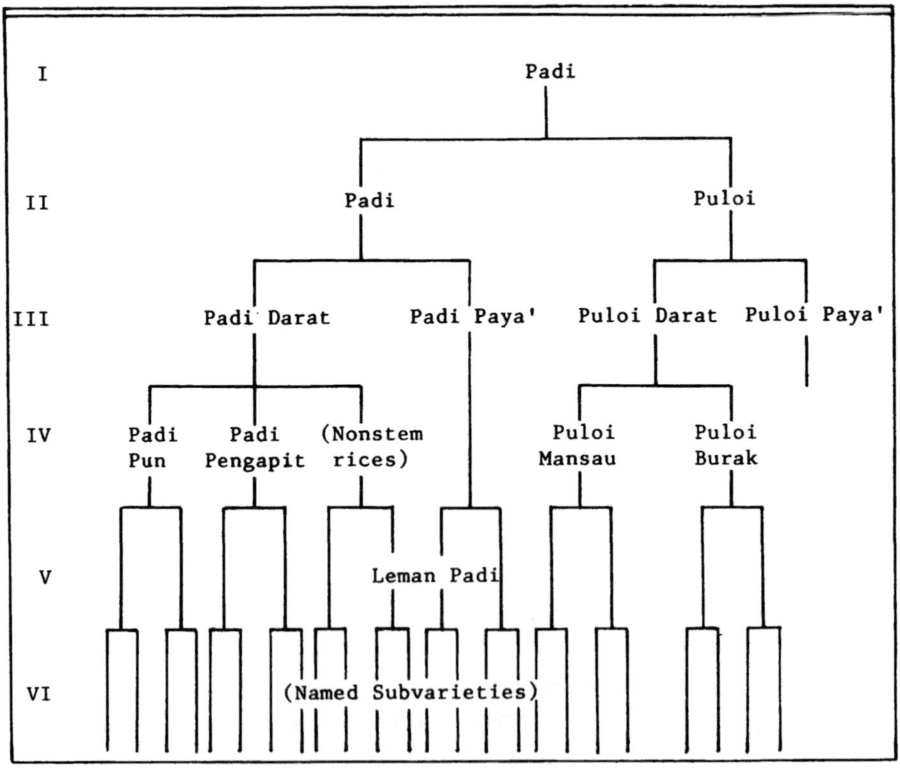

pengapit 'next-to-stem rices', the varieties of second and third greatest importance; and these three are contrasted to the ordinary varieties. (There is no similar differentiation on the basis of ritual priority or importance among the swamp rices, which is consonant with the lesser, overall emphasis on ritual in swampland as opposed to dryland swiddens.) When glutinous rather then nonglutinous varieties are involved in level IV, the color of the rice is at issue, the two varieties being mansau 'red' and burak 'white'. (The red/white contrast is not relevant to the glutinous swamp rices, because only a single such variety is planted by the Kantu'.) In level V, leman padi 'rice variety' is at issue. The Kantu' distinguish among rice varieties on the basis of differences in origin, in physical attributes, and in several lesser factors as well. These differences are reflected in the unique names given to each variety. With regard to origin, rice varieties may be named after (1) the place of origin (e.g., padi Lemanak and padi Nsana' originated on the Lemanak and Nsana' rivers, respectively); (2) the group of origin (e.g., padi Bugau and padi Singanan originated respectively with the Bugau, a Kantu' subgroup, and with the Singanan, downriver Islamicized Dayaks); or (3) the manner of origin (e.g., padi belabok is so named because the original handful of seed was taken by belabok 'stealth' from the swiddens of another longhouse). With regard to physical attributes, rice varieties are variously named. For example, puloi alus is so named because its grain is alus 'fine', while puloi sayap is so named because its sheath resembles sayap 'wings'. In the rare circumstance in which there is marked variation within a rice variety, each subvariety also is named. This is level VI in table 39. Thus, the padi Lemanak variety is subdivided on the basis of color into Lemanak burak 'white Lemanak' and Lemanak mansau 'red Lemanak'[4].

The usage of these different levels of terminology varies according to the context and need of a given conversation (much like the use of the differing levels of terminology for swiddens, as discussed in the Introduction). For example, a household rarely uses the varietal name for its padi pun 'stem rice'. Although there are several different varieties of stem rice in the longhouse Tikul Batu, any given household has at most one variety. Since the varietal name would differentiate nothing, a household usually refers to its stem rice as simply padi pun. The two padi pengapit 'next-to-stem rices' are with equal frequency referred to by this term and by their varietal names. The use of their varietal names is more common in interhousehold conversations, because there is great variation among households in the two rice varieties that are chosen for the next-to-stem rices, and the varieties chosen by one household may not be known by all of the other households in the longhouse. The nonstem rices are referred to in almost all contexts by their varietal names. There is in any case no term in the Kantu' language that means 'nonstem rices'. Swamp rices may or may not be referred to by varietal names. If a household plants only a single swamp rice variety in a swidden, as is often the case, they likely will refer to it simply as padi paya' 'swamp rice', since nothing would be differentiated by the varietal name here. Red and white glutinous rices usually are referred to by these terms, except when one or the other is represented by more than one variety in a single swidden. The overall importance to the Kantu' of each category is outlined in table 40, based upon the distribution of seed rice amongst the different categories in a sample of swiddens[5].

TABLE 40
Relative Importance Of Different Categories Of Rice

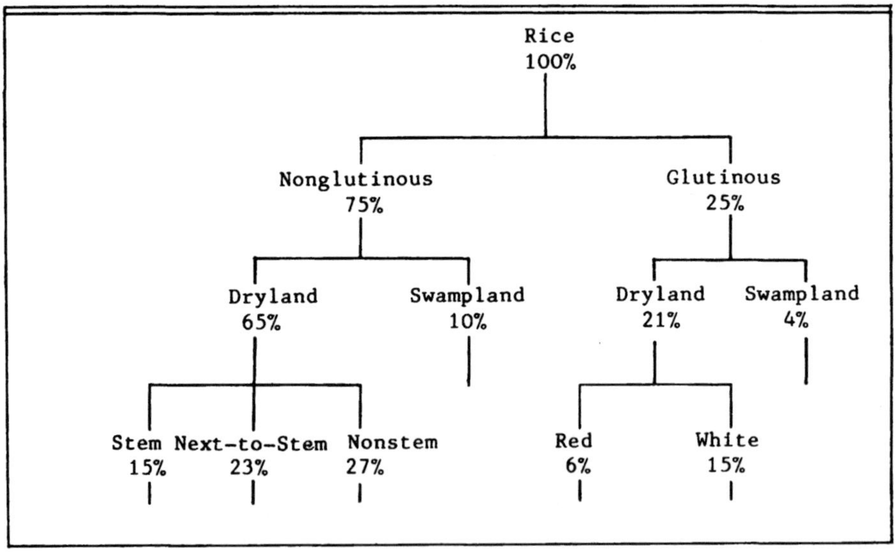

The great diversity in the Kantu' rice stocks is a function of several different factors, the most important of which is the even greater diversity and uncertainty of the local environment. The characteristics of different varieties suit them to different aspects of this environment, hence by planting many of them the Kantu' maximize the amount of the environment that they can exploit, while also minimizing their risk. One example is the local meteorology, specifically the great fluctuation between drought and rainfall. The Kantu' have adapted to this by developing and planting some rice varieties that require drought during their most critical stages of growth and some that require rainfall. Another example is local variation in forest age, referring here to the distinction between memudai biak 'young secondary forest', memudai tuai 'old secondary forest' and kampung 'primary forest'. The Kantu' say that all of their rice varieties grow equally poorly in the first of these forest types; many grow well and somewhat fewer grow satisfactorily in the second; and all grow equally well in the third[6]. A third relevant aspect of the environment involves drainage and elevation, with respect to which the Kantu' distinguish lempa' 'floodzone', mungo' 'highland', and bukit 'mountain'. They say that while most dry rice varieties grow well in the floodzone, some few do poorly; and while most varieties will grow satisfactorily in the highlands, some few will do better or worse. They say that all varieties grow equally well in the mountains. A final example again involves drainage, but here the distinction is between paya' 'swampland' and darat 'dryland'. With respect to this distinction, all rice varieties are clearly differentiated. The Kantu' never plant in swampland the varieties customarily grown on dryland, nor do they ever plant on dryland the varieties customarily grown in swampland. The Kantu' say that the distinction between dryland and swampland varieties is based on a difference in root length. They maintain that the longer root of the swamp rices enables them to survive the chronic inundation to which such land is subject, inundation that would kill the shorter-rooted dry rices. The swamp rices, in turn, cannot be grown on dryland because of higher moisture requirements, at least during the later stages of growth. These varieties often are planted initially in dryland nurseries, and then only in midgrowth are they transplanted to swampland, where they then bear their grain without any problem. However, if they are not transplanted, but raised to maturity on the dryland, they do not bear any grain.

The planting of multiple rice varieties, to maximize the Kantu' exploitation of their environment, is supported by certain ritual prescriptions. Whenever stem rice is planted in a swidden, there is a ritual prescription that it be accompanied by the 'first next-to-stem rice' and the 'second next-to-stem rice', which makes for a minimum of three varieties in all. There is another prescription that applies to all swiddens whether they contain stem rice or not. It stipulates that a minimum of one glutinous variety and one nonglutinous variety be planted in each swidden. The explanation for doing this is so that the antu padi 'rice spirit' and antu puloi 'glutinous rice spirit' always will have someone with whom to talk. In the life of rice, as in the life of humans, the Kantu' view conversation as one measure of a normal, social existence. Finally, there is a ritual proscription against planting a single rice variety in more than one swidden of the same household in a given year. Thus, when a given household makes multiple swiddens in a given year, the minimum number of rice varieties that it must plant is the sum total of the minimum numbers in each of its swiddens. The minimum number is two (viz., one glutinous and one nonglutinous variety) in any swidden in which stem rice is not planted and four (viz., one stem rice, two next-to-stem rices, and one glutinous variety) in any swidden in which it is planted. The Kantu' explain this proscription by saying that a given variety of rice will react vengefully if it is divided among two or more separate swiddens. The empirical value of this proscription to the Kantu', given that the average household makes two or three separate swiddens each year, is to ensure the multiplicity and hence diversity of the household's rice stocks[7].

These prescriptions and proscriptions were violated in none of the 1975 or 1976 dryland swiddens, but they were violated in many of the swampland swiddens. This is another example of the less ritualized character of swamp rice cultivation. This fact aside, it is generally true that fewer rice varieties are planted in swampland than in dryland, the mean being 3.5 varieties per swidden in the first case and 11.5 varieties per swidden in the latter. The lesser number of varieties planted in swampland is related to the lesser number of varieties available. Of the forty-four distinct, named rice varieties at Tikul Batu, thirty-nine are dry rices and only five are swamp rices. This means that a household does not need to plant anywhere near as many varieties in a swamp swidden as in a dryland swidden in order to exploit the same proportion of characteristics of extant varieties.

The principle of agricultural diversity, as expressed in the planting of multiple rice varieties, ultimately is opposed by a principle of economy or efficiency that is expressed in the planting of fewer varieties over greater areas. The existence of this latter principle is reflected in the manifest dislike of the Kantu' at being forced - by an insufficiency of seed - to plant 'too many' rice varieties in a given area of swidden. The line between too many and too few varieties can be inferred from the actual, average densities of rice varieties in 1975 and 1976: an average of 3.5 and 9.5 distinct varieties were planted in each hectare of dryland and swampland, respectively. As swidden size and the importance of economies of scale increase, these densities decrease; and as swidden size decreases and the importance of diversity increases, these densities increase.

Acquisition of Rice Stocks

Seed rice is first acquired, by a given household, when the household is created through the partition of its parent household. In household partition, both eating rice and seed rice is divided on an equal, per person basis among the resulting households[8]. The only exception is the seed of the ritually preeminent stem rice. A newly created household usually is not prepared to perform the elaborate and costly sacrifices involved in its cultivation. After a new household has been on its own for some years, and is richer in both labor and material resources, then, the Kantu' say, is the ideal time for it to acquire stem rice. For example, the last household in Tikul Batu to acquire stem rice (at the time of this research) was household #13 in 1975, approximately ten years after it was created through the partition of household #4.

When a household is ready to acquire stem rice, it must go to the household from which it originated. It cannot take stem rice from any other household. For example, among the Tikul Batu households that had acquired and were cultivating stem rice as of 1976 (viz., #1, 2, 3, 4, 5, 6, 7, 8, 9, 11 and 13), household #1 obtained its stem rice from household #2, 7 from 6, 11 from 1, and 5, 6 and 13 from 4 (figure 4). This same principle will apply when the households that do not yet have stem rice decide to acquire it. Thus, the Kantu' say that household #12 must obtain it from household #7, 14 from 2, 15 from 1, and 16 and 17 from 8. (Following a household's initial acquisition of stem rice, if it ever thereafter needs to replenish its seed stocks [e.g., following a failed harvest or accidental spoilage of stored seed], it also must turn to its household of origin.) It sometimes happens that the household of origin is unable to provide seed rice. In that event, the junior household can turn to any other household in its line of origination or keturun 'descent'. The households within a single descent line are said to be sebenih 'of the same seed rice', referring specifically to the stem rice. Within Tikul Batu there are three such descent lines, a separate one for household #9, one for household #2 and its lineal offshoots (viz., #1, 11, 14, 15) and one for households #3, 4 and 8 (which share a common ancestral household) and their lineal offshoots (viz., #5, 6, 7, 12, 13, 16, 17). The actual acquisition of the seed of the stem rice is a simple matter. The recipient household enters the swidden of the donor household at harvest time and harvests sufficient grain for seed for the coming year. The households are prohibited from giving or receiving any compensation for this grain.

Following a household's initial aquisition from its parent household of the seed of a number of different varieties of rice, it is likely to acquire additional varieties from other sources as well. One such source lies within the household's own swiddens. Occasionally, that is, a new rice variety simply appears in a swidden, which the Kantu' call Datai kediri' 'Came by itself'. In some such cases, the Kantu' observe that the new variety was sown in one swidden and then belaki 'took a husband' in the adjoining swidden (viz., through a process of hybridization). The Kantu' also recognize cases in which a new variety is created by the apparently spontaneous alteration of an extant variety. The most common alteration of this type is known as madi 'become nonglutinous rice', in which a recognized strain of glutinous rice takes on the characteristics of nonglutinous rice, especially with respect to husking characteristics. Other common alterations involve increased difficulties with respect to threshing, increased brittleness of the grain, and a general degradation of hardihood and productivity.

In most cases, however, a household acquires new varieties of rice from the swiddens of other households, typically following a failure of the harvest in its own swiddens. The frequency with which such interhousehold transfers occur among the Kantu' can be measured as the generational age of the rice varieties held by particular households (table 41). If a given variety was acquired after the marriage of the couple belonging to the youngest generation in the household, I assigned it a value of zero, signifying that it was acquired during the current generation. If a rice variety was acquired before the marriage of this couple but after the marriage of the couple from the next elder generation in the household, it was assigned a value of one, signifying

TABLE 41
Age Of Household Rice Stocks

Household	Number of Rice Varieties (as of 1976)	Mean Generational Age of Varieties (as of 1976)
#3	14	1.9
#4	13	2.1
#16	13	1.5

that it was acquired during the first ascending generation, and so on. In the three-household sample evaluated in table 41, the generational age of individual rice varieties ranged from zero to four, with a mean of about two, as shown.

There are several different methods by which one household can acquire seed rice from another. One exceptional method is called belabok 'take by stealth'. The Kantu', when traveling in another region and seeing a promising looking stand of rice, delight in pocketing a handful or two for planting in their next season's swidden. If the variety thrives, all that is harvested is set aside for seed, and in time it can become one of the household's principal varieties. Although only a small fraction of extant rice varieties now are known to have been brought to Tikul Batu under these circumstances, the fact that many varieties are named after distant river systems (e.g., the varieties named Rajang and Sambas are named after rivers distant by 150 and 270 kilometers, respectively, from Tikul Batu) suggests that many other varieties originally were brought into the Empanang valley in this way. A more common arrangement for the interhousehold transfer of seed is ngari 'pass the day'. Under this arrangement, workers from the recipient household assist in harvesting the swiddens of the donor and take their wages in seed rice. Wages usually are calculated as a percentage (viz., one-third) of the amount harvested. A third arrangement is called beganti-besilih 'exchange-replace'. This involves either of two kinds of exchanges of seed rice between two households. Beganti refers to a direct exchange of seed between two households, one of which has a surplus of seed of one variety of rice and a shortage of another, and the second of which has a shortage of the former and a surplus of the latter. Besilih refers to a transfer of seed that is not reciprocated until some time in the future. Here one household borrows from another the seed that it needs, in the expectation of being able to return it after the next harvest. In either type of transfer, the seed involved is taken from and returned to the donor's storage bins (not swidden), and the amount returned is identical to the amount taken. A fourth, distinct arrangement for transferring seed rice is called meri' 'give'. This involves the gift of seed from one household to another, without either immediate or future compensation either being made or stipulated. (However, such gifts do create an obligation on the part of the recipient to treat the donor similarly when the latter experiences a similar need.) Under this arrangement, the seed rice sometimes is taken directly from storage and sometimes reaped by the recipient in the donor's swidden. The fifth and final arrangement under which seed is transferred between households is called nyuruh nanam 'ask to plant'. Under this arrangement one household asks another to plant its seed rice for one year. This is done only when a household skips a year in the swidden cycle. Household #12 skipped the 1974 swidden year, for example, because its solitary male went off on a long trading expedition. In his absence, the household members made only one small swampland swidden, giving all of their dry rice seed to other households. This was necessary to ensure the availability of viable seed for the next season. If this dry rice seed simply had been stored in the rice bins until the next season (viz., for a total of eighteen months), it no longer would have been able to germinate. The households most willing to plant another household's seed like this are those whose own supply of seed rice is for some reason inadequate. In exchange for maintaining the donor household's seed stocks, the recipient household gets the use of the seed. No other compensation is made to either household, and the amount of seed returned to the donor household is identical to that originally taken. The recipient household takes the seed not from the donor's swidden but from its rice bins, and it returns it there as well. The relative frequency of each of these four ways of arranging interhousehold transfers of seed - with the exception of the last mentioned, which is very infrequent - is presented in table 42[9].

The principal factor that determines which one of these arrangements prevails in any given transfer is whether the two households involved belong to the same longhouse or not. This can be demonstrated by first ranking the four different arrangements in terms of the directness and promptitude of compensation or reciprocity. Wage payment is the most prompt and direct, gift giving is the least, and exchange-replacement

TABLE 42
Methods Of Acquiring Seed Rice

Theft	10% of varieties
Wage payment	30% of varieties
Exchange-Replace	20% of varieties
Gift	40% of varieties
Total	100% (40 varieties)

falls in between. The theft of seed rice involves no compensation at all, but nor does it represent part of a pattern of generalized reciprocity, as does gift giving; so theft should be seen as even further removed from gift giving than is wage payment. Looking at these four types of arrangements as a scale, then, interlonghouse transfers of seed cluster at one extreme, while intralonghouse transfers cluster at the other (table 43). The patterns observed here reflect a pervasive difference between interlonghouse and intralonghouse

TABLE 43
Socioeconomic Character Of Interlonghouse Versus Interlonghouse Transfers Of Seed Rice

		Relationship between households	
		Same longhouse	Different longhouse
Type of arrangement for transfer of seed	Theft	0	4
	Wage payment	1	11
	Exchange-Replace	5	3
	Gift	10	6

N = 40 cases of interhousehold transfers of seed rice in the histories of household #3, 4 and 16.
$X_c^2 = 8.42$
$P < .05$

relations. The latter are characterized by more generalized reciprocity, whether of goods or labor, and the former by less generalized reciprocity (or none at all, as in the case of the theft of seed rice).

For similar reasons, a household needing seed rice will seek it from another household in the same longhouse whenever possible. This not only avoids presuming upon weaker social ties, but it minimizes problems of transport as well. But in a very bad year, every household in the longhouse may be badly off, and all will be forced to seek seed at other longhouses. This is what happened in 1974, in which fully eight out of thirteen households at Tikul Batu did not harvest sufficient rice for the next year's seed, much less sufficient to eat during the coming year[10]. The five remaining households shared the year's misfortunes to the extent that they were unable to provide seed to any of the eight needy households (with a single exception). Poor harvests also prevailed in the three Kantu' villages located within one-half hour of Tikul Batu, and no seed could be obtained from them either. The eight needy households were forced to obtain seed from four longhouses, belonging to other subgroups of Kantu', located at distances of one hour's to one day's travel from Tikul Batu. Each of the eight households went to an average of two of these longhouses, obtaining seed from a mean of 2.2 separate households in each of the longhouses visited. The most important factor in the selection of these donor households and longhouses was proximity of kin ties to the needy households (aside from the initial prerequisite of a sufficiency of grain).

It also is possible for a needy household to obtain rice seedlings from another household, either within the same longhouse or in another longhouse. There is only one arrangement wherein this is possible. The recipient household secures the permission of the donor household, goes to the latter's swidden and uproots, prepares and transports the desired number of cuttings, and then makes a ritual payment to the donor of a piece of iron, called urat besi 'iron root'. This is the only form of compensation permitted, and its ostensible purpose is to prevent the donor's rice spirits from taking offense at the transfer. This is a source of concern because these are transfers not just of seed, but of the living rice plant. Payment of the 'iron root', the Kantu' say, will make the donor's remaining rice plants as strong and healthy as iron and so mitigate any deleterious effects of the transfer. Payment of the 'iron root' may have a more mundane function as well, as economic compensation for the transplants given by the donor and as a means of limiting the demands for transplants of the less favored households on those more favored[11]. This interpretation is suggested by the fact that although failure to pay the 'iron root' is mali 'taboo', recipient households often are slow to pay. And when some do pay, they try to make a token payment of an iron nail. The Kantu' frown on such attempts, however, saying that anyone who makes such a payment will become sick with pains as sharp as a nail. The importance of the intrinsic value of the iron also is suggested by the fact that the iron root does not need to be paid when two households exchange transplants, although the potential offense to their respective rice spirits might be presumed to be just as great.

Households that need rice seedlings can obtain them from any other household, but there is a tendency to obtain them from the households with which they have the closest kin ties. Among the sixteen households at Tikul Batu in 1976, there was a statistically significant tendency (x^2_c = 9.3, $P < .005$) for households linked by direct consanguinal or affinal ties to cooperate in the transfer of rice seedlings[12]. There was no similar tendency for households in need of seed rice (the seed of stem rice excepted) to obtain it from the households with which they had the closest kin ties. This difference reflects the fact that the interhousehold transfer of living rice seedlings is a weightier matter, involving greater risks to the donor household, (viz., in terms of offense to its rice spirits), than is the interhousehold transfer of seed alone. A household is not likely to assume these risks for another household unless they are closely related.

Whatever the origin of a given variety, whether swamp rice or dry rice, a subsequent process of human instigated selection determines whether the new variety will flourish or die out. All of those varieties for which the household has relatively little seed, including not only newly acquired varieties but also varieties whose previous harvests were poor, are planted together in a single stand. The Kantu' call this nanam campur 'mixed planting'. The relative success of the different rice varieties in such mixed stands is closely monitored by the household. After one or more years, if a given variety in a mixed stand appears to have grown favorably, its seed rice is harvested separately, to be followed by planting in a larger, monovarietal stand in the next season. On the other hand, if a variety in a mixed stand does not grow favorably, it continues to be harvested and planted in a mixture, or it may even cease to be planted altogether. This practice constitutes a rigorous test of the relative merits of the two or more varieties planted together, rigorous in that extraneous variables that might affect harvest success (such as type of soil, drainage, and so on) are held constant. With the vagaries of time, each of the household's rice varieties eventually will be planted in mixture with one or more other varieties. Thus, this process of mixed planting and selection continuously functions to maintain and increase the overall quality of the household's rice stocks, by increasing the number of rice varieties that grow favorably relative to other varieties, while decreasing the number of varieties that grow unfavorably relative to others.

This process may be termed the gross selection of rice varieties. There also is a process of finer, intravarietal selection that consists in selecting the finest grains from each stand of each rice variety for the seed that will be planted the following year. This selection necessarily is based on visible characteristics, and these are two in number. First is the <u>pemadas</u> 'goodness' of the grain. The Kantu' look for a full, whole grain, as opposed to empty, withered, or otherwise damaged kernels. The second characteristic is the <u>belayan</u> 'abundance' of the grain, pertaining to the number of grains per panicle, and the number of panicles per plant. At least one more characteristic is critical to Kantu' selection for or against a given rice variety, but it is not visible during seed selection: the threshing and husking qualities of the grain. Because this characteristic is not visible in the swidden, any problem with the threshing or husking qualities of a given variety necessitates grosser selection at the varietal level, meaning elimination of the entire variety. The inability to deal with this problem through finer selection at the intravarietal level (viz., during seed selection) explains why problems with threshing and husking are the ones most often cited by the Kantu' as the reason for eliminating a variety.

Nonrice Cultigens

The nonrice, swidden cultigens include both ritual and nonritual plants. The former category, called <u>ngkayu</u> 'relishes', comprises at least thirty different, named types[13]. Almost half of these types have named varieties. For example, <u>lingkau</u> is one named type of relish, and it is represented in the Kantu' swiddens by two varieties, lingkau keribang 'maize' and lingkau lesit 'job's tears. The origin of most of the relishes is similar to that of the rices. The Kantu' say that most of the relishes were given to them by the spirits, with just two or three originating from other regions or from the government.

Each year the average household plants twenty to twenty-one different relishes in its swiddens. There ae several reasons why the Kantu' plant such a large variety. First, they value the variety that these different relishes add to their basic diet of rice. Also important is the fact that there is great variation among the relishes in the length of the maturation period. Accordingly, by planting as many of these relishes as possible, each household can ensure that a minimum of several different types will be mature and available for consumption during every month of the year (see chapter 10). A final factor in the planting of so many different types of relishes is the diversity in the local environment. Table 44 shows what percent of all relish types 'favor' or grow well in each local microenvironment, based upon Kantu' assessments of each type. It is apparent that there is little differentiation

TABLE 44
Adaptation Of Different Relishes To Different Microenvironments

Microenvironments	Percentage of Relish Types Favoring Each Environment	
	Do not favor	Favor
Dryland	4%	96%
Swampland	96%	4%
Floodzone	6%	94%
Highland	56%	44%
Mountain	8%	92%
Young secondary forest	92%	8%
Old secondary forest	72%	28%
Primary forest	68%	32%

among relish types with respect to dryland, floodzone or mountain land, on the one hand, or with respect to swampland or young secondary forest on the other: almost all types can live in the first three microenvironments, and very few can live in the latter two. On the other hand, there is considerable variation with respect to the highlands, and also with respect to old secondary forest and primary forest. This differentiation among relish types in environmental needs and tolerances is one explanation for the number and variety of the relish types themselves, given a varied local environment and the Kantu' need to adapt to it[14].

The ritual nonrice cultigens are called utai 'things'. There are twelve major varieties, all of which are planted in the 'place of the seed rice' in the swidden of most ritual importance - the stem swidden for households that plant stem rice - to each household each year[15]. They are planted to succor and protect the rice spirits that are presumed to dwell in this place. For example, kunyit 'turmeric', ncekur 'East-Indian galanggale', and lulut are planted in the belief that they will have the same salubrious effect upon the health of the rice spirits as they have upon the health of humans, by whom they are regularly used medicinally. This emphasis on giving medical aid to the rice spirits is reflected in the names of two other utai: manang padi 'rice shaman', and pencelap 'that which makes cool' ('cool' here meaning 'healthy'). The attribute of pencelap that gives it this value in the eyes of the Kantu' is its own extraordinary fertility. They say that if just one leaf of pencelap falls to the ground, it will take root and grow, a quality that the Kantu' would like all of their cultigens to possess. The good health of the rice spirits also is at issue in the planting of tembako 'tobacco' and tubai[16]. Tobacco, in addition to being smoked by the Kantu', is used as a poison against certain plant pests and diseases, and tubai is used as a fish poison. The two plants are planted in the 'place of the seed rice' expressly because of their toxicity, in the hope that it will keep predators away from the rice spirits. Two other utai, the flowering duyung and serai 'lemon grass', are planted there in the belief that their appearance or bouquet is attractive to the rice spirits. The attraction of jeranang and ngkerebai seems to be that both yield a blood-red dye. The most important utai of all, singkenyang, is valued chiefly for its augural properties (see page 290), which are related to the irregularity of its flowering.

The source of seed for the relishes is the harvest from the previous year's swiddens, some portion of which always is set aside for this purpose. The source of cuttings are plants still growing in the previous swiddens, with the exception of chili peppers, which are grown from seed first in a nursery. There rarely is any difficulty in obtaining seed or cuttings for these cultigens, either from the household's own swiddens or from those of other households. More problematic are the _utai_, the seeds or cuttings of which cannot easily be transferred from one household to another. For example, the interhousehold transfer of turmeric is completely prohibited. Any household wishing to obtain or replenish a stock of turmeric must seek out a plant growing wild in the forest. The cultigens known as _pencelap_ and _singkenyang_ can be obtained only from a household that is senior to the recipient household in its own descent line. The two primary sources of _singkenyang_ in Tikul Batu are households #1 and 8. Household #11 is the only other household in the longhouse that plants this. It obtained the plant from household #1 at the same time that it obtained its stem rice. Household #15 also could obtain _singkenyang_ if it wished, from #1, and households #16 and 17 could obtain it from #8. The other households in the longhouse, however, not being junior to either household #1 or #8 in their respective descent lines, can obtain this plant only with difficulty. _Singkenyang_ can be brought into a new descent line only by staging a major ceremony involving the sacrifice of one or more pigs. Such a ceremony does not appear to have been held among the Kantu' in the Empanang valley within the past several decades.

TECHNOLOGY

Techniques of planting vary according to whether it is dry rice, swamp rice, or nonrices being planted.

Dry Rice

There are two distinct techniques for planting the dry rices: one is employed during the initial planting, and the other is employed later as a redressive measure.

Initial Planting

Planting is a two-step operation, the first of which is called _nugal_ 'use a dibble stick' or 'dibble'. The _tugal_ is the dibble stick. At the beginning of each day of planting, each participating male cuts a new dibble stick in the forest adjoining the swidden that is to be planted. They select species of trees with dense wood for this, both so that the stick's tip will be slow to wear or fray during the day, and so that the stick as a whole will be sufficiently heavy to facilitate penetration of the earth. The shaft of the dibble stick, excepting only twenty to twenty-five centimeters on either end, is shaved down to a diameter of about four centimeters, using the brush sword. This is done to eliminate unnecessary weight from the stick. Both ends of the stick are left approximately 2.5 centimeters larger in diameter. At its upper end, this larger diameter enables the worker to grip it with his hand in a more open position, which can be maintained with greater ease for longer periods of time. At the lower end of the dibble stick, the larger diameter adds weight to the working or 'free' end of the stick. This increases the ease and accuracy with which the dibbler can swing the stick about over the earth before him, picking out, often in difficult terrain, the spots of tillable ash among the rocks, unburned timber, and unburned humus. In addition, the greater diameter at this end also is needed to make a dibble hole of sufficient size. Finally, the bottom four to five centimeters of the stick is shaped, again with the brush sword, into a sharp, conical point. This point then is hardened by scorching it for some ten minutes with fire. Periodically during the day, as this point frays and blunts with use, it is reshaped with the brush sword. At the end of the day the dibble sticks are discarded. An occasional male makes a dibble stick with a permanent, detachable tip. This tip is made from one of the hardest of the forest trees, the ironwood tree, and since it wears but little during use, it can be used and reused. A tip of this sort is stored in the swidden house or longhouse between uses, only the shaft being made anew each time that it is used.

The dibble stick is held alternately in one hand then the other. Shifting the stick from hand to hand minimizes fatigue, and it also facilitates the dibbling of holes to either side of the worker. Very occasionally, an exceptionally energetic or hard pressed farmer, or a youth anxious to demonstrate his planting prowess, uses two dibble sticks, one in each hand. The physical strain of this dissuades the average worker, however. In all cases the intent is to make an acceptable hole, of sufficient depth and definition, with just a single stroke of the dibble stick. When dibbling in moist and sticky earth, the stick is given a twist on the upstroke, to free it cleanly from the earth and leave behind a well defined hole. When unburned or poorly burned brush or branches obstruct the work, they are slashed with brush swords (providing that the area involved is sufficiently well burned to be planted otherwise). Dibbling usually is done in groups, in which the dibblers work abreast of one another in a line running perpendicularly to the direction in which the work is proceeding. The dibblers space themselves from two to four meters apart, depending upon both the number of dibblers and the size of the swidden.

The dibblers make their holes five to seven centimeters deep, at a density ranging from 3.6 to 12 holes per square meter, with a mean of 7.3/square meter (standard deviation 1.7). The ideal density is considerably greater than this mean. When one or two workers plant alone, free of any other constraints, they average 11.8 holes per square meter. The much lower mean density is due, first, to the use of large, reciprocal work groups, in which the attention to the work is lower, especially when large amounts of rice beer are being consumed as well. A second factor that produces lower planting densities is the presence on the ground of either relishes (planted right after the burn, as much as several weeks before the rice planting) or large unburned logs. Relishes reduce the amount of nutrients available for the rice plants, and unburned logs reduce the amount of ground area available. Only in primary forest, however, is unburned timber of a size to cause any significant reduction in arable area.

The second step in planting is menih 'use seed' or 'sow'. The seed is grain that has been threshed and winnowed but not husked. It is prepared for planting, first, by being direndam 'soaked' to stimulate germination. Once this has been done, the seed must be planted within three days, at the latest. After that, the Kantu' say, Benih dipajoh petok 'The seed rice is consumed by the germ sprout'. After (or before) being soaked, the seed of two or more rice varieties that previously have not been planted together may be mixed together. This is done with new varieties that the household wishes to introduce into its swiddens on an experimental basis, or with old varieties that either are not producing well or are in short supply, as discussed previously. Not all rice varieties can be mixed together. For instance, varieties that differ in size cannot be mixed, because the mixture could not be milled properly. Neither would a glutinous variety be mixed with a nonglutinous one, because there would be ensuing difficulties with cooking. Two other rice varieties that would not be mixed are one that stores well for a long period of time and one that does not. Similarly, a variety with a short maturation period is never mixed with a variety having a long one, because a stand of this mixture would have to be harvested twice, each time selectively. Finally, two varieties with different environmental needs and tolerances, such as a swamp rice and a dry rice, also are not planted together.

The last stage in the preparation of seed rice for planting, undertaken usually on the morning of the planting day, consists in mixing it with the seed of various relishes. Two factors determine which relishes are mixed and planted with the rice. One factor is compatibility. For example, the Kantu' say that they cannot mix seed maize and seed rice, because the former would dominate the latter if sown into the same dibble hole. The second factor that determines whether a relish is planted with the rice is its ability to reproduce from seed. Many important relishes, including sugar cane, bananas and all three tubers, are reproduced only vegetatively by the Kantu'. Given these constraints, the relishes whose seed is mixed with the rice typically include only _lingkau lesit_ 'Job's tears', _retak_ 'long bean', _kawa_ 'pumpkin', _gemalai_ 'wax gourd', _genok_ 'bottle gourd', _pusut_ 'dish cloth gourd', and _ntimun_ 'cucumber'.

The seed is carried in _raga' pemenih_, and _takin pemenih_. _Raga'_ and _takin_ are the general terms for two types of round, plaited, square-bottomed baskets. _Raga' pemenih_ and _takin pemenih_ refer to the variant, in each category, that is made specifically for _pemenih_ 'sowing'. The _takin pemenih_ is unadorned, but the _raga' pemenih_ typically is finished with geometric designs on the exterior, done in fine strips of bamboo in colors of white, red and black. The _takin_ is the larger of the two, averaging thirty-four liters in size. It is used to carry the seed rice from the longhouse to the swidden. The _raga'_, averaging just five liters in size, is filled from the _takin_ and then carried by the sower during the actual planting of the swidden. The amount of seed carried in the average _raga'_ can be sown by the average worker in one hour and twenty minutes. This basket is worn on the sower's left hip, with her left arm and shoulder passing through a cord attached in a loop to two holes on one of its sides. The sower's left hand often rests on the basket rim, maintaining it in an upright position while she bends and straightens at the waist in the course of the work. The right hand is invariably the sowing hand. With this hand, the sower takes one handful of seed at a time from the basket and uses it to sow as many holes as possible before reaching in again. The hand is held in a closed position, with the tips of all four fingers resting against either the thumb or the palm, to keep the seed from falling out. A portion of each handful is released into each hole by raising the thumb from its position against the index finger.

Thus, seed is released across the middle joint of the index finger for the most part. Seed is released only after the body has bent at the waist, putting the hand nearer the ground - from midcalf to knee height - to maximize the accuracy of the throw. The sowing rate averages 61.5 liters per hectare.

The accuracy of sowing is high but far from perfect. A significant percentage of each throw of seed typically misses the intended dibble hole. The Kantu' implicitly accept this as the price of a more erect and less fatiguing sowing stance and a higher rate of sowing. They compensate for the surety of some seed missing the holes by throwing a relatively large number of seeds - the average in a small sample was twenty-three - towards each hole[17]. In addition, the sowers correct their worst misses by scooping missed grains into the hole with their feet or, though much less frequently (because more awkward), with their hands. In addition to having to cope with missed throws, the sowers also must cope with missed dibbles. They frequently come upon patches of ground that the dibblers in front of them have somehow missed. If the patch is of considerable size, they call back a dibbler to go over it. But in the more frequent case where only one or two dibble strokes were missed, the sowers make the holes themselves - using their big toes - and then sow them.

Sowing is done, like dibbling, by a line of people working abreast of one another. The line of sowers follows in the wake of the line of dibblers. Because sowing is slower work than dibbling (viz., forty-nine holes versus eighty holes per minute, respectively), the sowers will fall behind an equal group of dibblers on each pass from one side of the swidden to the other. If the sowers fall more than approximately five minutes behind the dibblers in a single pass, then some of the male dibblers cease dibbling and join the sowers. Even young bachelors will do this, without evincing any of the embarrassment over sowing that is reported from many other Dayak groups[18]. On the other hand, if the sowers are working right at the heels of the dibblers, the latter are apt to get agitated and order any males who are sowing to cease and return to dibbling. While males may join or leave the sowers to maintain a balance, no female ever joins the dibblers. Some degree of imbalance between the dibblers and the sowers appears to be desirable. Thus, if the dibblers finish a pass from one to five minutes ahead of the sowers (with four to five minutes being the average), there is no transfer of dibblers to the ranks of the sowers. Rather, the dibblers use these minutes to sit at the edge of the swidden and pehaps to smoke or resharpen the tips of their dibble sticks, but mainly to rest. The sowers, in contrast, usually do not rest when they finish a pass. Rather, at that point the dibblers get up and begin again, and the sowers follow immediately after them. Rarely during the day do the sowers sit down in the swiddens (viz., other than during formal breaks to eat and drink).

The desired imbalance between dibblers and sowers is reflected in the composition of the planting groups. Given the difference in rates of work, the planting group need contain only 6.1 dibblers for every 10 sowers, but in fact the ratio in the average group is 8.5/10. Given this latter ratio, the dibblers in the average planting group need work only forty-three minutes for every hour that the sowers work, the remaining seventeen minutes being available for their rest[19]. This normative imbalance between dibblers and sowers, in providing the former with more rest time, implies a cultural recognition of dibbling as a more physically demanding task than sowing. This also is implied in the Kantu' rationalization for minimizing the length of each pass across the swidden. The Kantu' say that the problem with longer passes - and hence with longer intervals between the rest periods at the end of each pass - is that this results in a fall-off in the quality not of sowing, but of dibbling. The relative need for rest periods aside, the greater physical demands of dibbling also are reflected in the fact that as males become older and pass their peak of health and strength, they customarily sow rather than dibble, dibbling only when the ratio of sowers to dibblers otherwise is too high.

There is some tension between the male dibblers and female sowers regarding this division of labor. The dibblers sometimes challenge the sowers to dibble, suggesting that they would find the physical demands of the task too great. The sowers, in response, jeer at the simplicity of dibbling and speak of the skill demanded for sowing. When the sowers fall far behind the pace of the dibblers, the latter (before they act to redress the imbalance) may jeer at the sowers, calling to them as they do to summon their dogs when hunting. The sowers, in turn, may criticize the thoroughness with which an area has been dibbled, the depth of the holes, and even the length of the males' dibble sticks. Conversely, if a sower feels well-disposed towards some particular dibbler, she may speak favorably of the length of his stick and the depth and close spacing of his holes. She may even say of his holes, <u>Nadai alah dibenih</u>, '[They] cannot [all] be sown'. Much of this praise and criticism is sexual. The length of a male's dibble stick refers to the length of his penis, and the spacing and depth of his dibble holes refer to the vigor of his sexual performance. The sexual overtones are attested to by the fact that when the female sowers come upon an undibbled area, they will shout to the dibblers, dapor 'hearth', which is a euphemism for the vagina (e.g., celap dapor 'cold hearth' describes a couple's abstinence from sexual relations). This sexual symbolism is clearest of all in the behavior of courting couples. Such couples try to align themselves within work groups so that the girl can fill with seed the holes that were dibbled by her suitor. In such circumstances the young male dibbles at a furious pace, in an attempt to demonstrate his vigor not only in work, but also in sexual activity.

The two lines of workers, the dibblers and the sowers, always begin work at the edge of the swidden. When the shape of the swidden varies from a square, the workers start from one of the long edges of the swidden, so that they move across its width rather than down its length (figure 23a). The explicit purpose of this pattern

FIGURE 23
Movement Of A Planting Party Through The Swidden

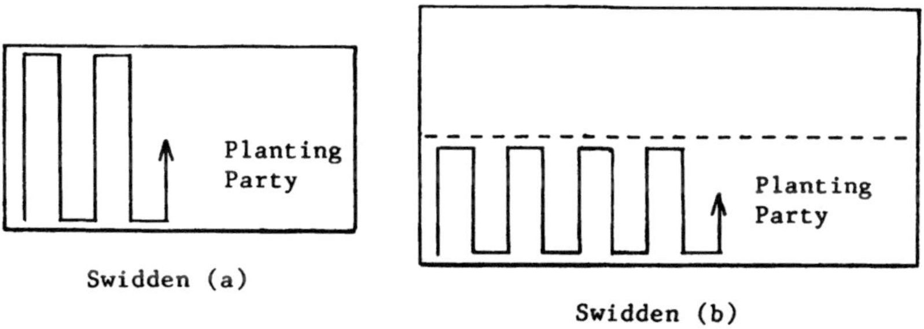

of movement is to minimize the length of each mwa 'face' or 'pass' across the swidden. The shorter the passes, the more frequent the rest periods of the dibblers between passes. The more frequent the rest periods, the higher the concentration of the workers and, as a result, the higher and hence better the density of planting. In a very large swidden, even the distance across the width of the swidden may be judged to be too long for a continuous pass. In this case the workers make their passes only halfway across the swidden, resting in the center (figure 23b).

When the lines of dibblers and sowers reach the edge of the swidden upon finishing a pass, they proceed (after resting first, in the case of the former) to work along the edge (dibbling and sowing the swidden margin all the while) in a direction perpendicular to that of the last pass, for a distance equal to the width of that pass. This positions them abreast of the next pass to be made (figure 24). The positioning of the workers

FIGURE 24
Prescribed Movement Of A Planting Party Along The Swidden's Edge

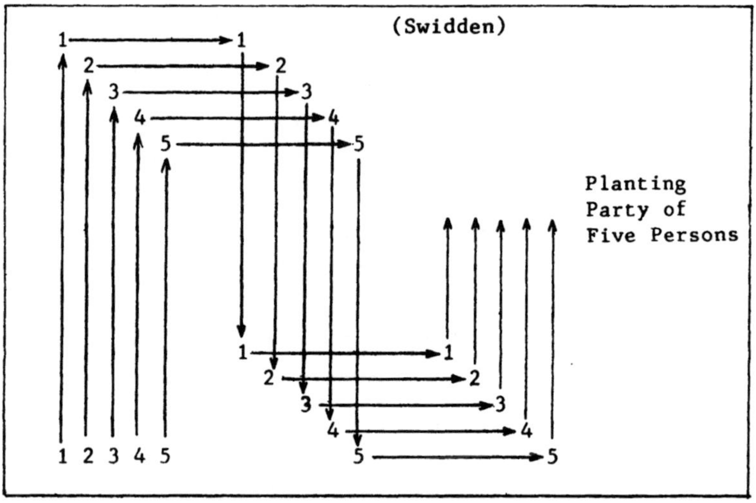

with respect to the cardinal directions is preserved throughout (e.g., in figure 24 worker #1 first crosses the swidden on the left side of #2, but then comes back on #2's right side, thereby remaining on the west side of #2 in either direction). The preservation of this positioning, at which the workers make a conscious effort, ensures that all workers walk the same distance along the edge of the swidden when moving from the point where their last pass was finished to the point where the new one will begin. The Kantu' consciously avoid the pattern that maintains the positioning of the workers with respect not to the cardinal directions, but to left and right (figure 25). In this pattern, after finishing the first pass worker #1 (e.g.) would have to walk the length of both that pass and the upcoming pass before being in position to work the second. Given a typical work group of twenty-five persons (dibblers or sowers) and assuming a spacing between workers of

FIGURE 25
Proscribed Movement Of A Planting Party Along
The Swidden's Edge

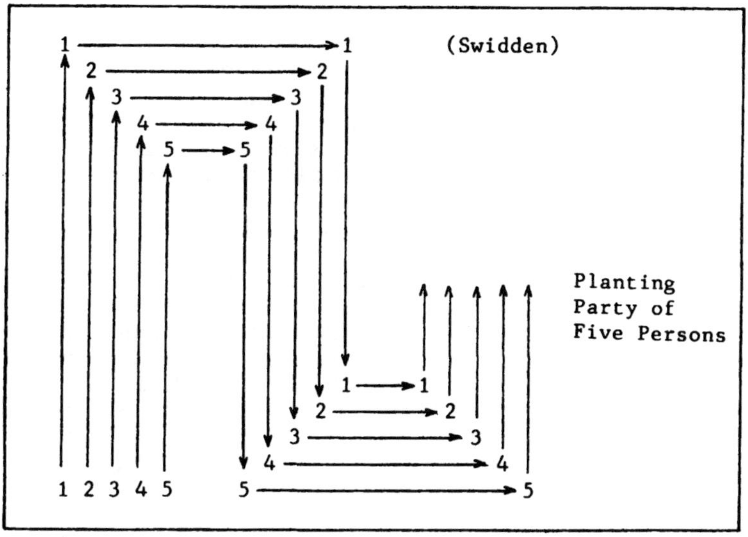

3 meters, this would mean a walk of 150 meters for this first worker in line, compared with a walk of just one-half that distance in the normal pattern (figure 24). As a result, this worker would be tired out just at the point of starting a new pass. In contrast, worker #25 (viz., still assuming a work group of twenty-five persons) would have virtually no distance to walk after the first pass, and so would have nothing to do but rest for the five to ten minutes that it would take worker #1 to walk (and dibble or sow) the 150 meters. (This would be in addition to any rest period that the group as a whole might take.) To a lesser extent, workers #2-12 also would have an extra distance to walk and workers 14-24 would have extra time to wait (their roles being reversed with the next pass). Since some workers always would have to wait for others, this pattern would increase (in fact, double) the amount of time taken by the work group to move from the end of one pass to the beginning of another (thus slowing down the overall rate of planting), despite the fact that the actual distance walked by each worker, overall, would be no greater than in the normal pattern.

This pattern of planting is interrupted in two instances. First, if the workers come across a patch of ground that did not burn either during the main firing of the swidden or during any refiring, they bypass it. Second, when planting a predominantly dryland swidden, any swampland that is encountered (e.g., along the marshy banks of watercourses branching through the swidden) also is bypassed. Since such swampland may cut across more than one mwa 'pass', and since it must be sown with different seed than that used in dryland, the Kantu' usually finish planting the dryland first and then plant the swampland, typically using a smaller work group.

Most rice varieties are planted discretely. The boundary between each variety is established by a single dibbler, who works in advance of the main planting party and marks the boundary with a line one or two dibble holes wide. If stem rice is being planted, it must be planted first, at the swidden's edge, followed by the two next-to-stem rices. Whether stem rice is planted or not, a glutinous variety cannot be the first to be planted (which is but one of many instances in which the ritual priority of the nonglutinous rice is marked). If the swidden being planted encompasses two or more different microenvironments, the Kantu' say that they usually will plant each rice variety so that it stretches across the microenvironmental boundaries (excepting only the critical boundary between dryland and swampland). For example, household #16 planted the five rice varieties in its 1976 Lubuk Menserai swidden so that each lay partially in and partially out of a strip of floodzone land that ran along one edge of this predominantly highland swidden. The household said that the purpose of doing this was Awak sama badas, sama jai', 'So that [all varieties] do alike well, [or] alike poorly', in that strip of floodzone land (figure 26). Implicit in this statement is recognition of the fact that this

FIGURE 26
Zoning Of Rice Varieties Within The Swidden

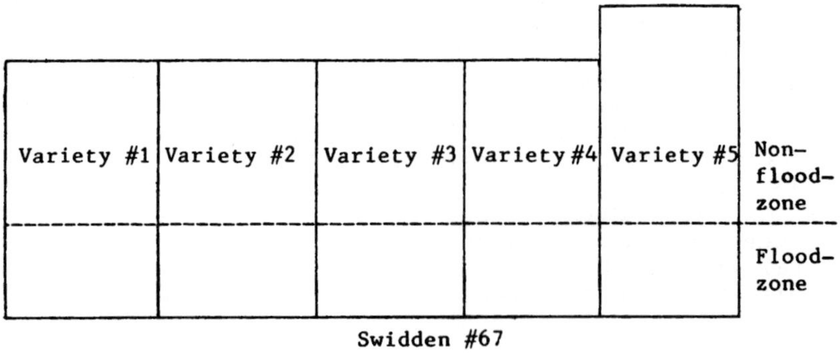

mixing of both rice varieties and microenvironments increases the odds that at least one combination of a particular variety and a particular microenvironment will be successful. For example, if household #16 had planted each of its five rice varieties either entirely within or entirely without the floodzone strip, it would have had just five different possibilities of success. As it was, the household had ten different possibilities of success.

This principle of mixing the microenvironments in which a single rice variety is planted holds within but not among swiddens [20]. For each variety the household selects that one swidden whose drainage, elevation and forest age best suit it. For their stem rice and next-to-stem rices, they select that swidden whose overall environment for growing rice is most ideal - meaning the swidden that is largest, has the best soil, and burned most thoroughly.

The workday never is cut short by omens during planting. If the Kantu' observe a minor omen during planting, they ignore it. If they observe a major omen, they schedule the sacrifice of a pig later in the year. The Kantu' explain this failure to observe omens on the basis of the great size of the planting work groups, which range up to sixty persons in size (compared with just three to four persons during slashing and felling). The cost of a disrupted day in lost man-days would be very heavy. Also important is the attendant cost in seed rice, which is soaked in water just before planting begins. Thereafter, delay in its planting by even one or two days would render it useless for planting.

It appears that the Kantu' have never observed bird omens during planting, but there is oral historical evidence that they formerly observed omens such as the breaking of the cord of a sowing basket (resulting in the spillage of seed rice), or the accidental extinction of the fire lit in the swidden every morning, and so on. Observance of these omens consisted in halting work for the rest of the day. In addition, the Kantu' formerly foreshortened the planting day for omens not observed but feared. In the morning they did not begin to plant as early as they now do, for fear of hearing either the insect _mpartikar_ or the _jani' apai aloi_ 'giant pig of Borneo'. In the afternoon they finished planting and left the swidden at the time of day known as _mata panas nutok_ 'the husking sun', so as not to hear either the giant pig or the _ngingit_ 'cicada'. In the event that they heard one of these creatures while engaged in planting, they say that the swidden would be _punas_ 'barren' as a result, although this outcome might be mitigated by means of a large offering to the spirits.

The 'husking sun' refers to the time of day at which the Kantu' usually husk rice. The Kantu' say that this also is the time of day when, after washing their work clothes, there still is enough time to dry them out before nightfall. They say of this time, Agi' tubah rangkai, 'Still plenty [of time to] dry'. Current observations suggest that this time is between 15:00 and 16:00. Assuming, then, that the Kantu' formerly ceased planting at approximately 15:30, which is two hours earlier than is currently the case, and further assuming a later start in the morning by a minimum of one hour, the total workday would have equalled a maximum of 190 minutes, or just over one-half as long as the current planting day[21]. Every planting day would have been foreshortened, given the Kantu' statement that they started late and finished early not when they heard the omen creatures, but rather in apprehension of hearing them.

The Kantu' at Tikul Batu abandoned this ritually foreshortened planting day (in addition to a variety of other ritual proscriptions) in 1970, through the paid offices of an Iban who was said to have had a jimat 'charm' that enabled him to abrogate such proscriptions. For the Kantu' at Tikul Batu, 1970 and 1971 were years of great trauma, during which their territory became the site of an armed conflict between the Indonesian government and a force of insurgents. As a result of the Kantu' proximity to this conflict, their swidden activities generally were circumscribed and rendered more difficult, and economic hardship became more common. In these changed circumstances, whatever value had once been attached to the shortened planting day was exceeded by its cost, and it was abandoned[22].

Redressive Planting

The principal problem with the initial planting of dry rices is failure to plant some parts of the swidden to the optimum density for rice plants. As mentioned earlier, this is associated with large work groups and heavy drinking. The range of densities in a typical swidden is examined in table 45. In a case such as this, the variation in planting density below the mean may be redressed by mubol 'filling in'. This operation is performed as often by males as females. Each worker carries both a sowing basket and a dibble stick. The worker first dibbles five or six holes in the ground immediately in front of him/her, stoops to sow them, then stands and moves on to another area. The Kantu'

TABLE 45
Density Of Planting In One Swidden (#44)

Mean Planting Density = 7.3 dibbled & sown holes/m^2.		
Standard Deviation = 1.7 holes.		
Percentage of Swidden Planted to Density that is:		
= Mean ± SD (viz., 5.6-9.0 holes/m^2)		- 66%
> Mean + SD (viz., > 9.0 holes/ m^2)		- 14%
< Mean - SD (viz., < 5.6 holes/m^2)		- 20%
	Total	- 100%

say that the seed rice used in a given area should be of the same variety as that planted there during the initial dibbling-sowing, with one or two exceptions. At the time that the Kantu' fill in a swidden, if they observe that a given rice variety is faring poorly, they may select a different variety for the filling-in. The selection of this second variety is based on whatever critical microenvironmental characteristics they deduce from the failure of the first variety. For example, when household #11 observed the failure of some stands of rice in their 1976 upstream Kenua' swidden, they concluded that some of the soil was too blai' 'wet' for dry rices and decided to fill-in with swamp varieties instead. A second and more common reason for filling-in with a variety different than that originally sown is the exhaustion of that seed during the initial sowing.

Filling-in results in the dibbling and sowing of an additional 4.6 holes per square meter, with a sowing rate of thirty-nine liters per hectare, on average. This density varies, however, according to the extant density of both rice and relishes in a given area. By the time filling-in is carried out, the principal relishes are planted and growing in the swidden. In one square meter containing one or more of these relishes, in particular maize or cassava, the measured density of filling in averaged 10 percent less than in one square meter containing no relishes. There also was an inverse correlation ($r=-.96$, $P<.01$) between the density of the extant rice plants and the density of filling-in (table 46)[23]. It was notable that even at the higher initial planting densities, filling-in - while performed at a decreasing rate - still was performed. This probably is due to the fact that seeking out the less densely planted patches of ground in a swidden is a taxing job. The Kantu' say that it gives them ligau-ligau mata 'bleary eyes' and pedih pala' 'an aching head', which is reflected

TABLE 46
Densities Of Planting And Filling-In
In One Swidden (#44)

	Dibbled and Sown Holes Per Square Meter							
Initial planting density	3-4	4-5	5-6	6-7	7-8	8-9	9-10	10-11
Mean filling-in density	7.2	5.3	5.1	5.1	4.4	4.3	3.6	2.4
Mean total density	10.7	10	11.1	11.1	11.6	12.7	13.2	14.4

in the fact that they customarily fill-in only for brief periods at a time. One consequence of the difficulty of this operation may be that areas where the planting density already is adequate or better than adequate are filled-in as well, without the intention or design to do so. This would explain why the total and presumably ideal planting density, which is attained in areas where filling-in is performed, is not a constant but varies somewhat, increasing gradually as the initial planting density increases (table 46).

In swiddens in which filling-in is performed, it is performed over only 18 percent of the swidden, on average. In many swiddens it is not performed at all. In 1976, it was performed in only fifteen out of thirty-five swiddens at Tikul Batu. Several factors determine whether or not it is performed. One is swidden type. Filling-in is associated with (dryland) secondary forest as opposed to primary forest swiddens (P=.007 for twenty-five 1976 swiddens). The almost total absence of filling-in among primary forest swiddens reflects the fact that the Kantu' perceive them as being generally less fertile - and hence less capable of supporting a higher plant density - than secondary forest swiddens. A second factor affecting the performance of filling-in is its chronology. If a given household is late in its inital planting, relative to the rest of the longhouse, then it may decide not to fill-in since any rice thereby planted would be attacked by a disproportionately large number of predators. Household #6 cited this threat of predation in explanation of its failure to fill-in its 1976 upstream Kenua' swidden, which was planted a full two weeks after the mean planting date for the longhouse. The third factor that affects whether or not filling-in is performed is the availability of labor. Competing demands on the household's labor force at this time of year include planting relishes, building the swidden house, going on trading trips and, especially, tapping rubber. The households with limited labor resources generally devote them to these activities. Only the households with more labor available perform some filling-in as well. This reflects a basic difference between the initial planting and any subsequent filling-in. The former is obligatory and is carried out over the entire swidden by interhousehold labor. It maximizes the amount of land planted per unit of labor and hence maximizes the return on this labor. Filling-in is optional and is carried out over just a small fraction of the swidden by household labor. It maximizes the amount of labor devoted to each unit of land and hence maximizes the return on this land. The relatively low, mean density of the initial planting produces a high return on labor, at the cost of high inputs of land. The higher density of filling-in produces a high return on land, at the cost of high inputs of labor. Consequently, filling-in is most attractive to labor rich households, for whom indeed it represents one of the few opportunities to intensify the utilization of labor in their swiddens.

Swamp Rice

The Kantu' plant some swampland using the normal, dryland planting technology. This is the relatively driest swampland, which they call tanah rangkai-basah 'dry-wet land'. This land will hold standing water after a rain, but will dry up during a subsequent period of dry weather. For the most part it consists in isolated bits and patches of land that rise slightly above the swampland within which they lie. The Kantu' distinguish this land from a second, wetter category of swampland, which is permanently covered by water, up to their ankles in depth. This second type is characteristically found on the outer margins of swampland, abutting on dryland. This sort of swampland is planted not by the usual method of dibbling and sowing, but by ngunsai 'scatter' or 'broadcast' sowing.

Broadcasting is done entirely by sowers, as often male as female. No dibblers are involved. Here too, a sowing basket is worn on the left hip, and handfuls of seed are taken with the right hand and released in small amounts by thumb and finger action. In broadcasting however, the release does not consist of a tight pattern of grain, thrown into one hole by a stationary hand attached to a body bent at the waist. Rather, the sower's body is upright, and the seed is released while the hand swings quickly through a 90° arc, alternating from the left to right quadrants in front of the body, so that the grain is thrown in a very loose pattern onto the water covered land. In planting on dryland, holes are dibbled at intervals of 36-38 centimeters and each one is sown with 20-25 grains of rice. In broadcasting on swampland, single grains are sown at intervals from less than one to more than several centimeters apart. While both techniques result in a square meter being sown with approximately two hundred grains, broadcasting may produce many more plants than dibbling-sowing. Each dibbled and sown hole produces one rice plant, regardless of the number of grains in the hole; whereas every single broadcast grain can produce a separate plant under optimal conditions. If any sizable percentage of the broadcast seed germinates, the resulting number of plants per square meter in fact is too great for optimal growth. The Kantu' say that such plants would _Ngai janga', ngai bebuah_ 'Would not want to branch, would not want to fruit'. Accordingly, the Kantu' usually thin out plants growing from broadcast seed, transplanting some to other areas of swampland before the close spacing can affect their growth.

The Kantu' sow this one type of swampland by broadcasting because it is economically attractive. First, the amount of seed needed to establish a single rice plant by broadcasting is potentially as little as 4 percent of the amount needed in normal dibbling and sowing. Second, the amount of planting labor needed - whether reckoned per plant or per hectare - also is very little, amounting to perhaps 10-20 percent of the amount used in dibbling and sowing. Third, thinning out the broadcast-seeded plants provides a source of transplants that is cheaper (in terms of labor costs) than any other source.

The majority of swampland in the Kantu' territory is covered by standing water more than ankle deep, and this can be planted only by the technique of _menih ukur_ 'chance sowing'. This technique is similar to broadcasting in that the ground is not dibbled before it is sown. But the way the seed is sown resembles broadcasting less than it does normal dryland sowing. Chance sowing consists in the release of a bunch of grains (approximately fifteen grains per bunch) in tight groupings, each grouping spaced from the others by the intended distance between the rice plants. Chance sowing differs from broadcasting, therefore, largely in its tighter bunching of grains, which represents an adaptation to the greater depth of water in the swampland in which it is practised. The Kantu' say that individually broadcast grains will not survive in water this deep, but tight bunches of grain may. Even then there is no certainty that the grains will succeed in producing a plant. The highly speculative nature of this planting technique is made explicit in its very name. Some Kantu' even go so far as to say that grain sown by chance sowing is dibuai 'thrown away'. The fact that chance sowing is practised nevertheless on most swampland is due to its potentially significant savings in labor. All areas in which chance sowing is not attempted (or in which it is attempted but fails) then must be transplanted, and this requires four to five times as much labor. This enormous difference in labor cost makes the Kantu' willing to risk labor and seed on chance sowing, in an attempt to minimize the amount of transplanting that they will have to do.

Among the three planting techniques used in swampland, dibbling and sowing is least susceptible to early failure. More susceptible is broadcast sowing, the seed from which is vulnerable to rice sparrows and other predators. Seed sown by the third technique, chance sowing, is most susceptible to failure. The Kantu' say that if this sowing is followed within five days by heavy rainfall, the swamp swidden will flood and the moving waters will dislodge the just rooted seedlings from the soil, killing them as a result. Such losses are not uncommon. In the average swamp swidden, between two-thirds and three-quarters of the seed sown by this technique dies in postplanting flooding. Regardless of which planting technique was used initially, if it is not successful the Kantu' replant with tambak 'transplants'. In 1976, some transplanting was carried out in twenty-five of the twenty-seven swiddens that contained some swampland. The Kantu' do no transplanting, after sowing failure, only in those cases in which the factors responsible for the death of the seed are expected to similarly affect any transplants. This was the case in two swiddens in 1976 (viz., #41 and 56). Their owners explained their minimal transplanting by saying, Kami meda' paya' ngai 'We saw the swampland would not [produce]'.

There are several different sources for transplants. If grain that was sown either by dibbling-sowing or by chance sowing survives and grows well, one-half of each plant can be uprooted and transplanted to another section of the swidden, in which the initial planting was less successful. The Kantu' say that this operation can be performed on rice plants the size of a mok 'condensed milk tin', which measures 7.4 centimeters in diameter. If plants of this size are not divided, in fact, the Kantu' say that they will not mature properly. Transplants also are taken from plants that were broadcast sown. The plants that grow from this seed are too small to be divided in two, but they grow so close together as to require thinning out for optimum growth. The most important source of transplants is the dryland terenak 'nursery'. The Kantu' plant these nurseries by the standard dryland technique of dibbling and sowing, the only difference being that the planting density is up to ten times higher. Indeed, the term for planting a nursery is nugal pisit 'plant closely'. The purpose of this high density is to minimize the amount of arable land that will be left unused when, after just one month, the nursery is emptied of its seedlings. The Kantu' always locate the nurseries on dryland, in order to avoid any threat to the germinating seed from either standing water or flood water. They prefer dryland that is relatively moist and is shaded from the direct rays of the sun, especially the afternoon sun. Often, therefore, nurseries are located along the western edge of the swidden, where the swidden meets the adjoining forest. The final factor in the location of the nursery is the distance to the swamp swidden to be transplanted. Since as many twelve thousand separate seedlings may be needed to transplant as little as one-tenth of a hectare, the cost of transporting seedlings from nursery to swidden becomes prohibitive over long distances. Whenever possible, therefore, the Kantu' locate their nurseries right at the edge of (if not within) the swidden for which its seedlings are intended.

In most cases, all transplants come from within the swidden being transplanted (whether this be from nurseries within the swidden, or from plants grown there from prior dibbling-sowing, broadcasting, or chance sowing). If the transplants needed cannot be found within this same swidden, the next most desirable source is a nearby swidden belonging to the same household. A swidden that is distant by up to fifteen minutes walking time is acceptable. In the absence of a proximate swidden, the Kantu' sometimes make a nursery on an isolated patch of ground near the swidden for which the seedlings are needed. More commonly, however, the needy household turns to another household with a nearby swidden, from whom it obtains seedlings either through payment of the urat besi 'iron root' or through exchanges. For example, households #2 and 3 exchanged rice seedlings in 1976. Household #3 took seedlings from household #2's swidden #40 to transplant its own swidden #42. Swidden #40 was closer to #42 by one-half hour than household #3's alternative source in its own swidden #41. Household #2 in turn took seedlings from swidden #41 to transplant its swidden #38. In this case swidden #41 was closer to

#38 by 1½ hours' travel time than was household #2's alternative source in its own swidden #40 (figure 2).

Whatever their origin, the procedures for preparing the seedlings are the same. First, any shoots that broke off from their roots during the uprooting of the seedling are discarded. Second, any earth clinging to the seedling is scraped off with the edge of a brush sword, making it lighter to carry and easier to push into the earth during transplanting. Third, the Kantu' cut off all but the upper twelve to fourteen centimeters of the root of each seedling. They say that this enables the transplant to take root quicker. Fourth, they cut off the upper ten to twenty centimeters of each seedling's leaves to prevent it from withering before it is replanted[24]. At this point the seedlings are ready for transportation and planting. They must be planted either the same day they are uprooted or the first, following day. By the second day following uprooting, they are likely to have died. In order to plant their seedlings in good time, the Kantu' even work on days that are ritually proscribed - one of the few occasions on which they do so.

The actual transplanting involves three steps. If the seedlings are from a nursery, the workers first divide each seedling with their fingers into two or three parts (the mean is 2.6), each one of which will be planted separately. Seedlings grown from broadcast seed are too small to require dividing, and plants grown from dibbling-sowing or chance sowing will have been divided earlier during uprooting. After being divided (or not), the seedling is gripped at its root between thumb and forefinger and then thrust into the earth. Alternatively, it is placed into a hole made with a dibble stick. The final step in this transplanting sequence is to uproot all weeds in the area being transplanted. This weeding is carried out simultaneously with the transplanting, and it typically is the only weeding performed in the swamp swiddens.

Seedlings are planted throughout the swampland swidden (or swampland fraction of a predominantly dryland swidden) until the overall density of rice plants reaches 14.1 per square meter, including any plants previously growing there from seed. As this latter number varies, the number of seedlings transplanted varies inversely. In most swamp swiddens some amount of seed is planted and survives, so that the swidden later requires only partial transplanting, which is called nyeleti' 'fill in'. Only in a minority of cases - nine out of the twenty-five 1976 swamp swiddens in which some transplanting was done - is the entire swidden transplanted, which is called nambak pengkal 'transplant from the beginning'. In some cases, this is due to the total failure of all seed planted there; but in other cases it is due to a conscious decision to plant the swidden entirely with transplants. The advantages of doing this, according to the Kantu', are that the survival rate of seedlings is much greater than that of seed, they are more competitive against weeds, and they produce a more vigorous plant [25]. However, the decisive factor is one of timing. If swampland cannot be planted until October or even November, only seedlings can be used, because seed will no longer thrive if planted at this late date.

The total sowing rate in swampland averages 146 liters per hectare. This figure includes seed sown in the swidden by dibbling-sowing, broadcasting and chance sowing. It also includes seed planted in an average-size nursery, for later transplanting into this hectare of swampland. It assumes that an average proportion of seedling will come from this nursery, with the rest coming from the seed sown in the swampland itself by any of the three techniques cited. There are several reasons why the sowing rate is so much higher in swampland than in dryland. First, transplanting usually is performed throughout as much of the swidden as necessary to fill it all in to the maximum planting density of 14.1 plants/square meter. In contrast, only 18 percent of the average dryland swidden is filled in to its maximum density of 11.9 plants/square meter, as noted earlier. Second, the maximum density in dryland is, in any case, lower than the maximum density in swampland. Third, the presence of standing water and/or flood water in swampland typically kills a large proportion of the seed initially sown there. The low rate of survival of seed sown in swampland compared with dryland is reflected in the fact (deduced from the respective sowing rates) that one liter of seed produces 1187 rice plants in the average dryland swidden and just 371 plants in the average swampland swidden [26]. The chances of seed surviving to produce a rice plant, therefore, are less than one-third as high in swampland as they are in dryland.

Nonrice Cultigens

The planting of nonrice cultigens or relishes in the swidden is called nanam. This term incorporates three different techniques, the most common of which is nugal-menih 'dibbling-sowing'. The seed of a few relishes is dibbled and sown with the rice. But for reasons discussed previously, many relishes cannot be sown into the same hole with rice and so must be dibbled and sown separately. Maize is probably the most important relish planted by this technique. Planting usually is done by adult females, each of whom works alone and carries both a basketful of seed and a dibble stick. They make several holes in succession and then sow them, with a varying number of seeds, depending upon the type of relish. A second planting technique is ngunsai 'broadcasting', which is most commonly employed to plant nsabi 'mustard', terong 'eggplant', jawa' 'Italian millet', arum and jejebung, two spinaches. If two or more relish types can grow together without difficulty, then they are broadcast together. Thus, mustard and spinach are sown together, but eggplant and Italian millet are sown separately. The third planting technique, nambak, consists in the planting of tambak 'seedlings' or 'cuttings'. This is vegetative propagation, whereas the first two techniques involve propagation from seed. The principal relishes planted by this technique are belaut 'sweet potatoe', keladi 'taro', rangki 'chili peppers', pisang 'bananas', ubi kayu 'cassava', and tebu 'sugar cane'.

Kantu' techniques of vegetative propagation can be illustrated in a description of the planting of cassava, which is the most important of this category of relishes. First, the planters take cuttings from cassava plants growing in the previous year's swidden. These plants, now approximately one year old, will have trunks approximately two meters tall. The cuttings are taken from this trunk. (When possible, the Kantu' take their cuttings from plants that have been uprooted by wild pigs in search of the subsurface tubers, rather than cutting down living plants still capable of producing leaves and tubers for human consumption.) After cutting away the horizontal branches and top 10-20 percent of the trunk, the planters cut the remainder into longitudinal sections averaging 15-16 centimeters in length and 1-2½ centimeters in diameter. They typically obtain ten to twelve sections from a single trunk. They prepare these sections for planting, first, by cutting the bottom end of each section (which goes foremost into the ground during planting) into a rounded point, called lancam, which they liken to the rounded point on the working end of a dibble stick. The Kantu' say that a rounded point will produce more roots and hence more tubers than would a diagonally shaped point. The top of each section is cut into a diagonal point, called kempal. This is done, the Kantu' say, so that the top of the cutting will leaf quicker and more profusely after planting. Once the cuttings have been thus prepared, they are put into a sowing basket. The planter, usually male, carries this on his back or on his left side. In his right hand the planter holds a dibble stick, with which he makes diagonal holes in the ground. The Kantu' say that cassava will not produce tubers if planted in a vertical hole. Finally, the planter thrusts a cutting half-way into each hole, then presses down the earth with one foot.

Each planting technique produces different spacing among the relishes. The plants of a given relish type are most widely dispersed when they are sown together with the rice, because the ratio of relish seed to rice seed is always very low. They are less dispersed when they are planted separately, either by dibbling and sowing or by the planting of cuttings. They are least dispersed and most concentrated when they are broadcast sown - which may place seeds within one or two centimeters of one another. The spacing between the relishes and the rice plants similarly varies. It is least when the relishes are sown into the same holes with the seed rice. It is greater when the relishes are broadcast sown. The broadcast seed of a relish may land right next to a rice plant or it may land as far away as one-half the distance to the next rice plant. Given that rice is sown over the entire surface of every (dryland) swidden at a mean distance of forty-two centimeters between plants, one-half of the distance between rice plants is twenty-one centimeters. Therefore, one-half of this latter distance, namely 10.5 centimeters, represents the average distance between relishes sown by broadcasting and the rice plants. The spacing between the relishes and the rice plants is greatest when the relishes either are dibbled and sown separately or else are planted by cuttings. In either of these cases, the relishes are planted equidistant between the nearest rice plants, meaning that the distance between relish and rice averages twenty-one centimeters, the maximum possible. These differences in spacing are an important determinant of which planting technique is selected for each relish type. The relishes with the most rigid spacing requirements tend to be planted by dibbling-sowing or by cuttings, while those with the least rigid requirements (at least with respect to the spacing between the relish and the rice plants) are planted in mixture with the seed rice.

The selection of planting technique also is influenced by broader spacing concerns. For example, the Kantu' try to plant cassava and maize in the center of the swidden as opposed to near its edges, in order to minimize predation by feral pigs and macaques. In addition, the Kantu' try to plant their maize in the best burned sections of the swidden (or near the sites of stack burning), and/or in parong 'ravines' where the soil is apt to be wettest. They try to plant dish-cloth gourds in the earth and ash between the buttresses of large tree stumps. The finer these spacing requirements, the greater must be the flexibility of the planting technique. Thus, while the Kantu' sow some dish-cloth gourds with the rice, they also sow some by separate dibbling-sowing so as to exploit the favorable micro-environment at the foot of tree stumps.

Certain swidden characteristics - well burned, distant from pest populations, and proximate to watercourses navigable by canoe - are preferred for all relish types. But many other characteristics - such as the age of the forest from which the swidden was cut or susceptibility to flooding - are desirable or not depending upon the type of relish in question. This is reflected in the fact that individual households tend not to plant all of their relish types in all of their swiddens. In the cases during 1975 and 1976 in which a single Tikul Batu household made two or more swiddens per year (each containing at least some dryland), an average of only 39 percent of all of the household's relish types were planted in all of the household's swiddens. The selectively planted types tended to be those with the least economic importance. The most important relish types - the one or two most important tubers, cereals and cucurbits - were planted in all swiddens.

In deciding where to plant their relishes, the Kantu' have the option of planting in another household's swidden. This is called numpang nanam 'join in planting', or 'guest planting'. If the household that owns a swidden is asked its permission and gives it, one or more other households can plant relishes there, using their own labor and seed, and later harvest the crop for themselves. In such arrangements no part of the crop is given to the owners of the swidden, nor is any other form of compensation made. The relish crops of the guest household(s) are kept distinct from those of the owner by being planted in a separate, discrete area of the swidden. Guest planting is a common practise: in 1976, eight of the sixteen households at Tikul Batu planted relishes as guests in eleven of the longhouse's thirty-five swiddens.

Some participants in guest planting say that its purpose is to enable more households to take advantage of the ideal growing conditions found in a few of the longhouse's swiddens. This proposition is examined in table 47, which shows the prevalence of three of these ideal conditions both in swiddens selected for guest planting and in swiddens in general. The prevalence indeed is greater in the former case, but the difference

TABLE 47
Guest Planting And Ideal Growing Conditions In The 1976 Swiddens

Ideal Condition	All 1976 Swiddens	1976 Swiddens Selected for Guest Planting
Primary forest	33% of swiddens	36% of swiddens
Floodzone	31% of swiddens	36% of swiddens
Thoroughly burned	Burn averaged 89% of swidden area	Burn averaged 94% of swidden area

is so slight as to suggest that access to ideal growing conditions is not the main purpose of guest planting. Stronger evidence suggests that the main purpose of guest planting is not for each guest household to plant its relishes in the best microenvironments possible, but for each household to plant them in as many microenvironments as possible. Thus, every household that plants relishes in the swiddens of another household invariably plants those same relishes in its own swiddens as well. In addition, among the households that plant relishes in the swiddens of other households, there are some in whose own swiddens the same relishes are planted by other households. For example, in 1976 seven different households planted maize in the swiddens of other households, and in three of these cases the household's own swiddens also were planted in maize by other households. There even were cases in which several households were linked together in a circle (figure 27). It is clear from figure 27 that household #12 (e.g.) planted relishes in a swidden of household #4 not because it was a better swidden but simply because it was different, given that household #12 and #4 together planted relishes in the same swidden of a third household, namely #13. The purpose of guest planting, therefore, is to increase the diversity of the microenvironments in which households plant their relishes.

Only the most important relishes - principally maize and cassava - are planted by guest planting. Maize is involved in 56 percent of all cases, cassava in 26 per-

FIGURE 27
Guest Planting Circles In 1976

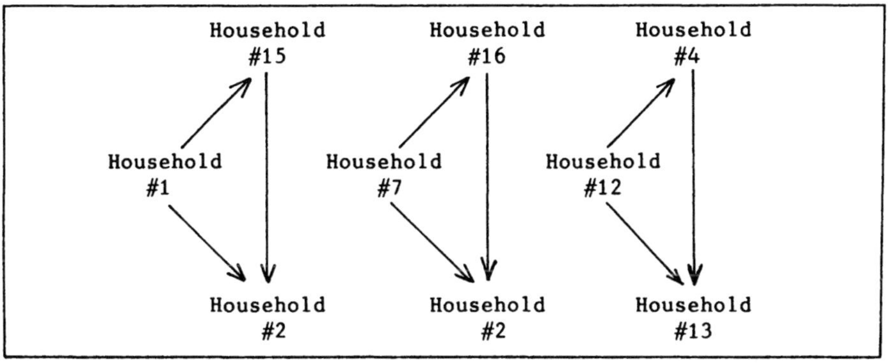

NOTE: The arrows point from the households requesting permission to guest plant and to the households granting permission.

cent, and the other tubers make up the remainder. The most important cultigen of all, rice, never is involved in guest planting, because the entire cultivable surface of every swidden invariably is planted in rice by the owning household. Any additional planting of rice would disrupt the optimal spacing of the rice plants, with deleterious consequences for the rice crop. In contrast, no household ever completely fills in its swiddens with relishes. There always are portions available (viz., portions planted in rice but with no relishes between the rice plants) in which additional relishes can be planted, with few if any ill effects on the extant relish or rice crops.

Although guest planting exacts little if any cost on the host household, permission to guest plant is not granted lightly. For example, it never is extended from a household in one longhouse to a household in another longhouse. Even within the longhouse, only the most closely related households request or grant permission to guest plant. Among the sixteen households at Tikul Batu in 1976, there was a statistically significant tendency ($X^2 = 5.44$, $P < .025$) for households linked by direct consanguineal or affinal ties to cooperate in guest planting.[27]

LABOR

The Kantu' devote a mean number of 12.9 workdays to the initial planting (viz., rice planting) of each hectare of dryland swidden. The individual work rates for dibbling and sowing average eighty holes and forty-nine holes per minute, respectively. These are the rates for male and female workers, respectively, who are capable of participating in reciprocal and wage labor arrangements. These rates apply to periods of intense work, carried out at normal speeds, but without interruption. Over the course of an entire work day, considerably lower holes/minute rates obtain. The rates also vary considerably from one worker to another and from one swidden to another, the observed range being from 67-106 holes/minute in dibbling and from 32-55 holes/minute in sowing. Several factors account for this variation. First, a steeply sloped swidden results in a slower work rate. So does a swidden that did not burn well and in which redressive burning operations were minimal. The large, unburnt logs that litter swiddens cut from primary forest actually obstruct the work less than unburnt brush and branches. The latter especially impede the sower, who must work either a part or the whole of her body into the brush in order to get her hand sufficiently close to the ground for an accurate throw. This is less often necessary for the dibbler, because the length of the dibble stick allows him to work with his hand some three-quarters of a meter above the ground. Work rates also are slowed, by as much as 30 percent, by ceremonial drinking, as discussed earlier. Finally, age differences among the workers, even among adults, are associated with some variation in work rates. For example, a male twenty-five years of age typically dibbles faster than a male of fifty, which reflects the fact that this task depends on body strength and endurance. In contrast, a fifty-year-old female typically sows faster than a female of twenty-five, which reflects the fact that the performance of this task depends more on experience and acquired skill.

Filling-in is performed at a mean rate of 15.5 holes per minute during uninterrupted periods of work. The decrease in the efficiency of labor, when dibbling and sowing are performed by the same person as they are here, is reflected in the fact that the same 15.5 holes could be dibbled and sown in just thirty-one seconds during the initial planting by separate dibblers and sowers[28]. The amount of labor devoted by each household to filling-in in 1976 ranged from zero to 5.2 workdays per hectare, with the mean figure being 1.8 workdays.

This mean figure represents the amount of labor necessary to fill-in approximately 0.18 hectare, or 18 percent of the swidden as a whole[29].

Labor inputs in swampland averaged 16.7 workdays per hectare for the combined operations of dibbling-sowing, broadcasting, chance sowing, and seeding a nursery of average size (viz., of the average size for transplanting one hectare of swampland)[30]. The preparation and transplanting of seedlings, and the associated weeding, consumed an average of 75.2 additional workdays per hectare (including all labor devoted to planting and uprooting seedlings outside of the swidden, that eventually were used within it). Variation from this figure is due to variation among swiddens in seed mortality and hence in the extent of transplanting that is necessary.

There is no single work rate for all of the different techniques of planting relishes. The highest or fastest rates are attained when they are dibbled and sown in a mixture with the seed rice. A somewhat lower rate is attained when they are broadcast sown. The lowest rates are attained when the relishes are dibbled and sown separately. For example, maize is sown at a rate of just 4.8 holes/minute. The work rates are equally low when planting cuttings. Thus, cassava is planted at a rate of 3.1 holes per minute. Overall, an average of nine hours per hectare were devoted to planting relishes[31]. The number of hours is not a perfect function of the number of hectares, however. The amount of labor devoted to planting each type of relish tends to be constant, regardless of the size of the swidden. This reflects the fact that the Kantu' demand for relishes is finite, unlike their demand for rice, due to the fact that the former are relatively less storable, less consumable, and less tradable.

Both the male and female members of the household work force participate fully in planting, with the participation of the former perhaps slightly surpassing that of the latter. Participation also is high among juvenile but able-bodied workers. Among workers incapable of reciprocal labor, meaning children, the aged and the infirm, participation in planting is negligible[32]. During other swidden phases, these marginal workers at least can make substantial contributions of labor to their own household's swiddens. But during planting even this contribution is limited by the fact that most swiddens are planted by large reciprocal work groups, which usually finish the work in just a day or two. While this day or two constitutes the total contribution of the marginal workers to planting, it represents only a fraction of the output of the able-bodied workers who - under the reciprocal labor system that prevails during planting - spend most of their time in the swiddens of other households, not in their own. This constraint does not apply to filling-in, transplanting, or the planting of most nonrice cultigens, however, and marginal workers can and do make greater contributions in these areas.

Of the sixty-nine swiddens made at Tikul Batu during 1975 and 1976, forty-six were planted entirely by reciprocal work groups and an additional four were planted largely by such groups. These groups ranged in size, in a loose association with the size of the swidden being planted, from eight to fifty-eight workers, including the workers from the owning household. The only swiddens in which reciprocal labor was not used were those composed entirely of swampland[33]. The multiplicity of techniques used to plant the swamp swiddens, coupled with their small size, make them less amenable to the use of large work groups. Because relatively so little land is involved, however, the overall percentage of swidden land planted by reciprocal work groups still reached over 97 percent for the years 1975 and 1976, which was much higher than for any other stage in the swidden cycle.

Virtually all households participate in reciprocal planting groups. Social disapprobrium is directed against any household that does not participate fully in the reciprocal labor system during planting, but rather milih 'chooses' among the other households, working with some but not others. Actual participation approaches the cultural ideal. For example, in 1976 there were sixteen separate households in the longhouse and a consequent maximum number of 120 possible labor relationships between any two different households, 100 of which actually were utilized. This rate of utilization of 80 percent compared with rates of just 17.5 percent during slashing and 19 percent during felling. During planting, each household joins with each of the others in the longhouse in reciprocal labor arrangements unless there is a good reason not to. Of the twenty instances in 1976 in which two households did not join in reciprocal labor, ten involved illness (viz., insanity in household #8), seven involved newness to longhouse life and custom (viz., on the part of the downriver Moslem heading household #17), and three involved interhousehold quarrels. The opposite principle prevails for households from different longhouses: they will not join in reciprocal labor groups unless there is a good reason to do so. In 1976, only households #5 and 9 participated in reciprocal labor planting parties with households from other longhouses. In both cases this participation was due to the proximity of both the kin ties and the swiddens of the households involved.

Participation in reciprocal work groups carries various costs. One cost stems from the lesser efficiency of labor in such groups, which is especially pronounced if much feasting and drinking is done. A second cost is that of the feasting and drinking itself. A household that plants a swidden with a reciprocal work group usually must call one work break at midmorning, to drink rice beer and/or arak, and to eat a special dish consisting either of cakes made from glutinous rice or a gruel made from this same rice, from millet, or from sago, garnished with trade sugar. At midday they must call another break to eat a meal that consists, in addition to a main dish of boiled rice, in side dishes of pork, chicken, or river turtle (the first two of which are sacrificed in connection with the planting ritual), as well as side dishes of swidden relishes. Sweetened black coffee is likely to be provided in addition to more rice beer and/or arak. The household that owns the swidden <u>nyangkung</u> 'provides' these items of food and drink to <u>all</u> workers from other households. The only exception to this is the main dish or staple of boiled rice. Each household provides its own during reciprocal labor. The owners of the swidden provide rice only to those persons from other households who are working on a <u>nolong</u> 'gift labor' or <u>nguli</u> 'wage labor' basis. In 1976 the market value of the food and drink consumed during the planting of an average household's swiddens was 2,548 rupiah. There also is a possible third cost to the use of reciprocal labor, involving a lack of fit between the size of the work group and the size of the swidden and a consequent loss of reciprocal workdays to the swidden's owners.

It is possible for a reciprocal work group to be too large. If the number of people in a work group exceeds the product of the swidden size (reckoned in hectares) and the constant of 12.9 workdays (viz., the mean workday input per hectare during planting), then the group will complete the planting of the swidden before the normal end of the workday at approximately 17:30. A short workday such as this represents a loss to the owners of the swidden, because thev still must reciprocate with full-length workdays in the swiddens of the other households. The swidden owners may incur a loss not only if the reciprocal work group is too large, but also if it is too small, such that they cannot complete planting on the first day but must reassemble on a second (or, rarely, third or fourth) day. On second days there again is a possibility of a shortened workday. In 1976, 47 percent of first planting days and 44 percent of second planting days were shortened.

Nine of sixteen Tikul Batu households experienced the loss of a short planting day in 1976. These tended to be households in which the ratio of workers (capable of reciprocal labor) to swidden hectares deviated from the longhouse mean. Table 48 shows that short first planting days were associated with worker/hectare ratios above the mean, while short second planting days were associated with ratios below the mean. This difference

TABLE 48
Shortened Planting Days And Worker/Hectare Ratios In 1976

		Mean ratio of household workers to swidden hectares
Swidden in which planting day was shortened on	1st day	3.6:1
	2nd day	1.5:1
	no days	2.2:1
All swiddens (total 27)		2.6:1

is based on the fact that the ability of a household to create debts in the reciprocal labor system is a function of the size of its work force, whereas its ability to incur debts is a function of the size of its swidden. Therefore, the larger the household's work force relative to the number of hectares in its swidden, the greater the relative amount of labor contributed by other households, and the greater the likelihood of a short, first planting day. Similarly, the smaller the household's work force relative to the number of hectares in its swidden, the smaller the relative amount of labor contributed by other households, the greater the likelihood that more than one planting day will be necessary, and the greater the likelihood of a shortened second day. Thus, this system of reciprocal labor penalizes deviation from the longhouse worker/hectare mean. The effect of the penalty varies according to the direction of deviation. In the case of deviation above the mean, which may reflect a relatively low intensity of labor among the household's workers, in the long run the penalty may stimulate greater efforts. In the case of deviation below the mean, which may reflect an unusually high rate of land use per worker, in the long run the penalty may dampen the household's efforts.

The threat of short workdays and the sundry other costs of reciprocal work groups are more than offset by their benefits. One benefit is aesthetic. The pleasures of ceremony, of feasting, and of working in a large, boisterous group, all make planting the most joyous of all swidden activities. More important than the aesthetics of labor, however, is the maximum and timely utilization of labor that is attained through the use of reciprocal work groups. The utilization of labor is potentially a problem during the planting stage because of the wide variation among households in the dates of planting. This variation is a function, in turn, of wide variation in the dates of burning as well as in the desired length of the interval between burning and planting. In the absence of the reciprocal labor system and assuming that each household planted its swiddens on its own, this variation in chronologies would result in a longer planting season within the longhouse and a lesser utilization of longhouse labor during that season. The actual reciprocal labor system cannot be used to make this comparison because there is no actual, nonreciprocal counterpart. I remedy this deficiency by constructing and comparing ideal reciprocal and nonreciprocal systems, as they theoretically would develop if labor were utilized to its maximum extent. Under the theoretical nonreciprocal system, assuming that all swiddens were planted using household labor only, the 1976 planting season at Tikul Batu would have spanned a minimum of thirty-six days, and the utilization of the total longhouse work force during that time would have equalled a maximum of just 35 percent[34]. Under the theoretical system of reciprocal labor, however, the 1976 planting season would have spanned a minimum of just twenty-six days, and the utilization of the longhouse work force during that time would have equalled a maximum of 51 percent[35].

In this theoretical comparison, therefore, the reciprocal system achieves a 28 percent shorter planting season and a 46 percent higher utilization of labor than the nonreciprocal system. This benefits the Kantu' in a number of ways. First, there are fewer days during the planting season on which longhouse workers are not engaged in planting. When labor is idled during the planting season, some of it inevitably is wasted because, while there are some alternate activities that can be engaged in during a one or two day lull, several of the most important activities cannot be. For example, a free day or two is not sufficient to start tapping a rubber grove or to go on a fishing or trading trip[36]. Under the theoretical nonreciprocal system, the average worker would have been idled for a minimum of twenty-three days during the 1976 planting season, compared with just thirteen days under the reciprocal system[37]. Thus, the potential for labor waste is 43 percent lower under the reciprocal system than under the nonreciprocal system. A second benefit of a shorter planting season and a higher intensity of labor is that this enables the Kantu' to concentrate more labor, sooner, in those swiddens that need to be planted quickly. As previously discussed, the secondary forest swiddens must be planted as quickly as possible after the burn, so that the post-burn growth of weeds is not given too much of a head start on the growth of the rice plants. Based on the preceding theoretical calculations, the reciprocal labor system would have achieved the planting of all secondary forest swiddens in 1976 within a minimum average of just 2.2 days of each burn date (which is very close to the ideal of one day), while the nonreciprocal system would have achieved a minimum average of 6.7 days from each burn date. The four to five day difference could cause significantly greater problems with weed growth under the nonreciprocal system. A third, possible benefit of a higher intensity of labor and shorter planting season is the minimization of losses to pests, especially the rice sparrow. The principal Kantu' strategy for dealing with the rice sparrows is to time the planting of the longhouse's swiddens so that the rice crops in all of them ripen at approximately the same time. This strategy is pursued more easily under the reciprocal system than the nonreciprocal system, because the former produces a much shorter planting season and, as a result, a shorter ripening season as well.

While most planting is done by reciprocal labor, a small percentage (usually less than 10 percent) is done by <u>nolong</u> 'gift labor'. The contribution that a donor household can make on any given day is limited by its total number of workers. However, a household can increase its contribution beyond this limit by earlier contributing labor (over and above its own reciprocal labor debts) to one or more other households, with the understanding that these latter will reciprocate this surplus labor in the swidden(s) of a designated household. For example, when the Kenua' swidden of household #8 was planted in 1976, three workers from household #16 joined in the planting, all contributing gift labor. However, the actual amount of gift labor contributed by household #16 was not three but ten days, because seven of the other workers in the planting group were there to reciprocate for seven workdays that earlier were contributed to their swiddens by household #16. Thus, household #16 used its labor resources earlier in the planting season to create labor debts that could be called in when its need for labor (to assist household #8) surpassed its own immediate resources. It also is possible for one household to give gift labor to another by giving reciprocal labor to other households to erase labor debts in the name of the recipient.

One household usually contributes gift labor to another only if the latter is in great need, due to either an overly small or a temporarily disabled work force. The work forces of the Tikul Batu households that received gift labor during the 1976 planting season averaged just 2.8 persons in size, compared with 4.5 persons for the rest of the longhouse. Household #8 received the largest contribution of gift labor. This was due not only to the fact that the household contained only a single worker capable of reciprocal labor (the smallest number of any household in the longhouse), but also to the fact that this sole worker was absent from the longhouse for most of the planting season (including the day his own swidden was planted) and so was unable to create many reciprocal labor debts by planting in the swiddens of other households. A large amount of gift labor also was given in 1976 to household #13, whose need stemmed not from the size of its work force (which, at six persons, was above average), but from the fact that it planted stem rice for the first time that year. This event places a major demand on the household's material and labor resources.

It sometimes happens that one household gives gift labor to another, even though the latter does not need it. This was true of the gift labor given to household #4 in its 1976 Serangir swidden and household #9 in its 1976 upriver Kantu' swidden. These gifts were due to the large amounts of resources (respectively third and fourth largest during the 1976 planting season) devoted to feasting in these swiddens. Thus, those households contributing gift labor did so in order to enjoy the food and drink, as well as the large crowd and festive atmosphere associated with a well-provenanced planting party. All of the households that contributed gift labor in these two cases came from outside of the longhouse Tikul Batu. Among households in the same longhouse, such a self-interested selection of households with whom to work (or not) would have been criticized by others.

The third and final type of extrahousehold labor used during planting is nguli 'wage labor'. This is the least important source of labor during this swidden phase, comprising just 1 or 2 percent of the total. Indeed some Kantu' even say, Kami nadai kala' ngupah urang nugal 'We have never hired people to plant'. This assertion refers to the fact that one household rarely makes a prior arrangement with another household for assistance, in exchange for a predetermined upah 'wage', in planting its swiddens. Such an arrangement is likely to be made only when two households have a close daily relationship, and when one of them has the ability to provide the goods while the other has great need for them. Of the five instances in which wage labor was used in planting a 1976 swidden, only one followed this pattern. This was the exchange of six kolak of glutinous rice by household #6 for three days of planting labor by household #16, which needed glutinous rice immediately to sponsor a curing ceremony. The Kantu' are likely to call such an arrangement 'wage labor' because of the prior settlement of terms. A second, more common way of arranging an exchange of planting labor for goods involves no prior discussion or settlement. Rather, on the morning of the day on which it is known that a given swidden is to be planted, workers from other households simply appear at the swidden on their own initiative and ask its owners for goods in exchange for their labor in the day's planting. If the swidden owners agree to the workers' request, the resulting arrangement also may be called wage labor, especially if the workers asked for rice in exchange for their labor, since this is the standard medium of exchange (far more common than currency) in all wage labor arrangements. The foregoing arrangement is less likely to be called wage labor if the workers asked instead for tembako-garam 'tobacco and salt' (also meaning trade goods in general), which are less common mediums of exchange. In a third and equally common type of exchange of planting labor for goods, the terms of the exchange are never set. The workers desirous of exchanging their labor for goods appear at a swidden being planted and simply join in the planting, without discussing their compensation beforehand with the swidden owners. At the end of the day, or on some subsequent day, the owners give them goods, the amount and type of which is at their own discretion. Because of the lack of specificity in this exchange, the Kantu' usually do not term it wage labor. As in the second type of arrangement, however, they are more likely to term it so if it is rice that is exchanged for the labor, as opposed to trade goods[38].

The households that exchange goods for planting labor are those that can most afford to do so. Of the five households doing so in 1976, three (viz., #2, 5 and 9) were heavily involved in trade and a fourth (#6) possessed the largest stores of rice in the region. The fifth household (#13) was of only average means and probably would not have accepted any wage labor but for the fact that it was planting stem rice for the first time, which obliged it to refrain from any ungenerous actions. The households that exchange labor for goods are those that most need to do so - in other longhouses. Whenever one household assists another household from within the same longhouse in planting its swiddens, its labor is reckoned and ultimately repaid within the reciprocal labor system. If a household contributes planting labor over and above its reciprocal labor obligations to another household in the same longhouse, the expectation is that this is - indeed, must be - gift labor. There is no such expectation when the labor comes from a household in another longhouse. When workers from another longhouse show up at a swidden that is about to be planted, the swidden owners will ask them if they have specific demands for goods in exchange for their labor (assuming that they do not present such demands on their own initiative). In response, the workers may present their demands, or they may indicate that the amount and type of goods are at the discretion of the owners, or they may say that they wish to plant on a gift labor basis. In all cases except this last, the swidden owners can then refuse the offer of labor if they wish.

CHRONOLOGY

The Kantu' believe that there is a fixed point in the solar year that is the best time to plant their swiddens. If they fail to plant at this time, they say, even the best omens will not ensure a good harvest: Asa dulau-dudi nugal, nadai termasuk burong agi' 'If planting is early or late, the omens no longer count'. The concern to plant at the proper time is reflected in the relatively high utilization of labor during the planting season: the figure was 43 percent for the longhouse as a whole during the 1976 season[39].

The Kantu' calculate this correct time for planting using a lunar calendar. As table 49 shows, this calendar

TABLE 49
The Kantu' Lunar Calendar In 1975

4th Moon	11 May - 5 June 1975
5th Moon	9 June - 5 July 1975
6th Moon	9 July - 3 August 1975
7th Moon	7 August - 1 September 1975
8th Moon	5 September - 1 October 1975
9th Moon	5 October - 30 October 1975
10th Moon	3 November - 28 November 1975
11th Moon	3 December - 28 December 1975

does not span a full solar year. The Kantu' say that they count the moons to mark time only while they <u>bumai</u> 'make swiddens', stopping as soon as the swiddens are completed and planted in rice. They say that they must not count the moons thereafter, while rice is growing in the swiddens (typically from September to March), because such measurement is offensive to the rice spirits. Instead, they mark the passage of time when the rice is in the ground by the successive stages of growth of the rice plant itself. The calendar includes a ninth, tenth, and eleventh moon because, while the Kantu' prefer to plant during the eighth moon, they sometimes are unable to plant until one, two, or even three months thereafter. The Kantu' say that the numbers assigned to each moon are a recent introduction or innovation. Formerly, the moons were distinguished by names (some of which are still in use, in addition to the numbers), each of which referred to some activity characteristic of its month. For example, the seventh and eighth moons were (and are) called the <u>bulan nunu</u> 'burning moon' and <u>bulan nugal</u> 'planting moon', respectively. The Kantu' have since numbered the moons in imitation of the twelve-month Gregorian calendar used by the national government. Because there are not 12, but approximately 12.4 moons in each 365-day year, however, the fit between the Kantu' and government calendars varies from year to year, with the moon numbers corresponding more closely to the month numbers in some years than in others. As table 49 shows, the correspondence was not all that close in 1975. A final, notable feature of the Kantu' calendar is the three-day gap between each moon. The Kantu' call this period, Udah bulan padam, umpai idup seria' 'After the moon is extinguished, before it lives again'. During these three days when the moon is not seen by the Kantu', they say that it shines in menua sida' sebayan 'the land of the dead'. These three days are not reckoned as part of either the preceding or following moon.

Each year the Kantu' determine the start of their lunar calendar with reference to the position of two constellations in the sky[40]: bintang tiga 'three stars' or Orion, and bintang banyak 'many stars' or the Pleiades (table 50)[41]. Indeed, the lunar calendar is important

TABLE 50
Correspondences Between The Kantu' Lunar And Stellar Calendars

Moon	Position of Orion	Position of the Pleiades
4th	--------------------	Disappear (at new moon)
5th	Disappears (at new moon)	Reappear (at new moon)
6th	Reappears	--------------------
7th	East of center of sky	--------------------
8th	Center of sky	--------------------
9th	West of center of sky	--------------------
10th	--------------------	--------------------
11th	--------------------	--------------------

to them only because it reflects these stellar cycles. It is the latter, according to the Kantu', that are the ultimate determinants of the proper planting time. Thus, the Kantu' say that rice planted during the ninth or tenth moons will not grow 'because' Orion and the Pleiades have passed their correct positions in the sky. This causal association derives from the Kantu' belief that one of the stars in Orion, Bunsu Bintang Tiga 'the youngest of the three stars', was their culture giver - teaching them how to regulate their social conduct and grow rice. As part of these teachings, Bunsu Bintang Tiga told the Kantu', Ti kami nyau melungan *kebong langit, bado' kita' nugal* 'Once we have passed the center of the sky, you stop planting'. In fact, this belief of the Kantu' in a causal relationship between stellar movements and correct planting time is based on a false inference. The stellar positions (like those of the moon) are merely an indicator of the correct time, not the cause. The true cause is the cycle of the earth's movement around the sun (which is reflected in the apparent movement of the stars) and the seasonal variation in a multitude of variables that is attendant upon this movement.

The good and bad times for planting rice, as reflected in the stellar positions and as these in turn are enumerated in the Kantu' lunar calendar, are shown in table 51. With respect to both the goodness of the individual

TABLE 51
The Moon Of Planting And The Success Of The Rice Crop

Planting Moon	Goodness of Grain	Abundance of Crop
4th	---------	---------
5th	---------	---------
6th	Average	Average
7th	Good	Good
8th	Good	Good
9th	Good	Bad (dryland) Good (swampland)
10th	Bad	Bad
11th	Bad	Bad

grain (*buah*), as well as the relative abundance of grain (*belayan*), the seventh and eighth moons stand out as the best for planting[42]. While the Kantu' believe that the ultimate cause of a given moon being good or bad for planting is the corresponding stellar positions, the proximate cause is believed to be pest predation. For three of four major rice pests (for which data are available), the amount of predation anticipated by the Kantu' varies according to the moon in which the rice

TABLE 52
The Moon Of Planting And The Extent Of Pest Predation

Planting Moon	Rodents	Rice Sparrows	Rice Bugs	Army Worms
4th	---	---	----	----
5th	---	---	----	----
6th	Bad	Bad	----	----
7th	Bad	Bad	----	----
8th	Bad	Good	Good	Good
9th	Bad	Bad ?	Bad	Bad
10th	Bad	Bad ?	Bad	Bad
11th	Bad	Bad ?	Bad	Bad

is planted (table 52). This covariation is based on two distinct principles of predation. In the case of the rice sparrows, the severity of predation varies inversely with the number of swiddens planted at the same time, because preying on each swidden also varies inversely with swidden number. Thus, predation by the rice sparrows is least in swiddens planted during the eighth moon only because most households and most long-houses plant their swiddens then; and their predation is greatest during the seventh and ninth moons only because fewer swiddens are planted then. On the other hand, the Kantu' maintain that rice bugs and army worms pose a problem in any swiddens planted in the ninth, tenth or eleventh moons, regardless of whether the number of swiddens planted in those moons is a majority or minority. Thus, whereas late planting merely concentrates the rice sparrow population, it results in an absolute increase in the rice bug and army worm population, by allowing several generations to develop[43].

The importance of planting during the correct moon, and so minimizing predation, is attested to by Kantu' explanations of their disastrous 1974 harvest. The Kantu' say that continuous rains postponed their burns until the ninth moon, so that they could not plant until the tenth moon. When they finally were able to burn, the swiddens all fired well and so, they say, the ensuing crop failures were due solely to the overly late timing of planting. Conversely, they attribute the very good harvest of 1975 to the timely planting of the swiddens during the seventh and eighth moons (viz., all planting was finished by 21 September 1974). The 1976 crop, which was better than the 1974 crop but not as good as the 1975 one, was planted earlier than in 1974 but later than in 1975 (viz., the 1976 planting did not finish until 2 October 1975). The Kantu' concern to plant at the right time is one of the most important determinants of the timing of the entire swidden cycle. It determines the prior timing of selecting, slashing, felling and (to some extent) burning, as well as the subsequent timing of weeding, guarding, harvesting and carrying.

While it is clear that it is bad for the Kantu' to plant during the late moons in their calendar, it is not clear that this is due solely to increased problems with pests, as they believe. Another likely factor, though not recognized as such by the Kantu', is the length of time that the seed rice has been stored. All seed rice for the coming year's swiddens is taken from the previous year's harvest. Therefore, the length of the harvesting-planting interval represents the length of time that the seed rice is stored before being planted. The Kantu' seed rice can be stored with safety for five to seven months, but after that its ability to germinate when planted begins to decline. Accordingly, the later the date of planting in a given year, and hence the longer the interval since the preceding year's harvest, the greater the danger of such a decline in germinability. The disastrous 1974 crop, for example, was not planted until the tenth moon in the Kantu' calendar. Assuming that the 1973 crop was planted and harvested at the normal times (August-September and February-March, respectively), then the seed used to plant the 1974 swiddens had been in storage for fully 8½ months, and hence had passed its peak of germinability. Similarly, the marked success of the 1975 swiddens may have been due in part to the use of seed rice at its peak of germinability. This seed had been stored for just 3½ months prior to planting, due to the late harvesting (because of the late planting) of the 1974 crops.

There is a second factor, also related to the germinability of the seed rice, and also not mentioned by the Kantu' themselves, that may help to explain why the later moons (and the earlier moons as well) are not as good for planting as the eighth moon. There is some evidence that the rainiest time of the Kantu' year (remembering that there are no distinct rainy or dry seasons per se) is this eighth moon. For example, during the eighth moon in 1976 the average daily rainfall was twenty-two millimeters, which was 90 percent higher than the average for the year as a whole. Given that some rainfall facilitates the germination and early growth of planted seed, it is possible that the high incidence of rainfall during the eighth moon has contributed to the Kantu' perception of this as the ideal planting moon.

Shorter-term decisions as to when to plant particular swiddens are determined by other factors, one of which is the timing of the burn. A continuing inability to burn a swidden (e.g., because of continuous rainfall) forces the postponement of planting, even if this results in planting later than the ideal time as set by the lunar and stellar calendars. Then, once the burn has been carried out, planting must follow within a certain period of time. As discussed in chapter four, the length of this period varies between primary and secondary forest swiddens. In the former this burning-planting interval must be long to dissipate toxic by-products of the burn, and in the latter this interval must be short to minimize weed growth. This distinction is reflected in the order in which multiswidden households burn and plant their swiddens. In 1976 eleven households cut two separate swiddens apiece (both at least partially on dryland) from forest of different ages, and while the older forest swidden was <u>burned</u> before the younger in 9½ of these cases, in 9 cases it was the younger forest swidden that was <u>planted</u> first. In both of the cases in which the swidden cut from older forest was planted first, the swidden cut from younger forest had not yet been burned and so could not yet be planted. The Kantu' desire to plant after a correct postburn interval ultimately is less important, however, than their desire to plant at the correct time in the lunar and stellar calendars. For example, in 1976 all eight primary forest swiddens were planted during the eighth moon. Among five of these eight swiddens, the burning-planting interval averaged forty-three days, which is ideal; but among the remaining three swiddens it averaged only seventeen days, which is far less than ideal. If the owners of the latter three swiddens had delayed planting by an additional twenty-six days, on average, they too could have achieved burning-planting intervals of forty-three days. However, this delay would have resulted in planting during the ninth moon. In order to plant during the eighth moon - which they regard as the best - as opposed to the ninth moon - which they regard as only the third or fourth best - these households settled for less than optimal burning-planting intervals.

Another factor that influences the chronology of planting in individual swiddens is the chronology of planting in other households. This influence derives from the fact that the longhouse planting season is short, intense, and concentrates the labor resources of the longhouse. Among households #3, 4 and 16, for example, the intensity of use of household labor throughout the thirty-four day 1976 planting season (viz., 30 August - 2 October) averaged 48 percent [44]. As a result of similarly high rates among the other households, one or more of the longhouse's swiddens were planted (by reciprocal work groups) on each of thirty days during this planting season. This tight scheduling is made possible by a careful, predetermined ordering of planting dates among all of the longhouse's swiddens. The position or order of each swidden in this schedule is fixed during nighttime meetings in the longhouse gallery, and it is determined by two factors, the first of which is the equitable distribution of planting dates among the longhouse's households.

This principle of equitable distribution operates to prevent the concentration of two or more planting dates, over a short period of time, within any one of the longhouse's households. There is a potential for such concentration, because most households require multiple planting dates each season (the mean is 2.3), as a result of making two or more swiddens and/or as as a result of making large swiddens that cannot be planted in a single day. The Kantu' make an explicit effort to ensure that when a household does require multiple planting dates, these dates are scattered throughout the planting season along with those of the other households and not concentrated at the most desirable point in the season, its beginning. The success of this effort is reflected in the fact that during the 1976 planting season at Tikul Batu, successive planting dates within the same household were separated by an average interval of seven days. This calculation includes all planting dates, including first, second and third dates in a single swidden. If the calculation is restricted to first planting dates (which are more important, because they place a greater demand on the longhouse's labor resources), an even more impressive figure of eleven days is obtained. This equitable distribution of planting dates formerly was supported by a ritual proscription. Namely, after a given household planted something (whether rice or relish) in one of its swiddens, it could not plant anything in any other of its swiddens for three days. This proscription not only supported the principle of equity, it also supported the principle of participation in the reciprocal labor system, since only by planting in the swiddens of other households could a household's workers have avoided being idle on its proscribed days.

The second factor that determines the distribution of planting dates among the longhouse's households is relative need, which principally varies according to the age of the forest in each household's swiddens. Since the ideal interval between burning and planting is long in primary forest swiddens and short in secondary forest swiddens, there is a strong tendency for the latter to be given precedence over the former in the longhouse planting schedule. During the 1976 planting season for example, each time that a secondary forest swidden was planted, it took precedence over an average of 5.5 primary forest swiddens that had been burned but in which planting had not yet begun (or if begun had not yet been completed), whereas each primary forest swidden that was planted took precedence over just 1.1 secondary forest swiddens. The tendency to give precedence to the secondary forest swiddens was even stronger than suggested in these figures because the data set contained 18½ secondary forest swiddens but only 8½ primary forest swiddens. If an adjustment is made for this difference, then it can be calculated that twelve primary forest swiddens were waiting to be planted each time that a secondary forest swidden was planted[45].

The timing of the initial rice planting determines when filling-in can begin. The Kantu' say that they cannot begin to fill-in a swidden until the rice initially planted there has grown tall enough such that the ground in which no rice was sown can be discerned easily. This height is attained an average of fourteen to fifteen days after the initial planting. The completion date for filling-in is determined by the celestial calendars. The Kantu' say that if they fill-in anytime after the full moon of the ninth moon in their calendar (which fell on 20 October in 1975), it will be too late in the year for the seed to live. The urgency to complete filling-in as quickly as possible is particularly clear in the case of the last swiddens in the longhouse to be planted, at the end of September or beginning of October, since filling-in could not even be started in these swiddens until as late as mid-October. Owing to these time pressures, filling-in usually is carried out relatively quickly, taking just three to four days from start to finish.

The dibbling-sowing, broadcasting and chance sowing

of swamp rices usually are carried out at the same time. If there is some dryland mixed with the swampland in the swidden in question, they all are performed on the same day as the dryland is planted. The nursery may be planted on this same day, but more often it is not begun, or if begun not completed, until five to ten days after this. The prolongation of this operation is due, in part, to the fact that it usually is carried out solely by each household's own work force. Thirty to thirty-five days after seed is sown by these four techniques, the seedling are ready for transplanting. If they are not transplanted within the ensuing ten days, they will become overly mature to be transplanted successfully. These time constraints, coupled with the relatively heavy labor demands of transplanting, sometimes make it difficult for the household to complete its transplanting on time. The Kantu' mitigate this difficulty by the particular timing of nursery planting. If a household sows their nursery some time after they sowed their swampland, they will complete it in just a day or two; but if they start to sow their nursery at the same time as they sow the swampland, they are likely to stretch this out over a period of many days. In either case, the timing ensures that the seedlings in the swampland and in the nursery are ready for transplanting at different times, which prolongs beyond ten days the period in which transplanting can be performed. By virtue (in part) of such timing, household #3 (e.g.) was able to carry out the transplanting of its 1976 swiddens over periods of twenty-three days and twenty-seven days, respectively, instead of just ten days.

The period during which relishes were planted, among the five 1976 dryland swiddens of households #3, 4 and 16, ranged from twenty-six to sixty-nine days; and the number of days on which relishes were planted during this period ranged from three to fifteen. In the one swidden in which each household concentrated its efforts, the time span averaged forty-nine days and the incidence averaged thirteen days. Compared to the dibbling and sowing of rice, therefore, there are two distinctive features of the chronology of relish planting. One is timing - relish planting typically begins before rice planting but does not finish until two months afterwards - and the other is intensity - relishes are planted on an average of just 20 percent of all days during their planting season.

The relatively low intensity of relish planting is associated with the fact that it is carried out (with the exception of the relishes sown together with the rice) by household labor alone. This makes a less intense pattern of planting both possible, since a household can plant relishes on any day on which at least one worker otherwise is unoccupied (whereas it can only plant rice on a day scheduled for it by the longhouse), and also necessary, since the labor input on any given day is limited to the household's own work force. In addition, during most of the period in which the relishes are planted, a part or all of each household's work force is unavailable for this work due to participation, under the reciprocal labor system, in the rice planting of other households. For example, on an average of 43 percent of the days on which households #3, 4 and 16 planted relishes, one or more members of the household's work force was absent, fulfilling reciprocal planting obligations elsewhere.

Many relishes are planted immediately after the burn (viz., the following day). The postburn residues that the Kantu' believe are inimical to rice are not believed to be inimical to the relishes (or not, at least, to all relishes). Relishes also may be planted long after the burn. The exact interval in each case varies according to the type of relish. Maize, for example, always is planted as soon as possible after the burn. This may be in part a function of its high demand for nutrients, given that the amount of nutrients present in the swidden is at its peak right after the burn and progressively declines thereafter. According to the Kantu', they plant maize early so that (with its short maturation period) it will be available as a supplementary food during the one or two month musim rapar 'hunger season' that often precedes the rice harvest. In contrast, there is no such urgency in the planting of cassava, which usually is the last relish to be planted. Its long maturation period precludes its use as a supplementary food during the upcoming famine season (although it can be used during the next year's famine season); and it also has low demands for nutrients, compared with maize.

The date on which relish planting must end is a function of the growth of the rice plants. The older the rice plants at the time that the relishes are planted, the greater the competition for nutrients that the latter will encounter from the former. For this reason, the Kantu' rarely plant any relishes more than two months after the planting of the main rice crop. (The longest interval I recorded was seventy days.) After that, the Kantu' say, Ngkayu alah oleih padi 'The relishes will be defeated by the rice'.

Chapter Six
Weeding

There are three distinct categories of weeds, each of which is dealt with by an equally distinctive weeding technique. The major category, rumput, comprises all undesirable noncultigens, whether growing from seed or from extant root systems, that are capable of being uprooted or severed from the ground[1]. The technique of uprooting or severing rumput is called mantun. The second category of weed, munti', refers to a single species of bamboo that commonly grows in swiddens. It usually represents the regrowth of bamboo that was cut down when the area was cleared for the swidden. The Kantu' remove this bamboo by a slashing technique called mumpun. The third category of weed, seringgir, refers to new growth on the stump of any tree that was slashed or felled during the making of the swidden. The Kantu' deal with this growth with a slashing technique called mantap.

The weeding phase consists, therefore, in uprooting grasses and seedlings, and cutting back bamboos and stump growth. Its purpose varies according to the age of the rice plants at the time it is performed. Weeding is termed nyiang kanong 'weed the pregnancy' when it is performed among rice plants still in their kanong stage, meaning before panicle emergence. The Kantu' say that the rice will nganong 'develop' only through this stage. After that, the immature grain will ripen but the rice plant will not otherwise develop or grow further. The performance of weeding up to the point of panicle emergence, therefore, is designed to enhance the growth of the rice plant - while it still is in progress - by eliminating competition from weeds. Weeding also may be performed after the rice has reached the succeeding stage, betangkai 'have panicles', when the panicles emerge. If so, it is termed nyiang tangkai 'weed the panicle'. At this stage, the uprooting and cutting back of weed growth is prompted not by the threat of competition, but rather by the threat of predation. The Kantu' say that the ripening rice plants are more vulnerable to rodents if there are weeds among the rice.

Weeds afford the rodents a means of climbing up to the rice ear and also provide them with protective cover from their own enemies (e.g., birds of prey). In addition, unchecked regrowth constitutes a potential obstruction to the coming harvest. Dense weeds hinder the movement of the harvesters through the swidden and, of greater importance, they make the panicles of grain more difficult to cut or even see. The purpose of weeding the panicle, then, is both to facilitate the activities of the harvesters and to hinder the activities of pests.

EXTENT OF THE WEED PROBLEM AND WEEDING

There is wide variation among swiddens in the type and extent of the problem with weeds, and hence in the type and extent of weeding required. For example, the Kantu' say that the felled trees in primary forest swidden typically produce seringgir 'sprouts' by weeding season, but the felled trees in secondary forest swiddens typically do not. Primary forest is cleared earlier in the year than secondary forest, by an average of eleven weeks or more, so there is a greater opportunity for stumps to produce new growth - by the time of the weeding season - in the former than in the latter. On the other hand, while stump growth is expected in any primary forest swidden cleared early in the year, most of it will die in a thorough burn. Thus, the Kantu' say that less labor has to be devoted to eliminating stump growth in a well-burned primary forest swidden than in one that has burned poorly. They also say that a poor burn will cause stump growth to become a problem even in secondary forest swiddens.

Bamboo growth also varies in severity from one swidden to another, according to whether or not bamboos were present in the forest originally cleared for the swidden. One determinant of this is forest age. The most problematic species of bamboo, munti', is never found in primary forest. Of the seven 1976 swiddens in which bamboos were slashed during the weeding stage, not one was a primary forest swidden. Finer distinctions between older and younger secondary forest also are important. The Kantu' say that there is more bamboo in the latter than in the former. This is one reason cited by the Kantu' for not farming very young forest. Drainage also is important in the distribution of bamboo. The Kantu' say that munti' grows widely only in the rich soil of the floodzone of the principal rivers, the Empanang and the Kantu'. Six of the seven 1976 swiddens in which bamboos were slashed lay at least partially in this zone. While preferring floodzone land, however, munti' does not thrive in the wettest parts of this, in the swampland. No bamboo were slashed in 1976 in any of the nine 100 percent swampland swiddens, even though all nine were located in the floodzone, and even though all nine were made in young, secondary forest, most of it just one to three years of age.

The growth of rumput 'grasses and seedlings' also varies from one swidden to the next, in part according to variation in the burn. In any swidden or portion of a swidden that does not burn well and in which post-burn redressive operations are not performed, grasses and seedlings establish themselves quickly (if indeed they were not established prior to the burn and have survived it intact). However, unburned portions of a swidden are not planted (unless redressive burn operations are carried out), and because they are not planted they are not weeded. This lack of association between burn success and weeding (as opposed to weed growth) is reflected in the fact that the Kantu' never attribute heavy expenditures of labor on weeding to a poor burn.

Weed growth also varies according to the age of the forest from which the swidden was cut. Specifically, the speed of weed growth varies inversely with forest age. In a secondary forest swidden two weeks following the burn, regrowth up to 20-25 centimeters in length can be observed in burned ground near any edge of the swidden bordering on secondary forest. In contrast, after a similar interval of time in similarly located burned ground in primary forest swiddens, no regrowth at all is observed. This is due to the greater prominence of herbaceous plants in secondary forest compared with primary forest. After secondary forest is burned, the surviving root systems of the herbaceous plants, as well as communities of herbaceous plants in nearby secondary forest, combine to quickly establish new plants in the swidden. And once established, these herbaceous plants grow relatively quickly. By comparison, after primary forest has been burned there are fewer herbaceous plants and more ligneous plants among those that initially establish themselves in the swidden; and this ligneous growth is much slower. The Kantu' say of such regrowth, Nadai tau' ngalah padi '[It] cannot defeat the rice'. As a result they rarely weed primary forest swiddens (table 53)[2].

The three secondary forest swiddens in table 53 in which no weeding was performed were cleared from grass-covered swampland. There are several reasons why regrowth

TABLE 53
Weeding In Primary Forest Versus Secondary Forest Swiddens

		Swidden Cut Primarily From	
		Primary forest	Secondary forest
Mantun 'weeding' performed	No	7	3
	Yes	1	24

N = 35 1976 swiddens.
P = .0001

is less of a problem in swampland, one of which is performance of the redressive burn operations nesap and mepat. These operations involve slicing, turning and treading the soil, and so they achieve near complete destruction of the root system of any grasses that covered the ground before slashing and felling. In a swampland swidden in which these two operations have been performed, therefore, the time needed for the grasses to reestablish themselves is increased greatly; and the need to carry out weeding is decreased correspondingly.

A second reason for the minimal weed problem in swampland is the use of the planting technique nambak pungkal 'transplanting from the start', which consists in planting a swidden entirely in rice seedlings as opposed to seed. When a swamp swidden is planted in seed, the rice begins on at best an equal footing with the weed growth. In contrast, when a swamp swidden is planted wholly in seedlings, the rice has a decided advantage over the regrowth. At the time of planting, the rice plants already will be 30-35 centimeters tall, whereas the weeds will consist in germinating seeds or at best roots. As the Kantu' say, there is little need to weed a swidden that has been planted completely with seedlings, soon after its clearing and burning, because the weed growth cannot really threaten the rice. When transplanting is not carried out until late in the planting season, however, the weeds will be better established and the competitive advantage of the seedlings will be correspondingly less.

The third reason for the lesser regrowth problem in swampland is the presence of standing water. A layer of water over the earth inhibits the initial growth of grasses in that earth, an effect that is recognized and utilized by the Kantu'. In most swamp swiddens, they take conscious measures to enhance this effect, one of which is the earlier discussed redressive burn operation nesap-mepat. This operation levels off high and dry spots in swampland, which otherwise are the likeliest areas of regrowth. In addition to lowering the level of the ground to ensure submersion of all swampland, the Kantu' may raise the level of the water. They achieve this by digging parit 'trenches' to nearby sources of additional water, which may be watercourses or simply wetter areas of swampland. Such trenches range from thirty to ninety centimeters in depth and no more than fifty centimeters in width[3].

There is one final factor that the Kantu' believe can affect the growth of weeds in some swiddens. This is the bird omen known as ketupong turun s'ari 'the Rufous Piculet gotten in one day'. This refers to hearing a Rufous Piculet on the very first day of seeking omens at the proposed site of a swidden. If this omen is heard, the Kantu' say, there are sure to be many weeds in that swidden. The Rufous Piculet otherwise is the most auspicious of all the omen birds. When heard during the initial omen taking at the site of any type of swidden, it constitutes assurance of a good harvest. There is an implication, therefore, that if the Rufous Piculet is heard on the first day it is sought, its benefit has been acquired with too little effort. There is a further implication that this imbalance of effort and reward is redressed by the weed problem that attends this first-day omen, for no such problem is associated with a Rufous Piculet heard after a longer period of searching.

TECHNOLOGY

There are two different techniques of weeding. One involves use of the hands only. With the fingers of one hand, each clump of weeds is grasped as near as possible to the ground and then pulled upwards. The intent is to remove from the earth not only the stalk and leaves of the plant, but its roots as well. The Kantu' recognize that the regrowth of weeds, after weeding, will be faster if the root system is intact than if it is not. The weeders work in a squatting position. (In damp or water-covered swampland, the weeding is done from an erect, but stooped, position.) Most often the weeders use just one hand for weeding, while the other hand grasps or lies across one knee, which helps to maintain balance. The weeders alternate use of the right and left hands for these two tasks.

The Kantu' prefer this technique of weeding because it is quickest. There are two restrictions on its use, however. First, the weeds must be sufficiently long to be gripped by the hand, and they must be sufficiently sturdy to bring up their root systems when yanked upwards. Second, the soil in the area to be weeded must be lemah 'soft' or 'loose'. Only in such soil is a pull on the above ground portion of the plant likely to result in the removal of the root system as well. These two conditions usually are present throughout swampland but not dryland swiddens. Some dryland weeds are short and grow close to the ground, making them difficult to grasp with the hand. There also are some species that separate too easily from their root systems. In addition, the Kantu' say that while some soil in a dryland swidden will be sufficiently lemah to permit weeds to be uprooted, there also is bound to be soil that is too kering 'hard' for this. In those portions of a swidden in which the weeds and soils are not conducive to the use of the hand in weeding, the Kantu' employ an alternate technique based on use of the bingkong, a short-handled, sicklelike tool.

This second technique consists in using the sickle to slice horizontally through the soil beneath each clump of weeds. Several such slices usually are made beneath each clump, two to three centimeters below the surface of the soil. A single worker invariably uses only a single sickle, wielded in the right hand. The blades are so angled as to make left-hand use impossible. The left hand usually is used to pull from the earth each clump of weeds, as it is sliced through by the sickle. As in weeding by hand, weeding with the sickle is performed from a squatting position; the only difference being that the position in this latter case is more upright, which is made possible by the length of the handle and shank of the sickle (which extend the weeder's reach by approximately fifteen centimeters).

After the weeds have been uprooted by hand or sliced through with the sickle, they are disposed of in any one of several ways, depending upon the drainage characteristics of that ground. In dryland lying above the floodzone, the weeds are ditabor 'scattered'. That is, they simply are dropped back to the ground. The Kantu' say that this can be done without fear of the weeds taking root again, because the heat of the sun will wither the weeds in a scant one-half hour. In floodzone land, however, the moist earth inhibits withering and promotes rerooting. As a result, the weeds in a floodzone swidden are ditugong 'piled' at intervals throughout the swidden. Weeds that have been uprooted or severed from the earth need access to both soil and sun to take root and grow anew; yet only the weeds at the bottom of a pile have access to the former, while only those at the top have access to the latter. No weed in a pile has access to both, as a result of which little rerooting occurs. Weeds also may be piled in swampland swiddens, where this piling serves a second function as well. The Kantu' say that weed piles in swampland quickly become burok 'rotten' and then feed and enhance the growth of the nearby rice plants. In other cases, the uprooted weeds in a swampland swidden simply are d'angkut 'carried' out of the swidden.

The Kantu' always begin weeding in that part of the swidden in which the weeds are longest and pose the most immediate threat to the rice plants. The work similarly progresses through the rest of the swidden, with the areas in which the problem is worst being weeded before other areas. Ideally, the entire swidden eventually is weeded. However, when the Kantu' say of a swidden, dibantun amih 'finished weeding', they do not mean that every single weed growing in the swidden has been removed. Rather, they mean that every single weed was removed from every part of the swidden at least once. After any given part has been weeded, however, weeds again will begin to establish themselves there. The Kantu' call these second growth weeds rumput baru 'new weeds', which they contrast to rumput lama' 'old weeds' or first growth weeds. While the Kantu' usually weed the old weeds, they usually do not weed the new ones, saying Rumput baru nadai alah dibantun 'New weeds are not possible to weed'. By this they mean that it is not possible to weed the new weeds during that stage of the rice plant's growth (viz., before emergence of the panicle) in which weeding can promote this growth. Indeed, most Kantu' households are hard pressed to rid their swiddens of first growth weeds before the end of this stage, much less the second growth weeds as well. If second growth weeds are weeded, this usually is done after the rice panicles have emerged, by households that were able to complete their weeding of first growth weeds during the preemergence stage (table 54).

The weeding of second growth weeds is never as extensive as the weeding of first growth weeds, because the new weeds do not pose as great a threat to the rice. Among twenty-five 1976 swiddens at Tikul Batu in which some weeding was carried out, in only five did this extend to second growth weeds. Moreover, in these five swiddens an average of only three workdays per hectare were devoted to weeding the second growth weeds, which compared with an average of forty-nine days per hectare devoted to weeding the first growth weeds. The weeding

TABLE 54
The Two Phases Of Weeding

		Associated Stage of Weed Growth	Associated Phases of Weeding
Stages of growth of the rice plants	Prepanicle emergence	Old weeds	'Weeding the pregnancy'
	Postpanicle emergence	New weeds	'Weeding the panicle'

of second growth weeds, therefore, is both minimal and highly discretionary. In the recent past, according to the Kantu', none of them ever weeded second growth weeds under any circumstances. The current practice represents an intensification of labor use in the swiddens and is perhaps one of the first signs of (and responses to) increasing population/land pressure in the Kantu' territory.

Among the twenty 1976 swiddens in which only first growth weeds were weeded, in fifteen swiddens this weeding was completed and in five it was not. In cases such as the latter, one of two distinctive weeding strategies is employed. First, the household may choose to complete weeding in as many rice varieties as possible, while doing no weeding at all in the remaining varieties. This strategy was followed in four of the five incomplete weedings in 1976. In three of these four cases, the varieties left unweeded were glutinous, reflecting the primacy of the nonglutinous rices in the diet of the Kantu'. The second strategy employed during incomplete weeding involves minimal weeding of all rice varieties and complete weeding of none. This strategy is based on the possibility that the farmer does not know, even by the time of the weeding season, which rice varieties are most likely to survive bad weather, pests and disease, and on the attendant possibility that the farmer does not know which varieties will be the best investment for weeding labor.

The failure to weed all first growth weeds before panicle emergence can be due to several factors[4]. One is simply the burden of the work. The Kantu' say that weeding, along with harvesting, places greater demands on their labor resources than does any other type of work. Measures of labor inputs support this view, yielding identical rates of forty-nine workdays per hectare for weeding and harvesting (in secondary forest swiddens). There also are subtler aspects of weeding that make it hard to perform in ways that are not true for other activities, including harvesting. For example, the Kantu' say that when they harvest they receive an asil 'reward' or oleih 'product', meaning the reaped rice. In contrast, they receive nothing when they weed. They say that the only product (as distinct from benefit) of weeding is the withered weeds themselves, which have no intrinsic value. Thus they call weeding, Kerja ti nadai isi' asil 'The work that does not have a product'. The effort put into weeding also can be affected by problems of personal security. The weeding season traditionally was the time for headhunting raids, to which anyone (especially a woman) weeding in the middle of an open swidden was vulnerable[5]. The Kantu' formerly dealt with this problem through the use of large work parties, the females in which weeded each swidden in turn, while the males guarded at the swidden edges. As the incidence of headhunting has declined over the years, the Kantu' have ceased to weed by these defensive work parties. In a limited sense, however, the need for the defensive work party has survived its demise. While endemic headhunting is no more, rare incidents do occur. Perhaps more important when discussing the lack of enthusiasm for solitary work in isolated swiddens, the traditional intertribal enmities still persist, along with the traditional fears. In any case, at the beginning of both the 1975 and 1976 weeding seasons, the Kantu' at Tikul Batu (along with tribesmen throughout the Empanang valley) were agitated with reports of roaming headhunters from Sarawak. During both scares many women refused to weed in the swiddens unless an adult male was present, and this reduced the overall weeding effort.

The impact of these several factors on the weeding effort varies somewhat from one household to another, in part according to variation in industriousness. Of the five households that failed to complete the weeding of first growth weeds in their 1976 swiddens, four had the reputation of being the least industrious in the longhouse. Their failure to complete weeding was seen by other households as 'in character'. This assessment is supported by the fact that the ratio of weeders to hectares needing weeding was 25 percent higher among those households that did not complete weeding in 1976 than among those that did.

The weeding workday closely resembles the slashing and felling workdays. For example, the workday is longer (averaging 350 minutes for household work groups) when the workers sleep in the swidden, as is commonly the case by this stage of the cycle. Like the planting workday, the weeding workday rarely is shortened by an ill omen. Whereas omens are not heeded during planting because this would be too costly (given the use of large groups), they are not heeded during weeding because they are thought to be of no consequence. Most ill omens are associated specifically with death as opposed to life, not failure as opposed to success. Accordingly, the Kantu' say that an ill omen can hardly have an adverse effect on their efforts, since their goal is to kill not grow the weeds. Further, because they have no projected use for the dead weeds, nor any intention to bring them back to their farmhouses or longhouse, it also is of little consequence if an ill omen becomes 'attached' to the weeds (which is the fear during the harvesting of the rice). Thus, many households pass through an entire weeding season without terminating a single day's work because of an ill omen. Exceptions are made only in the case of the less frequent, more potent omens, such as the call of the barking deer or an ominous dream.

The two techniques, _mumpun munti'_ 'slashing bamboos' and _mantap seringgir kayu_ 'slashing stump growth', are identical in technique. In each case the worker, using a brush sword, slashes through the offending vegetation at a point as near to its base as possible. The two techniques differ only in the thoroughness with which they are performed. Thus, the Kantu' do not necessarily slash all stump growth in a swidden, nor do they necessarily slash stump growth in every swidden in which it is present. In contrast, the Kantu' seldom allow bamboos to grow freely in a swidden, particularly while the rice panicles have not yet emerged. Indeed, the Kantu' typically clear the swidden of growing bamboo not once, but twice. That is, the Kantu' slash the first growth bamboos and later slash the succeeding second growth bamboos as well. The main reason for this is the striking rates of growth involved. The bamboo _munti'_ attains a growth rate of two or three centimeters per day, which far surpasses the rate of most other types of swidden weeds and presents the rice with formidable competition for available nutrients. The thoroughness with which the Kantu' slash the bamboos (viz., in comparison with the stump growth) also is facilitated by the fact that these bamboos grow in isolated clumps. In the average swidden in which bamboos constitute a problem, there may be only two or three small clumps per hectare. These clumps can be located and slashed with relatively little labor, compared with other, more widely scattered types of weeds.

LABOR

The amount of labor devoted to weeding - in those swiddens in which it is performed at all - averages forty-nine workdays per hectare. This figure is based on records of total weeding inputs in a sample of swiddens[6], as well as on direct measures of weeding in 9.3-square meter plots in a number of additional swiddens. The average rate in these plots was 1.7 minutes per square meter (assuming an uninterrupted period of work and the removal of all weeds from the ground), and the range was from 0.9 minute to 3.9 minutes per square meter. One factor in this variation is the extent of unburned timber in the swidden. Relatively much time is required to remove weeds growing near or under such timber, and in general unburned timber restricts the movement of the workers through the swidden. A second factor is the hardness of the earth and the related need to use a sickle versus one's hands alone. The greater the use of the sickle, the slower the work rate. A third factor that affects the rate is the number of weeds per measure of ground, which in turn is a function of the success of the burn, the drainage pattern, and the density of planted relishes. When bamboos are present and require slashing, the overall work rate still approximates forty-nine days per hectare, because the bamboo displaces the other weed growth. This is less true of stump growth, and so when it is present in a swidden the overall work rate may be somewhat higher. A final factor affecting the rate of weeding is the weeder. One relevant variable is age: children and adolescents often participate in weeding, but their work rates are considerably below the mean for adult workers. Also relevant is the gender of the weeder: the rates achieved by adult females average 10-15 percent faster than the rates achieved by adult males. This probably is a function of the fact that females do more weeding than males and consequently become more adept at it.

In forty-three out of fifty 1975 and 1976 swiddens in which some weeding was done, it was carried out largely or entirely by intrahousehold labor. Which household members join in the weeding and which do not is a function of age and gender[7]. The Kantu' do not regard weeding as an activity that either is or should be the primary concern of one gender or another; and indeed all adult males and females participate in weeding. However, while both males and females weed, the contribution of the latter is proportionately much larger. Among households

#3, 4 and 16 in 1976, each adult male participated in an average of 46 percent of all days on which any weeding was done in his household's swiddens, whereas the participation of each adult female averaged 83 percent. Gender differentiates among juveniles to an even greater extent. Juvenile, unmarried females make significant contributions to the household's weeding, but juvenile, unmarried males often do not. Most juvenile males participate, albeit minimally, in the weeding of their own swiddens; but without exception they do not participate in weeding other household's swiddens; and in a few cases they do no weeding at all. Once married, the young male's participation in weeding increases, but it still may not attain its maximum. If there are adult married males of two generations in a household, the male in the younger generation typically takes a smaller share of the weeding than the older male. For example, in household #3 in 1976, Angkol (in his twenties) participated in 30 percent of all weeding days while Gayan (his father) participated in 89 percent; and in household #4 the same year, Suring (also in his twenties) participated in 48 percent of all weeding days while Pagah (his mother's brother) participated in 60 percent. This apparent division of labor assigns the relatively undemanding task of weeding to the elder male, and frees the younger male for more physically demanding tasks.

The Kantu' say that weeding is not a reciprocal labor activity. Such labor was used in only nine of the twenty-eight 1976 swiddens in which some weeding was done (whether of weeds, bamboos, or stump growth). Even in these nine swiddens, an average of only 5 percent of the weeding was done by reciprocal laborers. Indeed, the Kantu' disagreed as to whether this was sufficient to justify the statement that reciprocal labor had been used at all. To the extent that the reciprocal system was used, much of the initiative came from maidens who desired one another's companionship during weeding. In order to work together, they used a variant of the reciprocal labor system involving the displaced repayment of labor debts. In a case (e.g.) in which maidens in households A and B want to weed together, but maiden A needs to weed for wages in a swidden of household C whereas maiden B does not, maiden B joins in the weeding in household C's swidden, but compensation for her work is paid by household C to maiden A. Maiden A then pays her debt to maiden B by weeding in one of the latter's swiddens.

Although reciprocal labor is used but little during weeding, there is reason to think that its benefits here would be as great as they are during planting, for example. As in planting, the use of reciprocal labor during weeding could be expected both to increase the overall utilization of the longhouse's work force and to quicken the rate of work. A shortened more intensive weeding season would benefit Kantu' because in weeding, as in planting, there are important constraints of time on the work. These arguments notwithstanding, reciprocal labor is used but little during weeding. One reason for this may be that until recently, weeding has not been a major concern for the Kantu'. Historically they favored the farming of primary forest over secondary forest, and in such forest no weeding is done except for the minor operation of slashing stump growth. This situation has changed just recently, as the Kantu' of Tikul Batu have finished clearing almost all of the primary forest in their territory. In 1975 primary forest still constituted an average of 60 percent of the hectares farmed by each household. In 1976, this percentage dropped to 36 percent, and by 1977 it had dropped still further to an estimated 25 percent. Thus, only now is weeding becoming as pervasive and important an activity as the other stages of the swidden cycle, such as planting. As weeding comes to equal planting in importance and as the potential benefits of reciprocal labor in weeding come to equal the benefits of reciprocal labor in planting, it is possible that the Kantu' will use increasing amounts of such labor in weeding their swiddens.

This prediction cannot be made unequivocally, however, for two reasons. First, weeding is regarded by the Kantu' as a somewhat discretionary activity, whereas planting is not. For example, when it appeared that household #8 could not muster sufficient intrahousehold or extrahousehold labor to plant its 1976 Kenua' swidden, the other households in the longhouse donated sufficient gift labor to ensure its completion. In contrast, when it appears that a given household cannot complete weeding in a swidden, no such offers of gift labor are forthcoming. Of the five household that could not complete weeding in 1976, only two received any gift labor from other households, and in neither case was the labor (just one or two workdays) nearly sufficient for completion. To the extent that weeding is thus discretionary, it is incompatible with the extensive use of reciprocal labor. For example, the use of reciprocal labor in planting is based on the assumption that each household will contribute sufficient labor to the swiddens of the other households to plant them to completion and also, when this labor is paid back, to plant their own swiddens to completion. This assumption is not valid in the case of weeding. Those households that do not contribute sufficient labor to their own swiddens to complete weeding, would not contribute as much labor to a reciprocal labor system as would be expected of them (viz., as would be expected of them by those households that do complete weeding in their own swiddens). The second reason why the use of reciprocal labor during weeding may not increase involves the extensive use of wage labor during this stage.

A greater percentage of weeding is done by wage labor than by reciprocal labor, gift labor, or even, in some instances, by household labor. The Kantu' say that time pressures create a greater need for wage labor during weeding than during any other swidden stage, with the exception of the harvest. The exchange of goods for labor during weeding always is based on a prior arrangement between the two households involved, in which the date and terms of the exchange are specified, and it always is called 'wage labor'. The rate of exchange during weeding follows the norm for wage labor: two kilograms of husked rice, or two kolak of unhusked rice, per worker per day. During the 1975 and 1976 weeding seasons, rice was given in 89 percent of all wage payments. An additional 9 percent were made in the currency equivalent (300 rupiah/day plus meals), and the final 2 percent were made in the equivalent in trade goods.

The ability to hire wage labor during weeding is a function of the amount of grain that the household has in storage, which in turn is a function of success either in swidden farming or in trading ventures. There is a particularly strong association between the latter and the extensive use of wage labor. The four trading households in Tikul Batu (viz., #2, 4, 5 and 9) used an average of twenty-three days of wage labor per hectare in weeding the 1975 and 1976 swiddens, which compared with an average of 2.4 days/hectare for the nontrading households. The extensive use of wage labor by the trading households appears to be less a function of their need to use this labor than of their need to pay for it. It often is difficult for the trading households to sell grain except as part of a wage labor arrangement; or, if sold on credit, it often is difficult for these households to collect what is due them except in the form of wage labor. The need to accept this labor (in effect, the low opportunity cost of this labor) enables the trading households to accept smaller returns per day of wage labor than the other households and results in generally higher overall weeding inputs per hectare in their swiddens.

The extent to which the trading households use wage labor during weeding varies inversely with the general success of the longhouse harvest. These households used an average of twenty-eight days per hectare of wage labor in 1975, following the poor 1974 harvest; but they used eighteen days per hectare in 1976, following the very good 1975 harvest. Following a poor harvest there are few households that can afford to hire laborers except for the traders, who have nonswidden sources of rice; but at the same time there are many households wanting to work for wages. Harvest success has an opposite effect on the use of wage labor by the nontrading households, whose only means of amassing rice is through their own farming. Thus in 1975, following the generally poor harvest of 1974, only one nontrading household (viz., #6) was able to hire any wage labor for weeding. But in 1976, following the very good 1975 harvest, seven of twelve nontrading households hired wage laborers for weeding. The 1975 harvests of these seven households met 107 percent of the year's needs, on average, compared with a figure of just 73 percent for the harvests of the five nontrading households that did no hiring.

For households whose most recent harvests were failures, wage labor weeding represents the best opportunity to acquire grain before the next harvest. After the weeding season, there is no further opportunity to perform wage labor before the harvest itself. Before the weeding season, the last major opportunity for wage labor is during felling; but this falls so early in the year that even the households whose harvests were poor may still have some grain left and, hence, would not yet need to perform wage labor. The relationship between prior harvest success and the performance of wage labor weeding was evident during the 1975 and 1976 swidden years. Following the poor harvests of 1974, longhouse members performed 108 days of wage labor in weeding the 1975 longhouse swiddens. But following the good harvests of 1975, longhouse members performed just seventy-six days of wage labor in weeding the 1976 swiddens.

A household's ability to perform wage labor during the weeding season is a function of the demand for weeding in its own swiddens and the timing of this demand. Occasionally the timing is such as to obviate any conflict. For example, the 1976 Kenua' swidden of household #12 was weeded later in the year than the 1976 swiddens of the other households. This allowed household #12 to first weed for wages in the swiddens of other households and then later complete the weeding in its own swiddens (assisted by wage laborers hired with the proceeds of its own, earlier wage labor). More often there are time conflicts between weeding one's own swidden and weeding for a wage in the swiddens of others. If a household performs the latter at the expense of the former, it decreases its chances for a good harvest and increases the possibility that it will need to perform wage labor during the following weeding season as well. Accordingly, if a household reaps a poor harvest, which they know will not satisfy their needs through the coming year (and assuming that they have no other source of rice, such as trading), they are likely to minimize the demands for weeding in their swiddens during the coming year by locating them in primary forest. The extent to which this is done is reflected in differences, in the ratios of weeders to hectares of secondary forest, between households that do and do not perform wage labor weeding. In 1975 the ratio for households that performed some wage labor weeding was 11.9 weeders/hectare (which is a light weeding burden), while the ratio was 3.5 weeders/hectare (which is a heavy weeding burden) for households performing no wage labor.

The impact of good and bad harvests on the hiring and performing of wage labor weeding produces a chronic imbalance within the longhouse in the supply and demand for such labor. After a poor harvest there are many households that want to weed for wages, but there are few households that can afford to hire them. Similarly, after a good harvest there are many households that can afford to hire wage labor weeders, but there are few households that need to be hired. Following a poor year this imbalance is partially corrected by the above average use of wage laborers in weeding the swiddens of the trading households, as already discussed. In addition, some of the households needing to earn rice can work as wage laborers for households in other longhouses. The variation among longhouses in harvest success is such that a longhouse invariably can be found whose previous harvest was good enough for the hiring of wage laborers. Following a good year the imbalance in supply and demand also is corrected by recourse to other longhouses. The households wanting to hire wage laborers seek out other longhouses whose previous harvests were poor enough such that their members are open to wage labor opportunities. Thus, following the poor harvest of 1974 at Tikul Batu, only 30 percent of the wage labor used in weeding the 1975 swiddens was contributed by households from other longhouses; but following the good 1975 harvest, this percentage perforce increased to 76 percent during the weeding of the 1976 swiddens. Such extensive use is made of extralonghouse laborers only when the number of willing wage laborers within the longhouse is insufficient to meet demands. If intralonghouse laborers are available, they are given preference in hiring.

CHRONOLOGY

In 1975 the average secondary forest swidden was weeded over the fifty-eight day period between 18 October and 15 December[8]. Weeding has no absolute chronology of its own. Rather, like all of the other swidden phases, its chronology is a function of the chronology of planting and burning. The chronology of burning determines when weeding can begin, and the chronology of planting determines when it must end. The burning of the swidden typically destroys or severely inhibits all extant plant growth, so weed growth usually dates from the time of the burn[9]. Some weeds break ground within just one or two days of the burn, but the Kantu' say that weeding cannot begin until the weeds are sufficiently long to be grasped in the hand and then plucked from the ground (assuming use of the hands and not the sickle). The time needed for weeds to attain this length averages six weeks from the date of the burn, but it varies considerably due to variation among swiddens in the conditions (e.g., a poor burn or location in the floodzone) that favor and accelerate weed growth. Once the weeds have attained the minimum necessary height, weeding is begun without delay. Any delay is inimical to the household's ability to completely weed the swidden while the rice panicles still are in their preemergent stage and still can benefit from the weeding. Panicle emergence occurs, at the latest, approximately fourteen weeks after planting. Given that the planting date determines when weeding can finish and the burning date determines when weeding can begin, the maximum average duration of the weeding season equals eight weeks (viz., fourteen weeks minus six weeks) plus the variable X, where X is the interval between burning and planting.

The relative demand that weeding places upon the household work force can be calculated as the ratio of the minimum possible duration of weeding to its maximum possible duration. The former is calculated as the product of the total number of hectares to be weeded by the household and the constant weeding input of forty-nine days per hectare, divided by the number of weeders in the household. The ratio of this figure to eight weeks plus X averaged 0.45 for the households of Tikul Batu in 1976, meaning that the average household had more than twice as much time available to it as was minimally necessary to complete its weeding. Nonetheless, this represents one of the greatest felt demands upon household labor resources in the course of the swidden cycle, as is reflected in the Kantu' statements that their need for supplemental wage labor is greater during weeding than during any other swidden stage, except for the harvest. This perceived pressure on labor is partially due to the demands for labor during this season for nonweeding activities as well. These include demands from other swidden activities, in particular transplanting swampland and constructing farmhouses. Some demand for labor during the weeding season also comes from cash earning activities such as rubber-tapping or wage labor in the swiddens of other households.

Chapter Seven
Guarding

Between the time that a swidden is planted and the time that it is harvested, it can tend towards one of two possible states with regard to predation. It can be angat 'hot', meaning that the crops are heavily preyed upon by pests; or it can be celap 'cold', meaning that the crops are relatively free of pests. The threat of the first state is an important determinant of human activity in other phases of the swidden cycle. As noted in chapter one, this threat can partially determine where swiddens - in particular primary forest swiddens - are located. As noted in chapter five, the threat of crop losses to pests also influences the chronology of planting and hence of the entire cycle. Finally, the chronology of harvesting can be influenced by the pest threat, being started or finished abnormally early in order to reduce losses. This manipulation of temporal and spatial variables is an indirect way of dealing with the threat of predation. Direct measures, called nginang 'guard', are used as well. Their importance is illustrated best by what transpires in their absence. During 1972 and 1973 the Kantu' were prevented by governmental decree from spending the night away from the longhouse, so they could not guard their swiddens to the normal extent[1]. Because of this, the Kantu' say, the harvests during these two years were very poor.

A subsidiary purpose of guarding is to supplement the diet with wild game. The two forest animals most valued as food - the bearded pig and sambar deer - also are major crop pests, and the Kantu' capitalize on the attraction of their crops in hunting these animals. During 1974-1976, 46 percent by weight and 62 percent by number of all deer and pig taken in hunting were taken in or near the swiddens.

The numbers and types of pests vary greatly from one swidden to another and from one year to the next, and there is corresponding variation in the extent of guarding. The Kantu' do not guard all of their swiddens all of the time. They guard only in response to prior evidence of losses to pests. Close monitoring of conditions in the swiddens is critical, therefore. During this interval between planting and harvesting, household members usually sleep in the longhouse rather than their swidden (except during the weeding season in secondary forest swiddens). They sleep in a swidden only if it is experiencing severe predation, otherwise relying on periodic daytime trips - called ngabas 'inspect' - to assess the pest situation and perform any needed operations. The average primary forest swidden is visited in this manner once every three weeks, commencing several weeks after planting and continuing until the harvest. The average secondary forest swidden is visited once every four days, beginning within several weeks of the completion of weeding and continuing until the harvest. The greater frequency of visits in secondary forest swiddens appears to be a function of their customarily richer harvests. In neither case do visits continue after the main harvest of the rice crop. Nonrice cultigens are present and harvestable in the swidden for many months after the rice harvest, but they are not of sufficient importance to merit continued guarding of the swidden for their sake alone.

INSECTS

With the exception of human sorcerers, all predators of rice and other crops are termed ama' 'pests'. The Kantu' differentiate the many types of pests into three main groups, one of which is termed ino' utai 'female thing'. The Kantu' use this term to refer to most insects, especially flying insects (regardless of their gender). There are many types of insects that can threaten the swidden crops. Of all of these, the Kantu' regard five - all predators of rice - as the most serious. These are the mpangau 'rice bug', mpit 'mole cricket', buntak bukit 'grasshopper', ulat jingkal 'army worm' and ulat tusok 'stem borer'.

Predation by these insects varies to some extent according to the dryland/swampland distinction. The most serious insect pest from one year to the next, the rice bug, always is a greater problem in swiddens made in swampland. Other insects, too, generally pose more of a problem in the swampland swiddens. The Kantu' attribute this to the fact that the rice plants germinate later in the swampland swiddens than in the dryland swiddens, which causes an increase in the relative pest burden of the former compared with the latter. This burden is an indigenous concept for the amount of predation in a given swidden, relative to all other swiddens. This amount is relatively small in a swidden planted at approximately the same time as most other swiddens in the longhouse. At any point during the growing season in that swidden, it is one of many susceptible swiddens and consequently it experiences only a small fraction of the total predation, assuming finite pest populations. The relative pest burden is greater in a swidden planted either well before or well after most of the other swiddens. During the growing season in this swidden, it will be one of but a few swiddens susceptible to predation, and consequently it will experience concentrated predation by a relatively large proportion of the pest population. Since swamp swiddens typically are not planted at the same time as dryland swiddens, they experience an unusually high pest burden (with respect not only to the rice bug, but also to other insect and noninsect pests). This burden falls on the swamp swiddens rather than the dry swiddens, because of the relatively much smaller area devoted to the former. Each year an average of 90 percent of all hectares farmed by the longhouse are in dryland. If all this 90 percent is planted at the same time, the pest populations will be dispersed over a relatively large area. In contrast, only 10 percent of each year's hectares are in swampland, so even if all of the swamp swiddens are planted at the same time, the pest populations will be concentrated in a relatively small area.

Increased insect predation results from deviation not only from the intralonghouse norm of planting timing, but from the interlonghouse norm as well. Just as most households within a give longhouse tend to plant at the same time, so too do most longhouses in a given region tend to plant (in more general terms) at the same time, namely from the sixth through the eighth Kantu' moons. Due to death, illness, or misjudgement of the rains, an individual longhouse sometimes is forced to plant later in the year (viz., during the ninth, tenth or eleventh moons); and this is associated with increased predation by insects, especially the rice bug and the army worm (cf. table 52). For example, most of the swiddens at Tikul Batu were planted during the eighth moon in 1976 and during the seventh and eighth moons in 1975, but most were not planted until the tenth moon in 1974, and both the rice and maize crops failed. The Kantu' attributed these failures to predation by the army worm, in part, and they attributed the unusual abundance of the army worm to the lateness of planting.

The Kantu' recognize this association between deviant

planting timing and increased predation, and they consciously try to prevent the latter by regulating the former. Each household tries to plant at approximately the same time as the others in the longhouse, and the longhouse as a whole tries to plant at approximately the same time as the other longhouses. These are the most important means by which the Kantu' attempt to forestall predation by insects. Once such predation is in progress, the Kantu' have but limited means of redress - most of which involve the use of poisons. There are three traditional poisons that the Kantu' use against insect pests, in particular the rice bug. One is *kayu garu* 'aloes-wood'. When burned in the swidden, this wood emits an odor that, while agreeable to humans, is believed to be anathema to all insect pests. A second poison comes from the *rengas* tree, the sap of which is a severe, contact irritant. The Kantu' burn this too in their swiddens, believing that the smoke will drive off insects. A third poison comes from a *raung* 'frog' that is known to have poison ducts in its skin. The Kantu' kill numbers of these frogs and scatter them in their swiddens. The rice bugs are attracted to the corpses, alight on them, and reportedly then are killed by the toxins in the animals' skins.

In recent years, while not relinquishing these traditional poisons, some Kantu' have pioneered the use of new ones as well. One such poison, used as yet by only a few households, is derived from the standard flashlight battery (a major trade item from the coast). The Kantu' cut open spent batteries, empty them of their zinc, and burn this in the swidden. The Kantu' say that the resulting fumes are noxious to insect pests. A second poison, pioneered by household #9 in 1974 and now widely used, is the mosquito coil (an occasional item of trade). Knowing that these coils drive away mosquitos, the Kantu' burn them in their swiddens in the hope that they will have a similar effect on the crop pests. The most important new poison is a commercial preparation of DDT, which the Kantu' sometimes buy from touring government officers and sometimes are given free. It is intended for use against the insects that prey upon pepper, but the Kantu' use it against the rice bug. They mix the poison with water and apply it to their rice plants using a simple sprayer made from bamboo. The Kantu' say that they invariably spray any swamp swidden being attacked by the rice bug. No Kantu' ever has used commercial poison in a dryland swidden, however. Some of the noncommercial poisons are used in the dryland swiddens, but not as frequently as in the swampland swiddens. The Kantu' say that the dryland swiddens (averaging 2.9 hectares in area) are simply too large to be treated with poisons, while the swampland swiddens (averaging only 0.4 hectare in area) are not. They also say that, as mentioned earlier, insect problems tend to be greater in swampland swiddens. Since yields tend to be much greater as well in these swiddens, these represent further reasons for the greater use of poisons, since the greater the amount of grain at risk, the greater the potential returns on labor and money devoted to protecting it[2].

The poisoning of insect pests is never more than partially successful in the swamp swiddens and is even less effective in the dryland swiddens. Ultimately the Kantu' rely more on the nyingkelan umai 'swidden sacrifice' and bwaya tanah 'earth crocodiles'. The swidden sacrifice is performed sometime after the planting and before the harvest in each swidden in which stem rice is planted. The owners of each swidden first prepare a pegela' 'offering' of rice beer and food, including one or two chickens and occasionally a pig as well. The owners invoke the pests, telling them that the offering is being made so that they (the pests) will not devour the rice crop. Part of the offering is reserved for the pentik, the offering post erected in the swidden during planting. The owners tell the pentik that it is given this offering so that it will be diligent in chasing pests out of the swidden. At the base of this post they bury a piece of iron, in the hope that their rice plants will acquire its qualities of strength. Finally, the swidden owners catch one of each type of insect pest, wrap them all in a leaf, and take it to the banks of the Empanang river. There the packets of every household in the longhouse are placed together on a small raft, and enjoined to go downriver to Java[3].

There is some variation from one year to the next in the performance of this ritual. In 1974, for example, it was completely omitted by many households in Tikul Batu. After observing how poorly the rice was growing that year, they say that they were too 'angry' to make any efforts at pest protection. In 1976 the rice grew well and all households performed the ritual, but separately and on varying dates. In 1975 the rice grew very well but was threatened by serious pest problems, so all of the households in the longhouse chose to perform the swidden sacrifice collectively, on a single day (viz., 25 September 1974). The Kantu' regard a collective performance as most potent[4].

If the swidden sacrifice is deemed incapable of dealing with the pest problem in a given year, the Kantu' have recourse to the 'earth crocodiles'. These are a pair of life-size images of crocodiles, one male and one female, that the Kantu' of Tikul Batu have constructed from clay. The Kantu' believe that if they make offerings to them, the earth crocodiles will devour all of the pests in the longhouse's swiddens. However, the Kantu' rarely make these offerings (no more than once or twice each decade) because they are extensive and costly, and because they require the entire longhouse to observe unusually strict and onerous prohibitions for eight days following the offering[5].

RICE SPARROWS

The second principal type of crop pest is the pipit 'rice sparrow', a tiny Munia that is popularly represented as the size of a man's thumb. The rice sparrows travel in large flocks and prey on the ripening grains of rice in the swidden. In the typical swidden the rice becomes vulnerable to their predation approximately sixteen weeks after planting, and it remains vulnerable through the harvest. The time of year during which the rice is in this stage often is called musim pipit 'the season of the rice sparrows', so fixed is the association between this stage of growth and predation by this pest.

The extent of predation by the rice sparrows is determined, in part, by deviant versus normative swidden timing, just as in the case of insect predation. The Kantu' say that within a single swidden, any rice variety that matures in advance of the majority of varieties will suffer from rice sparrow predation. Similarly, they say that any swiddens that are planted before the majority of the longhouse's swiddens will experience heavy bird predation. In 1975 not all households at Tikul Batu planted at the same time: many were forced to plant relatively early, and a number of these households later reported serious problems with rice sparrows. On the other hand, in 1976 no household reported a serious problem with birds, which the Kantu' attributed to the relatively closer timing of planting in all swiddens. Finally, just as deviation from the longhouse planting norm results in heavy predation in a single swidden, so does deviation from the interlonghouse norm result in heavy predation for the entire longhouse. The interlonghouse planting norm, with respect to rice sparrows, is the eighth moon (cf. table 52). The Kantu' of Tikul Batu say that if they plant during the eighth moon, they usually will not have a problem with the rice sparrows. If they plant earlier, during the sixth or seventh moons, they say that they will have a problem - even if all of the households in the longhouse plant at the same time.

A second determinant of the extent of predation by the rice sparrow is deviation in productivity. For example, the Kantu' say that predation by the rice sparrows (like the insects) generally is worse in swampland swiddens than in dryland swiddens, which is in accordance with yields that average over twice as high in the former as in the latter. They also say that predation is worse in any swidden whose productivity has not been diminished - before the rice sparrow season begins - by attacks from other pests. Thus, the members of household #14 noted that the rice in their 1976 Setungkus swidden suffered heavy, early losses to the rice bug and, as a result, did not suffer from later predation by rice sparrows.

In the event of serious predation by rice sparrows, there are several means of defense available to the Kantu'. They possess one technique for capturing the birds involving use of a glue boiled from the latex of their rubber trees. The Kantu' smear this glue onto branches and tree trunks lying on the surface of the swidden. Alternatively, they smear it onto five or six lengths of bamboo that are fixed perpendicularly (in a circle) to the top of a six-meter bauer kepulut 'pole for glue', which then is erected in the center of the swidden. When rice sparrows, feeding in the swidden, alight on the glued branches or poles, they become stuck and eventually die.

The Kantu's most effective means of defense against the rice sparrows consists not in killing them, however, but in scaring them away. They accomplish this by stringing lengths of liana through the swidden, attaching them at intervals to tall poles of bamboo that have been stuck in the ground. They call these tali buru 'chase lines'. All lines are led to the farmhouse. If this house does not command a good view of the swidden, or if there is no house in the swidden, a small shelter is built especially for use at this time. One or two persons can protect an entire swidden by sitting in this shelter during the day and working the chase lines. When a flock of rice sparrows alights in some part of the swidden, the watchers yank several times on the appropriately located line. The movement of this line, and especially the more exaggerated movement of the poles to which it is attached, causes the flock to take flight. The chase lines do not require unremitting attention, and so people usually do other work (e.g., plaiting mats or baskets) as they guard. Once harvesting begins, the harvesting itself usually suffices to scare off the rice birds without the additional use of chase lines.

Chase lines are not erected in all swiddens, but only in those to which the rice sparrows pose a distinct threat. They were erected in thirteen of thirty-four swiddens in 1975 and in only one of thirty-five swiddens in 1976. These data illustrate the marked difference from one year to tne next in the extent of bird predation. Data from a single year illustrate the difference in predation from one swidden type to another. In 1975 there was less usage of chase lines in dryland than in swampland swiddens ($X_c^2=7.66$, $P<.01$ for the 34 swiddens), and there was less usage in primary forest than in secondary forest swiddens ($P=.0043$ for the 34 swiddens). In both cases it was the most productive swidden type (viz., swampland and secondary forest) in which predation and guarding were most extensive.

ANIMALS

The third category of crop pest involves the broader grouping called jelu 'animals'. This grouping encompasses most nonhuman animal life, excepting the insects and birds already mentioned, and also excepting ikan 'fish'. Domesticated animals are subsumed under the subgrouping jelu laman, laman referring to the cleared or open space immediately surrounding the longhouse. Of the numerous named jelu dwelling in the territory of the Kantu', six are crop pests of importance: two even-toed ungulates, the rusa' 'sambar deer' and the jani' 'bearded pig'; two primates, the nyumoh 'pig-tailed macaque' and the kera' 'long-tailed macaque'; and two rodents, the cit 'rat' and the tupai 'squirrel'.

Deer and Pigs

Both deer and pigs pose a serious threat to swidden crops. The former eat the stems of the rice plants, and the latter eat the grain as well as the tubers of any root crops, especially cassava. Because of this difference in food preference, deer predation and pig predation occur at different times during the swidden cycle. Deer predation begins an average of thirteen to fourteen weeks after the rice has been planted and continues through the rice harvest. Pig predation usually does not begin until the rice harvest is under way, but may continue long after it is finished, so long as there are cassava tubers in the swidden.

Variation in Predation

The extent of predation by deer and pigs often varies considerably from one swidden to another. One factor in this variation is the distance from the longhouse to the swidden. It is apparent from table 55 both that there is less pig predation than deer predation, overall, and also that the former is influenced by the swidden-longhouse distance while the latter is not. The Kantu' say that pigs generally are more wary than deer of humans and human habitations and are more sensitive than deer to human smells and sounds. Thus, pigs only enter the vicinity of the longhouse to feed in the middle of the night, after all human sounds have ceased. Proximity to humans aside, the extent of pig predation in the vicinity of the longhouse also is affected by the lack of proximity to primary forest, which - as the Kantu' say - is the pigs' proper habitat[6]. This is not a factor in deer predation because their proper habitat is the very mixture of secondary forest, fruit grove, brush and open field that is most abundant in the vicinity of the longhouse.

TABLE 55
Swidden-Longhouse Distance And Predation By Deer Versus Pigs

	Swidden-Longhouse Distance	
	< 30 minutes	> 30 minutes
Percentage of 1975-1976 swiddens containing defensive measures against deer	46%	46%
Percentage of 1975-1976 swiddens containing defensive measures against pigs	23%	34%

NOTE: The presence or absence of predation is deduced from the presence or absence of defensive measures.

A second swidden characteristic that influences the extent of deer and pig predation is the degree of insulation from the surrounding forest. When deer and pigs enter a swidden, they tend to feed only in the areas nearest the swidden's edge, from which the flight distance into the adjoining forest is shortest. The larger the swidden, the smaller the proportionate area comprised by this strip along the perimeter in which deer and pigs will feed. In a one-hectare swidden, for example, 36 percent of the total area lies within ten meters of the swidden's edge and adjacent forest; but in a two-hectare swidden, only 26 percent of the total area lies within ten meters of the edge. If a given swidden is adjoined on one or more of its four sides by another swidden, it is totally insulated from the forest and from deer and pig predation on that side or sides. The Kantu' say, indeed, that a swidden that is tengah urang 'between other peoples' [swiddens]' never requires defensive measures against deer and pigs (or any other animal pests). These benefits of insulation can be obtained from more distantly removed swiddens as well. For example, in 1976 and for some years prior to this, the members of a nearby Iban longhouse made swiddens higher and higher up the big Kenua', a stream running along the southern border of the territory of Tikul Batu[7]. In so doing, they progressively cut off the Kantu' territory from a large tract of primary forest running higher up into the nearby foothills. This tract is the closest pig habitat of any extent to the Kantu' territory. As the Iban have progressively interposed their swiddens between this tract of forest and the Kantu' territory, the Kantu' have noted a progressive decline in pig predation in their territory, which they attribute to this interposition.

Deer and pig predation also may vary greatly from one year to the next. Some of this is due to yearly variation in the productivity of the longhouse's swiddens relative to the interlonghouse norm. For example, the data on use of defensive measures in the swiddens indicate that there was deer and/or pig predation in 65 percent of the 1975 swiddens at Tikul Batu, in 54 percent of the 1976 swiddens, and in less than 50 percent of the 1974 swiddens. This corresponds to the fact that the 1975 harvest was well above the interlonghouse norm, the 1976 harvest was close to it, and the 1974 harvest was far below it. The extent of deer and pig predation varies, therefore, according to the attractiveness of the longhouse's crops relative to the crops of other nearby longhouses. In addition, the extent of predation varies according to the attractiveness of the crops of all the longhouses relative to the wild fruit crops in the forest.

Once each year, from October through March (but concentrated in December and January), the forest trees bear and drop their fruit. This is the time of year that the Kantu' call musim buah 'the fruit season'. This fruit season is an important factor in the behavior of certain animal populations, principal among which is the pig, but including macaques and, to a lesser extent, deer. At the beginning of the fruit season, the Kantu' say, large numbers of pigs migrate into the Kantu' territory from the Kapuas Lakes region, a distance of at least thirty statute kilometers[8]. At the end of the season, they return downriver. Some feral pigs live in the Kantu' territory year-round, but they are fewer in number. The Kantu' call these latter jani' ubi 'cassava pigs', suggesting that the only pigs that live in the territory year-round are those that can be supported by the cassava crops. This seasonal and territorial variation in the food supply of the bearded pig is reflected in Kantu' hunting patterns. They say that they only hunt during the fruit season, when the migrant pigs are abroad in the territory. In fact they hunt throughout the year, but they make two-thirds of their kills during the six-month fruit season[9].

The fruiting of the forest trees, and the effect of this on pig movements, is not constant from year to year. At intervals of usually four years, all of the trees bear heavy crops of fruit, resulting in greater numbers of pigs feeding within the territory[10]. The Kantu' say that when the fruit crop is heavy, huge bala 'troops' of pigs come to their territory from as far away as the east bank of the Kapuas river (a distance of at least thirty-five statute kilometers), crossing this broad river in a single mass. (Such crossings of the Kapuas river are a matter of historical record among the Malay peoples living along its banks.) The increased numbers of pigs in the territory also is due to increased local breeding. Indeed, the Kantu' regard increased breeding as a forecast of a heavy fruit crop. They say that if the sounds of pigs mating can be heard at the longhouse, then the fruit crop of the coming season is certain to be heavy. Pigs always mate at the time of bungai timul 'the flowers appear [on the fruit trees]', and the offspring from this mating are born at the time of buah labuh 'the fruit falls [from the fruit trees]'.

Although more pigs feed in the Kantu' territory during a year in which the fruit crop is heavy, their attraction to the fruit actually results in a decrease in their predation on the swidden crops, according to the Kantu'. This decrease in predation and in attendant crop losses is marked, because there also is a decrease in the efforts devoted to protecting the swiddens during such years. Instead of protecting their swiddens, many Kantu' go off at the height of the fruit season (December and January) to camp in the forest and feed on the fruit themselves. During the heavy fruit season of 1976, for example, some of the members of Tikul Batu camped in the forest for as long as three or four weeks. Such expeditions would cost them dearly, in terms of crops lost to pigs from the unprotected swiddens, but for the fact that they are undertaken only during years of heavy fruit crops, when the wild pig population concentrates its predations on the forest fruits as opposed to the swidden crops.

The freedom to undertake fruit expeditions during heavy fruit seasons is important to the Kantu'. The forest fruits are an important, alternate source of food and trade goods during famine seasons. Famines occur when the crop from the past year's swiddens has been exhausted and the crop from the current swiddens is not yet ready to be harvested - a time of year that tends to coincide with the fruit season (table 56).

TABLE 56
The Relationship Between Swidden Harvests, Famine, Fruit Crops, And Swidden Predation

Year (1-3)	(1) Average Fruit Crop		(2) Heavy Fruit Crop		(3) Average Fruit Crop	
Calendar months Oct.-Mar.	(10-11-12-1)	(2-3)	(10-11-12-1)	(2-3)	(10-11-12-1)	(2-3)
Swidden crop		Poor		Good		Poor
Swidden related famine	No		Yes		No	
Fruit crop can offset famine	No		Yes		No	
Extent of pig predation permits fruit expeditions	No		Yes		No	

Famines tend to occur, therefore, during the fruit season following a year in which the swidden harvests were poor. The swidden harvests never are poor during a year in which the fruit crop is heavy. The Kantu' say that if the fruit trees in the forest bear a heavy crop, so too will the rice plants in the swiddens. This heavy rice crop precludes the possibility of famine the following year, but it does not preclude the possibility during the current year, during the heavy fruit season in which this same plentiful rice crop ripens. Famine during this current, heavy fruit season is related not to the goodness of the currently ripening swidden crops and coming harvest, but rather to the poorness of the preceding year's swidden crops and harvest. Since the rice crops tend to be good during a year in which the fruit crop is good, the rice crop during the preceding year is likely to have been poor in comparison. As a result, the Kantu' are likely to experience rice shortages during this heavy fruit season. They can partially redress these shortages through the exploitation of the fruit, which the Kantu' are free to gather because of the diminished need to defend their swiddens against pigs[11].

There have been historical changes in both the fruit crop and the pig population. The Kantu' say that now they rarely hear pigs mating, and when they do hear them, the fruit crop is never as heavy as it used to be. (They attribute this decline to the increasing incidence of sexual delicts in their territory.) The Kantu' note that not only has the productivity of the trees declined, but their total number has diminished as well, due to the clearing of primary forest for swiddens. Finally, the Kantu' note a decline in the pig population, which they attribute not to the diminishment in fruit trees and crops, however, but to the increase in guns and people. This attribution is problematic: if the increase in guns and people has had such an impact on the pig population, it should have had a similar impact on the deer population; but the Kantu' say that the present deer population (including barking deer and mouse-deer, as well as sambar deer) is as large as ever. I suggest, therefore, that the decline in the pig population is due to the diminishment of its preferred habitat and food sources. Conversely, the maintenance of the deer population at a constant level, despite the increase in people and guns, probably is due to the increase in its habitat - the secondary forest and farmland - and the food sources therein.

Guarding Techniques

The Kantu' use several types of wind activated devices to scare off deer and pigs. One type, the kelekitau (from kitau 'wave'), consists in a pole five to six meters long, one end of which is thrust diagonally into the ground, while from the other a winnowing tray is suspended. The tray moves about in the wind, scaring any animals that approach the swidden. A second device, the ketekong (from tekong 'hit'), scares by both sight and sound. This too consists in a winnowing tray suspended from a pole, but in this case a club of wood is attached to the tray, and a tin box is additionally suspended from the pole. (These boxes come into the interior as containers for various trade goods.) When the wind blows, not only do the tray and clapper move about, but the latter strikes loud notes from the tin box. A third device, less frequently used, consists in a framework of poles covered with a Kantu' sun hat and other materials to create a humanlike appearance. The device is suspended from an inclined pole, so that the wind can move it about and play upon its 'garments'. The Kantu' say that this device, and also the kelekitau first mentioned, are primarily directed not at the pigs but at the deer, which are more sensitive to visual stimuli.

Of much greater importance than scaring is fencing, of which there are four types. In increasing order of labor cost and efficacity, they are the tali jintang, lalau, perimpah, and pagar. The tali jintang 'stretched line' consists of a single strand of liana strung horizontally between tree trunks or driven posts, at a height of approximately one meter, along the perimeter of the swidden. It provides a minimal deterrent to deer - but none to pigs - that are threatening to enter the swidden. The lalau 'rail [fence]' consists of a single, horizontal rail laid onto forked posts - or lashed to nonforked posts - that are fixed vertically into the ground along the swidden's edge. The rails average six to seven meters in length, and the posts average one meter in height. The rails are taken from amongst the unburned timber in the swidden, and the posts (as with the posts for the stretched line) are cut from living trees at the swidden's edge. The rail fence also provides a deterrent to deer only. The third type of fence, which provides a greater deterrent to deer and a minimal deterrent to pigs, is the perimpah 'brush fence'. The Kantu' construct this by felling living trees and piling them, complete with limbs and foliage, into an unbroken barrier along the swidden's edge. In order to construct a brush fence, suitable trees must be available in the forest immediately adjoining the swidden. If the adjoining forest is too young, the Kantu' instead may construct it of bamboo. Living bamboo is cut, sharpened at both ends, and - again complete with limbs and foliage - thrust at both ends into the ground, in the form of an arch. Succeeding arches are overlapped, so as to form a continuous barrier. If the forest immediately adjoining a swidden lacks bamboo as well as large trees, the Kantu' will make do with a rail fence. The fourth type of fence is the <u>pagar</u> 'true fence'. This is a more certain deterrent to pigs as well as deer than the brush fence. The true fence is made of four horizontal rails lashed at even intervals to vertical posts. The posts are spaced approximately two meters apart and have an aboveground height of approximately one and one-third meters. The rails and posts are cut from living trees. The Kantu' say that trees of the requisite diameter and character are distributed unevenly through their territory, and the construction of the true fence is inhibited in those parts of the territory where such trees are absent.

These four fence types employ different principles of deterrence. The deterrent of the stretched line and rail fence is visual. Deer can easily crash through or jump over either type of fence, but if a deer has never (or but rarely) entered a swidden, the Kantu' say, it will be deterred by the sight alone of a stretched line or rail fence. (This will not deter pigs, which are said to be less sensitive to visual stimuli, as already noted.) However, after a given deer has entered a given swidden a number of times or after, as the Kantu' say, <u>Rusa' nya' nyau suah kemi'-bira' dapi' umai nya'</u> 'The deer is regularly urinating and defecating in the swidden', construction of this visual deterrent is ineffective. Even if the stretched line or rail fence is erected before any deer have entered the swidden, it may lose its effectiveness as the season progresses and it loses its abnormity. In either case, a physical deterrent or barrier then is needed and is provided by the brush fence or true fence. Based on these differences in the principle of deterrence utilized, the Kantu' construct some fence types in anticipation of predation and others only in response to predation. The stretched line and rail fence always are erected early in the season, in swiddens in which predation is anticipated but not yet experienced. In contrast, the brush fence and true fence are erected only later in the season, in swiddens that already have experienced predation.

Differences in labor costs also are an important

TABLE 57
The Labor Costs Of Different Types Of Swidden Fences

Fence Type	Labor Cost
Stretched line	3200 meters/day or 0.12 workday/hectare
Rail fence	700 meters/day or 0.57 workday/hectare
Brush fence	250 meters/day or 1.6 workdays/hectare
True fence	< 125 meters/day or > 3.2 workdays/hectare

factor in this timing (table 57). The relatively low labor costs of the stretched line and rail fence make it possible to construct them before the need and hence return on this labor is certain, whereas the relatively high labor costs of the brush fence and true fence make it more necessary to construct them only after this need and return is certain (table 58). For example, during 1975 and 1976, deer and/or

TABLE 58
The Cost And Efficacy Of Each Type Of Swidden Fence

	Efficacy of Fency Type Dependent Upon Construction Prior to Predation	
Labor Cost	Yes	No
Low	Stretched line Rail fence	
High		Brush fence True fence

pig predation was a problem in forty-one out of sixty-nine swiddens. A stretched line and/or rail fence was first erected in twenty of these forty-one swiddens, although brush fences or true fences later had to be erected in fourteen of these twenty instances. (Brush fences and true fences were erected in nineteen additional swiddens in which they were not preceded by either the stretched line or rail fence.) In 70 percent of the instances in which stretched lines and/or rail fences were erected, therefore, they eventually proved insufficient to contain deer and pig predation and required replacement by the brush fence and/or true fence, representing a loss of the labor devoted to their construction. In the remaining 30 percent of these instances, however, the initial construction of stretched lines and/or rail fences sufficiently deterred predation to obviate any later necessity to erect brush fences or true fences. The savings of labor achieved in this minority of instances was sufficiently great to make the initial erection of stretched lines and rail fences a chance worth taking. The strategy of initially erecting stretched lines or rail fences, and later erecting a brush fence or true fence if the former fails, carries an average labor cost of 2.0 workdays per hectare, which compares favorably with the average labor cost of 2.4 days per hectare of straightaway making a brush fence or true fence[12].

In addition to scaring away and fencing out the deer and pigs, the Kantu' hunt and trap them. Three types of traps are in current use: the singong, the jerungkang, and the peti'. The singong 'snare' is made from a length of thick liana or steel wire. Only deer are taken with it. The Kantu' say that the body mass and strength of the pig is too great to be held by a snare. The jerungkang 'set-spear' consists of a pole stuck diagonally into the ground at one end and sharpened to a point at the raised end. Set-spears are placed behind fallen trees or specially constructed barriers of brush and timber. In either case, any deer passing through the area is forced to jump the barrier and impale itself upon the raised tip of the spear. Pigs never are taken with this trap. With their shorter legs and lower stature, they are both less capable of jumping obstacles and less in need of doing so. The peti' 'spring-spear' is by far the most important of the three types of traps. Like the set-spear, the spring-spear employs a sharpened pole. In this case, however, the pole is driven by a second pole, bent back under tension, and released by the tripping of a trigger (figure 28)[13]. A variety of the spring spear, occasionally used by the Kantu', is the peti' senapang. This consists of a loaded senapang 'shotgun', which fires when a line attached to its trigger is tripped. Both deer and pig can be taken with either variety of peti'.

The Kantu' set all traps along the edges of their swiddens, between the animal populations in the adjoining forest and the ripening swidden crops. They choose those points along the perimeter where heavy animal movement

FIGURE 28
<u>Peti'</u> 'Spring-Spear Trap' For Deer And Pigs

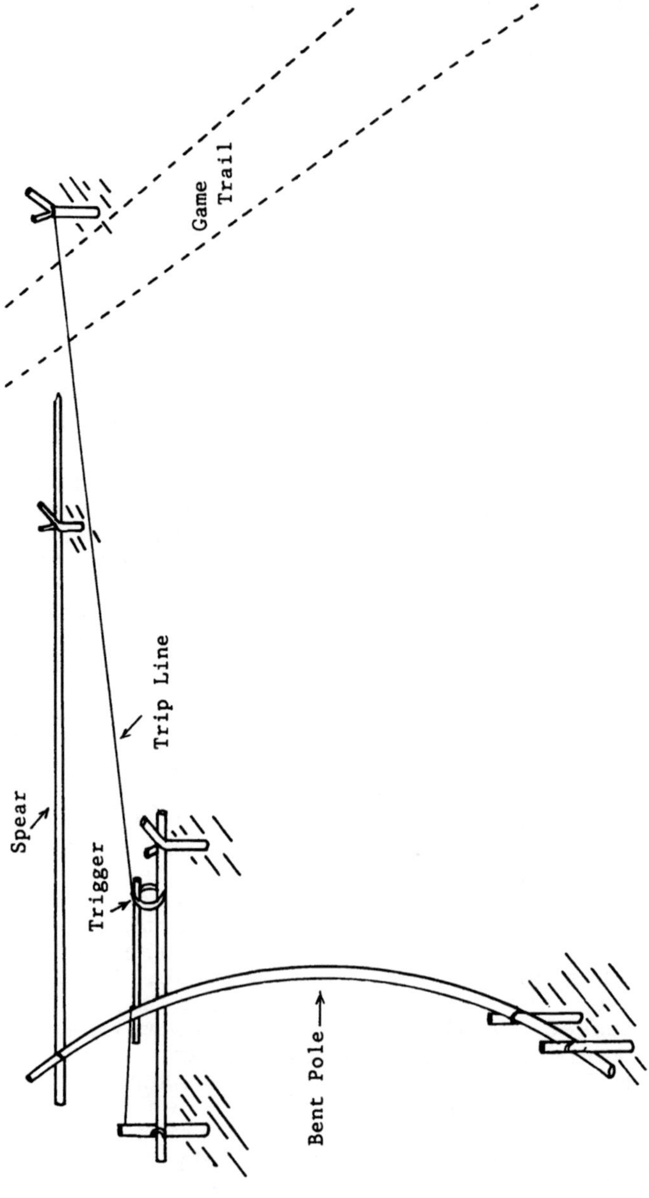

SCALE 1:20

either is anticipated (due, for example, to stands of cassava - a favorite food of pigs - near the perimeter) or already is in evidence (as deduced from recognizable game trails). The Kantu' further increase the likelihood that an animal crossing the swidden's perimeter will pass through a trap by the use of fencing. In some instances they erect rough barriers of timber and brush on each side of a trap, so as to direct passing animals into it. More often they make gaps in existing brush fences or true fences and set their traps there. They do this less often with the less effective stretched lines or rail fences.

The Kantu' hunt with either spear or gun. The sangkoh 'spear' consists of a hand forged iron tip fixed to a hardwood shaft two meters or more in length. The senapang patah 'breakable gun' is a single shot, breech-loading (hence 'breakable') shotgun, taking a twelve-gauge shell. Most guns are made entirely by the Kantu'. A few senapang kit 'cap guns' or muzzle-loading matchlocks (which utilize a percussion 'cap') are still in use. These are variously of indigenous or European manufacture. Black powder is obtained through trade and is used not only in the matchlocks, but in the shotguns as well. The Kantu' hand load their shotgun shells, reusing plastic casings of European origin. There are three techniques for hunting with spear and gun near the swiddens: ngipa', mujok and ngasu. In ngipa' 'ambush', the hunters lie in wait for passing deer and pigs, occasionally using a raised blind. In mujok 'coax', the hunters stalk the animals in the areas (viz., of the swidden) where they are known to be feeding. These two techniques are most often used at night, with the aid of flashlights. The third technique, ngasu 'hunt', always is used during the day. It involves using dogs to track down deer and pig, beginning at a swidden where the animals are known to have fed recently.

Hunting and trapping are less efficient means of protecting the swidden than are scaring and fencing. Hunting and trapping are commonly practised, nonetheless, because they yield a return not only in terms of swidden protection, but also in terms of food provision[14]. The flesh of the deer and pig is of both aesthetic importance to the Kantu' - they prefer it to that of all other animals, including the domestic pig - and quantitative importance - the Kantu' obtain more animal protein from deer and pig than from any other source, except for fish. Each year the average household consumes ninety kilograms (undressed weight) of deer and feral pig, not counting the smaller barking deer and mouse-deer. The importance of this food source is such that many Kantu' bemoan the historic decrease in the local pig populations, despite the fact that this has been attended by a decrease in their predation in the swiddens.

The Kantu' do most hunting and trapping of deer and pig in or near their swiddens. In 1975 and 1976, 58 percent of all deer and pigs taken within the Kantu' territory were taken with traps along the edges of swiddens. Some of the remaining 42 percent, taken by spear and/or gun, also were taken in the swiddens or were taken in nearby forest after being tracked out of the swiddens. The remainder were taken when feeding in groves of rubber trees or fruit trees. In all cases, therefore, the deer and pig were taken at or tracked from their food sources, where the likelihood of a hunter encountering an animal is greatest [15]. Comparing one food source with another, the likelihood of an encounter is greatest in the swiddens. The average swidden is larger than the average rubber or fruit grove, and it contains more food for a greater portion of the year than do either rubber or fruit groves.

The relatively high probability of encountering deer and pigs in the swiddens makes practical the use of traps there. Traps for deer and pigs are rarely set elsewhere, because at no other food source is this probability high enough to justify the labor cost of setting them [16]. Outside the swiddens, deer and pig are taken exclusively with spear or gun. Some of the deer and pig taken at or tracked from the swiddens also are taken by hunting as opposed to trapping, but the latter outnumber the former by at least four-to-one. This decisive preference for trapping as opposed to hunting, in circumstances in which the concentration of game is relatively high (viz., in and around the swiddens), is due to the much higher returns to labor of trapping. One hour devoted to hunting yields an average of 0.75 kilogram of deer or pig flesh, whereas the average hour devoted to trapping yields 3.45 kilograms of flesh[17].

Macaques

Among animal pests, the deer and pigs are followed in importance by the pig-tailed and long-tailed macaques. Both of these primates raid the swiddens in large troops and feed mostly on maize, although the pig-tailed macaque also may feed on cucumbers and rice. Because of this focus on maize, predation by macaques peaks early in the season and drops off once the maize has ripened and been harvested. Predation is intense in some swiddens and nonexistent in others, depending upon proximity to the macaque populations. The pig-tailed macaque's habitat is the primary forest in the hills on the south-west edge of the Kantu' territory, and the habitat of the long-tailed macaque is the secondary forest and fruit groves immediately bordering the two principal rivers in the territory, the Kantu' and the Empanang[18]. The macaques appear never to range beyond a distance of one or two swidden lengths (or approximately 350 meters, maximum) from their respective habitats (tables 59, 60). Their predation also is affected by the quadrennial fruiting cycle of the forest trees. The Kantu' say

TABLE 59
Distance From Primary Forest And Predation By Pig-Tailed Macaques

		Distance of swidden from habitat of pig-tailed macaques in primary forest	
		< 2 swidden lengths	> 2 swidden lengths
Swidden preyed upon by pig-tailed macaques	No	0	23
	Yes	6	5

N = 34 1975 swiddens.
P = .0003

TABLE 60
Distance From Major Rivers And Predation By Long-Tailed Macaques

		Distance of swidden from habitat of long-tailed macaques along major rivers	
		< 2 swidden lengths	> 2 swidden lengths
Swidden preyed upon by long-tailed macaques	No	0	40
	Yes	9	20

N = 69 1975 and 1976 swiddens.
P = .0002

that predation by macaques in the swiddens is minimal during a heavy fruit season and is heavier during those years when there is little fruit in the forest. This relationship is evident in the 1975 and 1976 data on predation by both pig-tailed and long-tailed macaques, although it is more evident in the case of the former than the latter (table 61). Between 1975, with its average fruit crop, and 1976 with its heavy fruit crop,

TABLE 61
The Extent Of Macaque Predation In 1975 and 1976

	1975	1976
Percentage of vulnerable swiddens preyed upon by pig-tailed macaques	54%	10%
Percentage of vulnerable swiddens preyed upon by long-tailed macaques	33%	29%

NOTE: 'Vulnerable swidden' refers to swiddens located within two swidden lengths of the preferred habitat of each primate.

the number of vulnerable swiddens that experienced predation by pig-tailed macaques declined by 82 percent.

The macaques cannot be fenced out of the swiddens, as can the deer and pigs. They can be scared away, but not by passive devices such as are used for deer and pigs. Only the presence in the swidden of dogs or humans has any effect. The long-tailed macaques especially are scared of dogs. If a dog is present in a swidden, the macaques will remain in the trees and not venture down to the ground to feed. They are less afraid of humans. Long-tailed macaques will cluster in trees on a swidden's edge and make forays into it even in the presence of workers. If directly approached by a worker, they will flee to the top of the nearest large tree. In the habitat of the long-tailed macaques along the banks of the main rivers, there are many huge, old fruit trees. Some were planted by the ancestors of the contemporary Kantu'. Some are wild and, by virtue of their location on the river's edge, have not been felled during the making of swiddens. At the top of these trees, the macaques are safe from the Kantu' on the ground below. Even if the Kantu' are carrying shotguns, the macaques are protected by their height, by the intervening foliage, and by the fact that they are, in any case, a poor return for a spent shotgun shell. The presence of these trees has enabled the long-tailed macaque populations to survive in their present habitat along the banks of the main rivers. Since the Kantu' locate their longhouses along the banks of these same rivers, hunting pressure (macaque meat also is esteemed as a food) long ago would have exterminated or driven away the macaques, if they did not have these trees as refuges.

In comparison with the long-tailed macaque, the Kantu' say the pig-tailed macaques are less afraid of dogs and more afraid of humans. They will not enter nor even approach a swidden in which humans are present. Accordingly, the Kantu' often stay in a swidden that is being plagued by these macaques solely for the purpose of keeping them off. The Kantu' will shoot them if given the opportunity, but this rarely happens. Not a single pig-tailed macaque was shot by the inhabitants of Tikul Batu during 1975 or 1976, nor were any taken in traps. In this respect, the Kantu' say that it is the smartest animal in the forest.

There are two types of traps for macaques. One is the <u>ringkap</u> (figure 29). The second type, and the one

FIGURE 29
<u>Ringkap</u> 'Box Trap' For Macaques

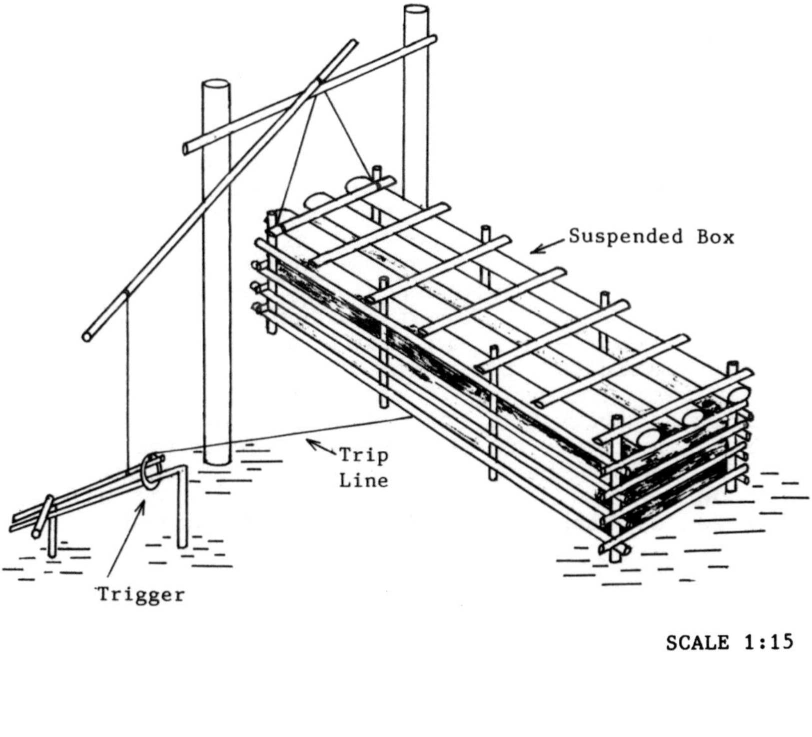

more commonly used, is the belantik. This is much like the spring-spear used for deer and pigs except that it is smaller and the sharpened pole is set not horizontally, but vertically, with the tip pointing towards the trip wire below it on the ground[20]. Both types of traps are set on the swidden's edge, but without attention to game trails and without constructing lateral fencing or barriers of brush. The macaques do not make regular trails and they are undeterred, and undirected, by vertical barriers. As the Kantu' say, Nyumoh nadai nemu pagar 'The pig-tailed macaques do not understand fences'. Rather, the Kantu' try to draw the macaques to their traps by baiting them, usually with ears of maize. At least eighteen traps were set in seven separate 1975 and 1976 swiddens that experienced heavy predation from the pig-tailed macaques, and at least thirteen traps were set in nine separate 1975 and 1976 swiddens that experienced heavy predation from the long-tailed macaques. The latter are said to be far less wary than the former, and they are taken with some regularity in these traps.

Rats and Squirrels

Of the six animal pests, the Kantu' regard the two rodents - rats and squirrels - as the least worrisome. Their attacks are concentrated just before and during the weeding season, when they utilize the dense stands of weeds as cover from the eyes of both humans and birds of prey. After the weeding has been completed, the lack of this cover restricts their predation to the edges of the swidden, close to the adjoining forest cover. Because these rodents make use of weeds as cover and because they otherwise confine their predation to the periphery of the swiddens, they pose the greatest problem in secondary forest swiddens - in which there are more weeds - and in small swiddens - which have a higher ratio of perimeter to area. Their predation does not vary according to any deviation in productivity or timing (table 52), nor does it vary according to the fruiting cycle of the forest trees. The only means that the Kantu' have for defending themselves against these two pests is trapping, which they do with small snares and sliding door traps. Both are baited.

HUMANS

The *ama'* 'swidden pests' do not include *mensia* 'humans'. Nor indeed do the Kantu' ever experience any loss of standing crops to theft by humans. However, the Kantu' believe that humans can and do threaten standing crops by *nepang* 'cursing'[21]. A crop can be cursed if someone skilled in sorcery stands before it and says *Semua padi to' mati* 'All of this rice will die'. Bountiful crops are especially likely to be cursed. The Kantu' believe that a crop cursed in this manner will die unless it has been protected by a *teruit* 'that which has barbs'. *Teruit* are made at the beginning of planting, from the variety of bamboo known as *temiang*[22]. The Kantu' stick one end of a length of this bamboo in the ground and cut the other end into five or six separate barbs. More potent *teruit* occasionally are constructed, in which the barbs are made of porcupine quills. The quills are partially inserted into a gourd with a menacing face carved on one side, or into a carved wooden image of the *burong kenyalang* 'Rhinocerous Hornbill'. *Teruit* of one type or another are constructed in all swiddens in which stem rice is planted. While the *teruit* can thwart nonhuman swidden pests, their principal function is to oppose human sorcerers. The Kantu' say that anyone who attempts to curse a swidden protected by a *teruit* will be devoured by its barbs.

LABOR

Some defense of the swiddens against crop pests incidentally results from the presence of humans in the swiddens for other reasons. The amount of labor that the Kantu' specifically devote to this defense varies greatly, from less than one workday per hectare to as many as fifteen to twenty workdays, according to not only the severity of predation but also the swidden type. The labor input averages one workday per hectare in primary forest swiddens, four workdays/hectare in secondary forest swiddens, and eleven workdays/hectare in swampland swiddens. The return on this labor consists in the increased percentage of the crop that survives to be harvested, in addition to a return - in some instances - in the form of edible animal flesh.

Chapter Eight
Harvesting

The Kantu' regard harvesting as the termination of the swidden cycle. When ngetau 'rice harvesting' is completed in a swidden, they no longer call it umai 'swidden' but rather memudai 'secondary growth'. They also say Udah ngetau, amih taun kitai 'After harvesting, our year is finished'. The next year begins - signified by the starting of a new lunar calendar - once they begin to slash the next year's swiddens. The importance of the harvest is reflected in the numerous pantang 'proscriptions' that circumscribe behavior for its duration. It is prohibited, for example, to wash rice in water, discard this water and then cook the rice in fresh water. Instead, the rice must be cooked in the same water with which it is first mixed[1]. Once the rice has been cooked, there is a prohibition against audibly scraping it with the ladle from the cooking pot. These and other, similar prohibitions are concerned with any behavior deemed disrespectful of the rice. Such behavior is prohibited throughout the year, but the prohibitions are particularly strong during the harvest. The Kantu' believe that if they show any disrespect towards the rice at this critical time of year, the harvest will be jeopardized. Thus, the Kantu' say that scraping a ladle against the rice pot sounds like the rumbling of empty stomachs, empty because of the crop failure that will follow this demonstration of disrespect towards the rice.

TECHNOLOGY

The harvest has three distinct phases: the early harvest of immature rice, the main harvest of mature rice, and the late harvest of slow ripening rice. There also is a distinct sequence of activities involved in processing the harvest.

Harvesting Immature Rice

The first activity of the harvesting season is matah padi 'the rice-breaking'. This begins with a search for propitious bird omens and the making of a food offering in the center of each household's principal swidden. Then one member of the household harvests several panicles of each variety of rice in the swidden. The manner in which these few panicles are harvested is distinctive in several respects, when compared with later harvesting in the swidden[2]. Before the harvester begins, a little bit of cooked rice is pressed onto the rim of the harvesting basket (at the point where its carrying strap is attached), which is said to cause it to fill quickly with rice. Then the harvesting is begun, using a sharpened piece of bamboo to sever the rice stalks. The harvesters do not use their fingers because, at this stage of the harvest, the rice stalks still are too green to be snapped in two easily. Nor do they use the finger knife at this stage, perhaps because of beliefs concerning the peculiar powers of iron. (For example, the Kantu' say that seed rice can be harmed if, while it is in storage, it is allowed to come into contact with a piece of iron; it may be fear of similar harm that causes the Kantu' to use a piece of bamboo instead of the knife during this first, ritually laden stage of the harvest.) Alternatively, the use of bamboo instead of the finger knife may be in acknowledgement of the former's precedence in the evolution of Kantu' swidden technology. The use of more traditional tools or techniques is a characteristic of all of the ritually important stages of the Kantu' swidden cycle. Another distinctive aspect of rice-breaking is the type of cut made on the panicle. Whereas during later harvesting each panicle is severed just below its lowest grain laden branch, during the rice-breaking the panicle is severed fully ten to fifteen centimeters below its lowest branch, which the Kantu' call ngetau panyai 'long harvesting'. The Kantu' are obliged to use this technique when harvesting immature rice, because such rice has to be threshed by a special technique requiring long stems on each panicle. As each panicle is severed by this long cut, it is placed into the harvester's left hand (the reaping is done with the right hand). When several panicles have been gathered in the left hand, they are transferred to the harvesting basket. Handfuls from different rice varieties alternately are coiled clockwise inside the basket, then counterclockwise, to keep them separate. When the reaping has been completed, the panicles of each variety are divided into thirds, and then thirds are taken from each type to make three mixed bunches. Each bunch then is tied up with ubong mansau 'red/ripe string'. (The Kantu' use this color of string not because mansau 'red' is the desired color of the rice in the swidden - kuning 'brownish yellow', the color of ripened rice, is its desired color - but because mansau 'ripe' is its desired condition.) One bunch is placed into a carrying basket set in the middle of the swidden; a second bunch is placed into an offering basket hung over this carrying basket; and the third bunch is taken to the swidden house and hung from the underside of the roof, directly over the spot where the rice storage bin is (or will be) located. The bunches are so placed to encourage the as yet unharvested rice to fill these baskets and bins as quickly and as high as possible.

After the panicles reaped during the rice-breaking have been disposed of properly, the household must diau 'abstain' from any further activity for the remainder of this day, and for the first and second following days as well. On the third following day, the household may carry out a second phase of the rice-breaking, depending upon whether or not it is cultivating the ritually important stem rice and whether or not it intends to participate in the year's end harvest festival. If both are true, the household then must reap the padi taun 'year's rice'. This resembles the earlier rice-breaking in all respects except that whereas the Kantu' only harvest several panicles of each rice variety on the first day, on this third day they completely fill their left hand twice with panicles of each variety. A second difference is in the harvesting on the third day of some job's tears and millet, in addition to the rice[3]. When the requisite two handfuls have been harvested from each rice variety, as well as from each of the two nonrice grains, the proceeds are taken to the farm house for storage for later use during the postharvest festival.

Participation in the two phases of the rice-breaking is determined not only by having stem rice and a role in the postharvest festival, but by other factors as well - in particular the material resources (especially the ownership of chickens) of the household. For example, because of its terrible 1974 harvest, household #3 (which has stem rice) was incapable of staging the rice-breaking in its principal 1975 swidden; but because its 1975 harvest was so good, it could and did stage the rice-breaking in its principal 1976 swidden. Nonmaterial factors also may be relevant. Household #5 was capable of staging the rice-breaking in its principle 1976 swidden, but it did not do so because its members were distressed by their recent involvement in an adat dispute.

The importance to the Kantu' of the rice-breaking has diminished considerably from historic times. The Kantu' say that rice-breaking used to be a gawa' raya 'major festival', which all households performed on the same day, by arrangement. Today there is no phase of swidden work that is begun by all households in the longhouse on the same day by arrangement. It may be that it is more difficult for the households of a longhouse to synchronize their swidden activities than it once was. Formerly, all or most households made their principal swiddens in primary forest. Today (viz., 1974-1976) the principal swiddens are divided almost evenly between primary forest (58 percent) and secondary forest (42 percent), which makes interhousehold synchronization more difficult because the swidden chronologies in the two forest types are not the same. Synchronization also could be more difficult today because of the greater size of the longhouse. The longhouse Tikul Batu (e.g.) is more than twice as large as many of the longhouses reported from the period prior to World War II. On the other hand, the diminishment in importance of the rice-breaking may be due to problems of resources. This period just prior to the harvest season is the most difficult time of the year to stage a major festival (which entails the feeding of large numbers of guests from other longhouses), because rice stocks always are at their lowest at this time of year, before the new harvest is brought in and after the past harvest has been depleted maximally by daily consumption. Therefore, the historic practise of staging a festival at this time of year implies that comparatively large harvests were being reaped by most or all members of the longhouse, with regularity. Today, the Kantu' rarely reap harvests sufficient for the staging (if they otherwise desired) of a major festival at this time of year. The Kantu' themselves maintain that their harvests have declined greatly from former times, an observation that is supported by this disappearance of the rice-breaking festival.

After the rice-breaking has been carried out in a swidden, the next harvest activity - or first activity, in the event that this earlier operation is not carried out - usually is nyuma'. The term suma' refers to rice that is not quite ripe and nyuma' refers to its reaping. The reasons for reaping immature rice during the rice-breaking are ritual in nature, but the reasons for reaping immature rice during nyuma' are purely material. The immature grains of glutinous rice types can be made into a much desired food, mpin. Every year, each household tries to harvest at least a few baskets of immature glutinous rice, expressly to prepare a small quantity of this delicacy. In 1976, twelve out of sixteen households harvested some glutinous rice for this purpose. Households with other demands on their labor forces must forgo this early harvest, however, because it yields a relatively small amount of grain for a relatively large amount of labor. The members of such a household may be allowed to join other households in making and consuming this dish. Alternatively, the households lacking mpin may be given some by other households. In this respect, mpin is like most other nonstaple, greatly desired foods, such as wild game. The households that have some share with those that have none.

There is a second reason to perform nyuma'. By harvesting at this early stage, the household can begin to draw upon its ripening crop at least one week before it would be able otherwise. This early access to the crop is important to those households that have exhausted their rice stocks from the previous year's swiddens, as is frequently the case: the 1974 harvest at Tikul Batu did not last until the 1975 harvest in ten out of thirteen households, and the 1975 harvest did not last until the 1976 harvest in eight out of sixteen households. For such households, the purpose of the early harvest is to provide not the delicacy mpin, but the daily staple padi pakai 'eating rice'. Accordingly they harvest not glutinous but nonglutinous rice, and they harvest larger amounts. For example, in 1976 household #3 took twenty-six kilograms of glutinous rice from its Kenua' swidden during the early harvest, for the purpose of making mpin; whereas household #16 took eighty-four kilograms of nonglutinous rice from its Lubuk Menserai swidden, for the purpose of basic consumption.

An early harvest is not performed by all households in need of eating rice. Of the eight households experiencing a shortage of rice at the start of the 1976 harvest, only three chose to perform an early harvest. The remaining five households chose instead to buy rice from traders and/or borrow it from other households, and to hold off on the harvest until their rice crops attained full maturity. Their aversion to performing an early harvest is based on several different factors. First, this harvest involves seeking out those panicles of rice that, while not yet fully mature, are still the closest to maturity of any rice in the swidden. As the Kantu' say, it is much more difficult to do this than simply to reap all of the panicles that come to hand. In addition, the immature grain is soft and must be subjected to a special baking process before it can be husked and eaten. Finally, the early harvest entails a cost in grain as well (although this is not mentioned by the Kantu'). Rice grain normally increases in weight until it attains full maturity, so if it is harvested beforehand its weight will be below its potential[4].

Because of these costs, households not suffering from a rice shortage never harvest immature eating rice. The one exception is when rice is suffering unusually heavy predation. For example, household #7 made an early harvest of one nonglutinous rice type in its 1976 Lubuk Menserai swidden, after the household members observed that this rice stand was being heavily damaged by an insect pest. As they said, it is better to harvest the rice in an immature state than to lose it completely to pests while waiting for it to mature.

Harvesting Mature Rice

The *ketam* 'finger knife' is the principal tool used during the main harvest. It is made from a rectangular piece of tin that is cut from a sardine can, flattened and fitted halfway into a wooden dowel so that the plane of the tin is perpendicular to the length of the dowel. The cutting edge of the tin is given an edge with a whetstone. There is evidence that this tool was adopted comparatively recently by the Kantu'. In the Kantu' language the term *ketam* is used to refer to a woodworking plane as well as to the harvesting tool. The plane's cutting edge is a rectangular piece of iron, in appearance identical to that of the harvesting tool. In the Malay language, which Kantu' in many respects resembles, the term *ketam* also refers to both a woodworking plane and a harvesting knife. The verbal form of *ketam* in Malay is *mengetam*, meaning 'plane [with a *ketam*]' or 'harvest [with a *ketam*]'. In the Kantu' language, however, *ngetam* means only 'plane [with a ketam]' and not 'harvest [with a *ketam*]'. In Kantu', *ngetau* means 'harvest [with a *ketam*]', but its nominative form, *ketau*, cannot be used to refer to the finger knife (the *ketam*). All of this suggests that *ketau* and *ngetau* are the original Kantu' terms, and that *ketam* and *ngetam* are more recent borrowings from Malay. This in turn suggests that the Kantu' were performing *ngetau* before they were using the *ketam*, meaning that the *ketam* or metal finger knife was preceded in the Kantu' swidden system by some other harvest tool. I suggested earlier that this prior tool may have been a sharpened piece of bamboo, like that still used today during the ritual rice-breaking[5].

The harvester holds the finger knife in the right

hand, with the wooden dowel vertical and the tin rectangle horizontal. The harvester wraps his/her four fingers counter-clockwise around the dowel, while wrapping the thumb clockwise (as viewed from above in both cases). The index and second fingers wrap around the dowel above the tin blade, with the side of this latter finger pressing against the upper surface of the blade. The third and fourth fingers pass below the blade, with the third finger pressing against its lower surface. The thumb is held above - but does not rest upon - the blade. The back edge of the blade (opposite the cutting edge) is pressed perpendicularly against the palm of the hand. Wilkinson suggested that the finger knife is held in this fashion so that the harvester's fingers hide the blade from the rice, which otherwise would take fright to see it[6]. My study of the Kantu' suggests, rather, that this distinctive grip is best explained in terms of the mechanics of severing the rice panicle from its stalk, to which task this grip is admirably suited indeed.

During this principal phase of the harvest, the Kantu' make a short cut on the rice panicle, just below its lowest grain bearing branch. The harvester first grips the panicle with the bottom tip of the thumb and the bottom and upper side of the last joint of the index finger. The thumb presses against the side of the stalk nearest to the harvester, while the index finger presses against its far side. The second, third and fourth fingers, passing around the far side of the panicle, pull the panicle against the blade of the finger knife. The bottom of the second finger pulls against the panicle above its contact with the blade, while the bottom of the third and fourth fingers pull against it below this contact. While the last two digits of the second and third fingers pull against the panicle, the blade of the finger knife continues to be held firmly between the knuckles of these two fingers. Finally, the right hand is pivoted slightly upwards at the wrist. This movement taughtens the rice stalk - now gripped in the fingers of the right hand - pressing the blade more firmly into it and ultimately severing it.

The finger knife always is held in the right hand. The left hand is employed in other ways. Sometimes it is used to gather the rice stalks close to the harvester's body and hold them while the panicles are severed. Alternatively, the left hand may have to be used to hold the rice panicles in an overly full harvesting basket. When the left hand is not being employed for one of these purposes, it too is used to sever rice panicles, pinching them off between the thumb's fingernail and the upper side and bottom tip of the index finger. An alternative technique, used on well ripened and brittle panicles not carrying too many grains, is manual stripping. This involves wrapping the four fingers of the left hand around the main stalk of the panicle, just below the lowest grain bearing branch, so that the tips of the fingers (and in particular the fourth finger) press against the palm. Then the closed hand is slid up the stalk to its tip, with the stalk passing under the thumb and between the curled fingers and the palm, which rips the grains off the panicle and catches them in the hand.

These manual reaping techniques are secondary in preference to the use of the finger knife during the early part of the main phase of the rice harvest. So long as the rice stalks still retain some of their moisture, elasticity and strength, the finger knife - with its sharp cutting edge - is superior to any manual technique. However, as the harvest progresses and as the rice stalks become drier and more brittle, and hence more easily severed or broken, the finger knife loses its comparative advantage, and the manual techniques initially confined to the left hand come to be used with the right hand as well. For example, in a harvesting party of nine persons in swidden #20, sixty-seven days after the commencement of the harvest, only one person was observed using a finger knife. The other eight persons were using their fingers alone.

As each rice panicle is severed with the finger knife, it is placed in a takin pengetau 'harvesting basket'. These baskets are very shallow in construction, some examples even being wider than they are deep. Their shallowness minimizes impediment to the worker's thighs and knees, and their breadth facilitates entry of the rice panicles. The baskets have an average volume of fifteen or sixteen liters.[7] The Kantu' also use the lanyi' 'carrying basket' during the harvest. This is the largest basket made by the Kantu', with an average volume of almost ninety liters. The lanyi' is carried on the back by means of a shoulder strap on either side, and a third tump strap that either is passed across the forehead or is held in one hand. One or more of these carrying baskets are placed in the swidden near the stand of rice currently being harvested. As the harvesting baskets are filled, they are emptied into the carrying baskets, which periodically are carried to the farmhouse and emptied into its rice bins. This minimizes the amount of time that the workers must take off from harvesting in order to carry rice back to the farmhouse. By returning to the farmhouse only whenever a carrying basket is filled, as opposed to whenever a harvesting basket is filled, the Kantu' reduce this time by over 80 percent[8].

The harvesting basket is carried in front of the body by a waist strap. At two corners of the basket, on the side carried closest to the body, two arches of rattan project above its rim. One end of a bark strap is tied to the arch on the harvester's left. To the other end of this strap is tied a wooden peg, some eight to ten centimeters in length. The strap and peg are passed around the harvester's waist and then through the rattan arch on the harvester's right. When this peg is righted, the basket is held fast against the harvester's body. When the peg is twisted down and out, a basket filled to a precarious height eight to ten centimeters above its rim can be unfastened and emptied with ease. The basket usually is worn slightly towards the left side of the body, which facilitates the use of the left hand in holding the grain in an overly full basket. In addition, this positioning allows the left thigh to be used, by periodically raising the left knee, to support the weight of the basket.

Cut panicles are placed in the harvesting basket without arrangement during this stage of the harvest. Formerly however, the Kantu' distinguished between those panicles on each rice plant bearing the smaller grains, termed <u>anak</u> 'children', and those panicles bearing the larger grains, termed <u>ino'</u> 'mothers'. The latter always were harvested and put into the basket after the former. The Kantu' say that by placing the 'mothers' on top of the 'children' in the basket, the children would be discomfited and would cry out to one another, <u>K'atau agi'</u> 'Higher still', urging one another to climb above their 'mothers' so as to escape their weight. As a result of the ensuing upward movement of the 'children' in the basket, the Kantu' say, the basket would fill more quickly with grain.

Not all cut panicles are placed into the carrying basket. When the Kantu' are harvesting glutinous rice and come upon a few panicles that are <u>matak</u> 'unripe', they sever them with a long cut and place them into their left hands. When they have a handful of these unripe panicles, they deposit them near the carrying basket or take them back to the farmhouse, for eventual use in making <u>mpin</u>. On the other hand, when the Kantu' are harvesting <u>not</u> glutinous rice but nonglutinous rice, and they come upon a few unripe panicles, they leave them to ripen for later harvesting. Unripe nonglutinous rice cannot be made into <u>mpin</u>, but only into <u>beras pansoh</u> 'bamboo rice' (viz., steamed in bamboo), which is not -

unlike mpin - a desired foodstuff. When the Kantu' are harvesting glutinous rice and come upon a few panicles of ripe but nonglutinous rice, again they sever them with a long cut and place them into the left hand[9]. (In this case, use of the long cut is dictated not by any special processing requirements, but solely by the need to carry the panicles in the hand.) When a handful of panicles of nonglutinous rice has been gathered, they are taken to the carrying basket or farmhouse, to be treated thenceforth just like all other ripe nonglutinous rice. In contrast, when the Kantu' are harvesting a stand of nonglutinous rice and come upon a few panicles of ripe but glutinous rice, they sever them with an ordinary short cut and place them into the basket along with the nonglutinous rice. The Kantu' explain this contrasting treatment by saying that glutinous rice is palatable if cooked in the same manner as nonglutinous rice, but the latter is not if cooked in the same manner as the former. Therefore, a little glutinous rice can be mixed with nonglutinous rice, but not the other way around.

The harvesters must contend with differences not only of ripeness and glutination, but of soundness as well. Some panicles, termed layo' 'withered', bear only flattened, empty, greyish-white grains. Other panicles bear grains that, while more rounded and normal in appearance, completely lack the inner endosperm. These are termed ampa' 'empty husks'. Still other panicles bear ampa' berat 'heavy empty husks', which differ from the normal grains by their pinched-in sides and their lighter coloration. Panicles falling into any of these three categories are rejected by the harvesters. Occasionally, a panicle is encountered whose lower half bears worthless grains, but whose upper half bears normal grains. In such instances, the harvesters sever the panicle in the middle and take the upper half alone. The harvesters also ignore panicles that have been knocked to the ground by wind or rain, unless some grain remains attached to them. They never attempt to retrieve individual grains that have shattered from such panicles, saying that to do so simply would consume too much time.

The progress of the harvest through the swidden is carefully ordered, both within the individual stands of each rice variety as well as from one stand to another. The main harvest in each swidden is begun in the one stand of rice that is quickest to ripen. Occasionally, a household cannot begin the harvest until well after many of the rice stands in the swidden have ripened, in which case it begins it in the one stand that has been quickest to overripen. As the Kantu' say, if the rice becomes too ripe, dry, and brittle, Ia ka' berayar 'It will want to shed'. When stirred by wind or rain, that is, lodging and shattering will occur. On the other hand, if the household discovers that pests are preying heavily upon one particular stand of rice, they may harvest that stand first to prevent any further losses, regardless of whether it is the ripest in the swidden (or whether it is even ripe at all, as noted earlier). If neither ripeness nor predation is a consideration, then the household will select one of the rice stands at either end of the swidden (figure 31a).

In each stand of rice, harvesting begins at the swidden's edge. If the harvesters instead began in a part of the stand away from the swidden's edge, they first would have to pass through the surrounding stands of unharvested rice. This is strenuously avoided by the Kantu' because so many panicles and grains get knocked to the ground. In addition to being most accessible, the rice nearest the swidden's edge often is in greatest need of harvesting, because damage from pests tends to be greatest there. For both of these reasons, harvesting begins at the swidden's edge and moves towards its center (figure 30a). Deviation from this pattern, called meling 'wind round' (figure 30b), formerly was ritually proscribed. This proscription has since been abandoned, due to an increasing necessity for irregular planting patterns in the Kantu' swidden system. One cause of this is the development of swampland cultivation, which

FIGURE 30
Movements Of Harvesters Within Individual
Individual Rice Stands

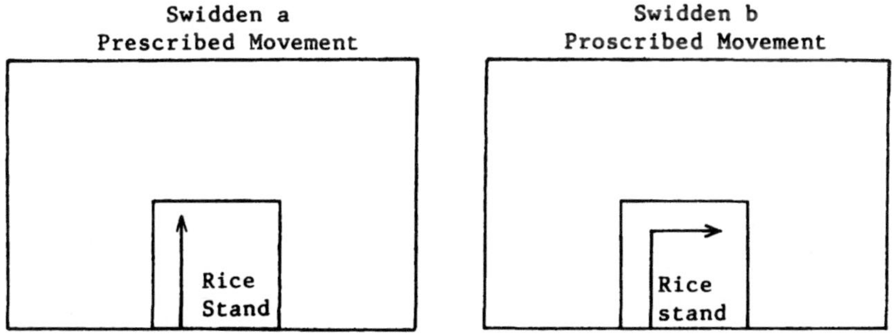

compels planters to follow the winding courses of the small streams along the banks of which much of the swampland is found. Another relevent development is the increasing deviation of overall swidden outlines from the preferred square or rectangular shape, which is due both to constricting and conflicting household rights to secondary forest and also to the related tendency to include small patches of secondary forest in primary forest swiddens and small patches of primary forest in secondary forest swiddens.

As the harvesters strip each rice plant of its panicles, they press its stalk to the ground with the sides of their feet. This makes it clear to all of the harvesters what portion of the rice stand has been harvested and what has not. This also means that after the main harvest has been completed in a given stand of rice, the only plants remaining upright should be those bearing unripe panicles - which can be found easily, therefore, during the succeeding and final phase of the harvest. The Kantu' try to ensure that only unripe panicles are left by monitoring and minimizing the distances between harvesters in the same work party. Whenever one or more harvesters chances upon <u>tai' pangan</u>' 'a friend's feces', the term given to ripe panicles overlooked by a fellow harvester, they cry out <u>Disintak tali takin</u> 'Tighten the waist strap on the harvesting baskets', meaning that the harvesters should decrease the intervals between them. The implication is that if the intervals are too great they will tax too greatly the harvester's powers of perception and concentration, resulting in overlooked panicles.

Not until all the ripe panicles in one rice stand have been harvested, according to a still extant ritual proscription, can the workers proceed to the next stand in the harvest order. By minimizing the passage of workers back and forth between stands of unharvested rice, this proscription minimizes the amount of grain knocked from unharvested plants by passing workers. This also is achieved by the practise, when the harvest in one stand has been completed, of moving to the immediately adjoining stand. This harvest order is called <u>betangkir</u> 'to adjoin'. The ideal harvest order consists in starting in a stand at one end of the swidden and progressing successively through the remaining stands to the other end (figure 31a). The value of this order is evident from a consideration of the consequences of the alternative. For example, the harvest order in figure 31b is problematic because, in order to reach stand #1 in which the harvest is to begin, the workers must pass through either the unharvested stands #3 and #4 or the unhar-

FIGURE 31
Movement Of Harvesters From One Rice Stand To Another

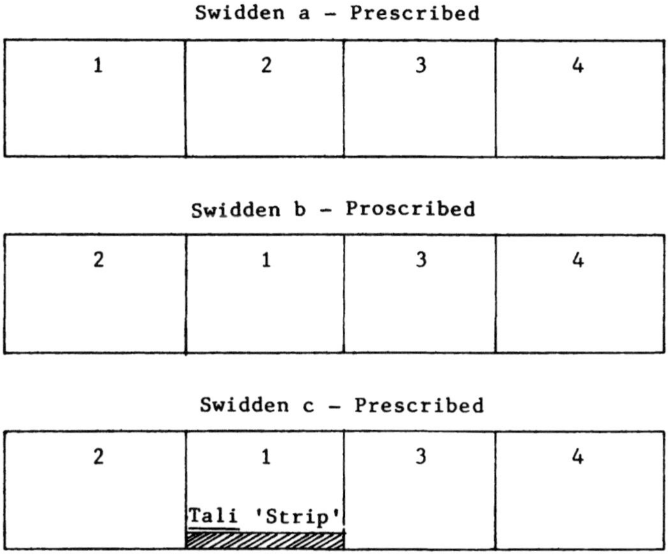

NOTE: The numbers denote harvest order.

vested stand #2. When the Kantu' pass through a rice stand in this fashion, they say of the rice therein, Amih kena' inik 'Finished by being tread upon'. Notwithstanding this threat, if a rice stand is suffering from severe predation, or if it is ripening overly slowly or quickly, the household may have to harvest it out of order. Assuming that this is the case with stand #1 in figure 31c, it is obvious that if this stand is harvested completely the unharvested stands #2, 3 and 4 must become separated. The Kantu' say that this would be inimical to the welfare of the rice, because Antu padi nadai tau' ngami' api, nadai tau' ngusung pangan 'The rice spirits could not obtain fire from one another, could not visit one another'. Therefore, stand #1 will not be harvested completely. Instead, a thin strip of rice will be left unharvested - which is called betali 'do a strip' - to provide a continuous link between the unharvested stands #2, 3 and 4 and a trail for any rice spirits wanting to travel between them. After stand #2 has been harvested and this strip is no longer needed, it too will be harvested. While the Kantu' use such strips out of express concern for the feelings of the rice spirits, the principle that is involved - that of maintaining an unbroken expanse of unharvested rice - is consistent with the aforementioned concern to minimize worker passage through - and hence grain loss in - unharvested stands of rice.

There is a final proscription governing the movement of the harvesters through the swidden: in the words of the Kantu', *Kitai nadai tau' bado' ditisi umai urang* 'We cannot finish [the harvest] at the edge of someone else's swidden'. In figure 32, for example, this proscription is honored in swidden a but violated in swidden b. The danger in the latter case, according to the Kantu',

FIGURE 32
Harvest Order Of The Rice Stands In Adjoining Swiddens

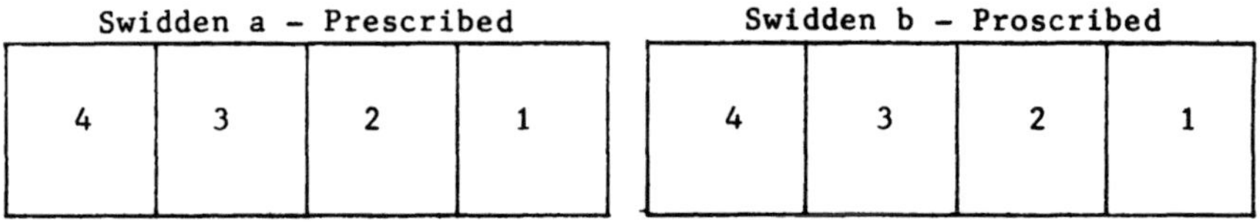

is that the rice spirits of swidden b might flee into swidden a. The Kantu' believe that the rice spirits will try to gather in any remaining stand of unharvested rice, even one in an adjoining swidden. However, they also believe that the rice spirits cannot cross over ground that contains no rice plants or only rice plants whose grain already has been harvested. Thus, the rice spirits clustered in stand #4 in swidden a cannot pass across the earlier harvested stands #1, 2 and 3 to swidden b. The retention of the rice spirits is a matter of importance to each household, because the presence or absence of these spirits is associated with the abundance or lack of abundance of the rice crop, respectively.

The harvest workday is similar in length to the slashing, felling and weeding workdays, except that it is more often lengthened as a result of sleeping in the swiddens (e.g., household members slept in the swidden houses as opposed to the longhouse on an average of 88 percent of all harvest days in the 1976 dryland swiddens of households #3, 4 and 16). The harvest workday also is similar to the slashing and felling (although not weeding) workdays in its susceptibility to foreshortening by ill omens. The Kantu' believe that it is especially important to heed ill omens during the harvest, because unheeded omens attach themselves to the rice being harvested; and when that rice is brought into the farmhouse for storage, the ill omen is brought along, posing a threat to the health and well-being of all those dwelling therein. In addition, the Kantu' believe that nonobservance of omens can cause the swidden to become ala' pungor 'burnt and desolate'. If a household is made aware of an ill omen, death or ritual sacrifice before the start of a harvest day, that day is proscribed; but if it is not made aware until after the start of the day, the day is just foreshortened. In a sample of households, harvesting was forestalled completely on 6 percent of all days during one harvest season; and among those days on which it actually was begun, it was foreshortened on a further 11 percent (table 62)[10].

TABLE 62
The Impact Of Omen Observance On The 1976 Harvest Schedule

Number of days in harvest season: 47 →

Proscribed for harvesting – Yes: 3
No: 44 →

Harvesting attempted – No: 17
Yes: 27 →

Harvesting foreshortened – Yes: 3
No: 24

Formerly there was an additional ritual proscription affecting the length of the harvest day. After completing the main harvest in each stand of rice, the harvesters had to quit work for the rest of the day. They could not begin on the next stand in the harvest order until the following day. Were this proscription in effect today, it would greatly increase the number of foreshortened days. For example, in the sample of swiddens analyzed in table 62, each of which contains an average of six different rice stands, this proscription would increase the number of days on which harvesting is foreshortened from three to nine. This would be a large number of foreshortened days, given the time pressure under which the contemporary harvest is carried out. The fact that this proscription once was honored suggests that formerly there were fewer time pressures associated with harvesting or that formerly there were fewer separate rice stands per swidden.

There is one final factor that may cause the harvest day to be foreshortened: rainfall. The Kantu' will perform most swidden tasks during a rainfall, if need be, but not harvesting. If the harvest is carried on during a rainfall, the product must be dried in the sun before it can be stored, even temporarily, or else it will spoil. This drying takes a day or more. When the harvest is interrupted by rain, therefore, the Kantu' prefer to stop for the duration of the rain and for one-half hour to one hour thereafter, by which time the standing grain will have dried, than to continue working and then have to spend a much greater amount of time drying the cut panicles.

In small stands of glutinous rice or swamp rice, the Kantu' may harvest without differentiating the padi pakai 'eating rice' that they will live off during the coming year from the benih 'seed' that they will use to plant the next year's swiddens. In a stand that has fared poorly and produced but little grain, they also may harvest without differentiation, of necessity setting aside the entire yield for the next year's seed. But in all other cases the Kantu' carry out separate harvests of eating rice and seed rice. Typically after completing 80-90 percent of the main harvest of eating rice in a given stand, the harvesters take time out to harvest its seed. The amount of grain harvested for seed usually is a little greater than the amount of grain needed to plant a stand of equal size, which allows for some loss of seed during storage and also allows for a possible expansion of cultivated area in the next year's swiddens. The type of grain selected for seed usually is the largest and healthiest that each stand has to offer[11]. By consciously selecting such grain from one year to the next, the Kantu' are able to maintain and improve the qualities of each of their rice varieties.

In mixed stands containing two or more different varieties of rice, the best panicles of each variety are selected for seed and then planted together again the following year. However, when the Kantu' observe that one rice variety is clearly outperforming the others in a mixed stand, they select the seed rice of that variety separately and then plant it separately the following year. This selection of monovarietal seed rice from a multivarietal stand is called muntih 'select', and it is the critical step in an ongoing process whereby superior rice varieties are introduced into the household's swiddens and inferior varieties are eliminated from them. <u>Muntih</u> can be performed in any mixed stand of rice except one containing the household's stem rice, where its performance is <u>mali</u> 'tabooed'. The Kantu' say that if a harvester performs <u>muntih</u> in a stand containing stem rice, openly rejecting this rice in favor of some other rice variety, the rice spirits will strike his/her eyes blind in revenge. Given a stand in which a stem rice is mixed with - and is being outperformed by - another rice variety, the household must <u>betakin dua</u> 'use two harvesting baskets' if it wishes to replace the former with the latter. Whereas <u>muntih</u> consists in selectively reaping the best panicles of one rice variety from a stand of two or more varietes, <u>betakin dua</u> consists in simultaneously selecting panicles from both of two varieties in a single stand, keeping them separate in two harvesting baskets carried at the harvester's waist. The basketful of panicles of the desired rice variety is kept as seed rice, and the basketful of panicles of the undesirable rice variety is eaten. According to the Kantu', the latter fate will not be evident to the undesirable stem rice at the time that it is reaped - it will be masked by the use of the two baskets - hence the harvesters will be spared the rice's ire over its rejection.

The ritual tension surrounding the selection of seed from the stem rice is characteristic of this phase of the harvest. For example, while the members of a household might harvest eating rice the day after one of them had an inauspicious dream, they never would attempt to harvest any seed rice. Further, if any household member observes an ill omen in the swidden while they are harvesting seed rice, not only must work cease for the day, but any seed already harvested must be dropped in the middle of the swidden. Not until the morning of the following day can this seed be brought into the swidden house for processing and storage. This observance is required once for each type of omen, even for those types already observed once during the harvest of eating rice.

Harvesting Late Maturing Rice

Like the main harvest, the late harvest consists in cutting ripened rice, using a short cut on the panicle. It differs from the main harvest in that it is performed on much sparser stands of rice. Hence, while the term for the main harvest is <u>ngetau</u> 'reap', the term for the late harvest is <u>ngega'</u> 'seek'. In 1976 households

#3, 4 and 16 harvested an average of just fifty kilograms of rice apiece during this late harvest. The late phase further differs from the main phase in the lesser importance attached to omens during it. The Kantu' say that omens need not be as strictly observed at this time because the panicles being cut are not the harvest's tangkai bungas 'first panicles', but rather its tangkai dudi 'last panicles'.

According to the Kantu', two types of rice are taken during the late harvest. The first type is padi matak 'unripe or late ripening rice'. Due to overall time pressures, the Kantu' usually cannot delay the main harvest in each stand until all of its rice is completely ripe. As a result some late ripening rice usually is left behind in each stand after the main harvest, and this is one object of the late harvest. The second type of rice taken during the late harvest is called smeli' or padi baru 'new rice'. This is the product of a second initiation of panicles on already harvested rice plants. In contrast padi matak is merely a late, first initiation of panicles. In either case, the panicles are fully ripened when they are reaped during the late harvest.

The late ripening rices (of whichever type) are not present in all rice stands in quantities sufficient to justify a late harvest. In 1976 for example, the late harvest was not performed at all in eleven out of thirty-five swiddens. In the remaining twenty-four swiddens, it was performed in an average of only one out of every two rice stands. Several factors determine whether there is sufficient rice for a late harvest in a given stand. One factor is when the stand is first reaped. If the stand is not harvested early, but is allowed to become very ripe first, there will be little unripe rice at the time of the main harvest and hence less need for the late harvest. The Kantu' also say that in such a case little 'new rice' will develop thereafter. The converse also is true. The earlier the main harvest in a given stand, the more unripe rice there will be at the time of this harvest and the more new rice afterwards, and hence the greater the need for a late harvest. Another factor affecting the production of the late ripening rices is rainfall and drainage. The Kantu' say that new rice will not develop if the main harvest either is preceded or followed by a period with unusually little rainfall. They also say that new rice is found more often and in greater quantities in swampland than in dryland. A final factor affecting the production of both new rice and late ripening rice is predation. Since the late ripening rices in any given stand or swidden are much less numerous than those ripening earlier, they bear a relatively heavier pest burden. A pest that is particularly burdensome in this regard is the rice bug, which often is credited with devouring all of the new rice in swamp swiddens.

Processing the Harvest

As each harvest basket is filled, it is emptied into one or more carrying baskets placed in the center of the swidden, as I noted earlier. When the harvesters return to the farmhouse at midday to eat, and at the day's end to eat and sleep, the males in the work party bring in the filled carrying baskets. If the panicles were harvested in swampland and had drooped into standing water, or if the panicles were rained upon while being harvested, they are sun dried upon removal from the carrying baskets. The Kantu' say that if wet panicles are threshed and stored, they will acquire a bau wakap 'bad odor' and will germinate while in storage. With the exception of these wet panicles, all panicles are threshed and stored directly upon removal from the carrying baskets.

There are two techniques of threshing: ngenas 'foot threshing' and ngerih 'scraping'. The former is used with rice that has been reaped when ripe, during the main or late harvests; and the latter is used with rice that has been reaped when still immature, during the early harvest. Panicles that are mature and brittle can be threshed only by foot. The second technique, which involves scraping the grain from the panicle using a sliver of bamboo, would be as likely to break the stems of these panicles as to separate them from their grains. Conversely, the elasticity of immature panicles renders foot threshing ineffective, at the same time as it renders hand scraping not only possible but necessary. The use of this threshing technique is facilitated by the long cut customarily used in reaping immature rice. Only the long cut leaves a sufficient length of stalk on the panicle to enable the worker to grasp it firmly during scraping. Because the Kantu' let most of their rice stands mature before harvesting them, the vast majority of the harvest is threshed by foot, using the ngenas technique.

Most foot threshing is done on a pelangkus, a rectangular grill constructed of thin lengths of bamboo. This grill, sometimes sided with wood planks, is laid onto a rough framework of bamboo that rests atop the pasah renas 'box for unwinnowed rice' in a corner of the farmhouse [12]. The rice panicles are emptied from the carrying baskets directly onto this grill. The thresher - rarely more than one at a single time - stands upright on the grill, using one hand to hang onto a strap or beam attached to the ceiling for support. The threshing involves using one foot to compress and roll over the rice panicles, pressing them against the instep of the other foot, alternating feet all the while. This causes the grains to separate from their stalks and to fall through the openings in the grill into the rice box beneath. After a pile of panicles has been threshed in this manner, the worker stops, stoops, and pulls the compressed mass of panicles apart. The panicles are shaken in the air over the grill and finally rubbed by hand back and forth against the grill itself, thereby freeing any remaining grain trapped among the rice stems or attached to their original stalks. After this hand threshing has been completed, the foot threshing is repeated a second time, and then the hand threshing is repeated again. After both repetitions have been finished, only rice stalks will remain atop the grill. These are discarded. Some bits of stalk with grains still attached inevitably will have passed through the grill into the rice box beneath. After each pile of rice panicles has been threshed, these bits of stalk are sifted by hand from the rice box and then threshed again, not atop the grill but on a kelaya' 'mat'. Working atop this mat, both foot and hand threshing are performed once more, and then the product is winnowed. The grain that remains after this winnowing needs only to be husked to be ready for cooking, and it is a standard source of eating rice for many households during the harvesting season.

Glutinous rices and seed rices usually are threshed from beginning to end atop a mat as opposed to the grill, so as not to mix them with the nonglutinous eating rices in the box below the grill. When threshing on a mat, as on the grill, foot threshing and hand threshing each are performed twice on each pile of rice panicles. In general the smoother surface of plaited mats is less suited than the grill to the rapid separation of grain from stalk, and so threshing on them is more time consuming. This is *why* the Kantu' only thresh on mats those few rices whose distinctiveness prohibits mixture with the others in the rice box.

The rice box consists of a lashed, rectangular framework of logs, the bottom and sides of which are lined with closely plaited mats or bark sheets. The length and width of the average rice box are 2.1 meters and 1.6 meters, respectively. The height varies, not only from one box to another, but also for a single box through the course of the harvest season. The rice box initially is built low to the floor of the swidden house. As it fills with threshed grain, the sides are built up to accommodate the increasing volume. This phased construction is due less to uncertainty over the final size of the harvest than to a desire to avoid any show of hubris in this regard. Were a household to build a tall rice box right at the beginning, this show of confidence in the size of the harvest to come might well offend the spirits, bringing ruin upon what otherwise could have been a bountiful harvest. While anxious to avoid any show of hubris, every household takes pride in a tall rice box. Thus, visitors to a swidden house always are invited to sit atop the rice box. The household members themselves typically sleep on top of their rice box at night (with the exception of young children who cannot control their bladder and bowels).

After threshing the harvested rice and placing it in interim storage in the rice box, the next phase in its processing is ngelekai 'sun drying'. If threshed grain is stored too long without being dried, the Kantu' say, it will be ravaged by a tiny, grain-boring insect. In addition, drying accentuates the difference in weight between whole grains and empty ones, and so it facilitates the subsequent operation of winnowing the latter out. Finally, drying reduces the weight of the grain and hence conserves labor when the grain later is transported from the farmhouse back to the longhouse.

Drying is carried out on the tanyo' 'platform' that is attached to all nighttime swidden houses. On the morning of a day chosen for drying, the platform is covered with closely plaited mats. Then the threshed rice is taken from the rice box and carried in baskets out onto the platform, where it is emptied onto the mats, into tall piles. If the sun is obscured by clouds, the rice is kept in these piles. But as soon as the sun breaks through, the rice is spread out to form a thin layer on the mats. This layer should be thick enough so that a goodly quantity of rice can be dried on the platform at one time, but not so thick that the sun's heat cannot penetrate to the bottommost grains. The piles of rice that are initially made on the mats always are too large to be spread out over the mats in a single layer of this optimal thickness. Rather, each pile is spread out in stages during the course of the day. After each layer has dried, it is covered by additional rice from the pile to form a new drying layer, and so on. Ideally, each pile of rice can be dried completely in this manner in the course of one day. The amount of rice that can be dried atop one kelaya' 'mat' in one day - called 'one mat' - is the unit used by the Kantu' to measure how much rice they dried on any given day.

In addition to this large-scale technique for drying grain, the Kantu' also have a smaller-scale technique, called ngancau. The Kantu' distinguish between the two by saying that ngelekai involves large piles of rice that are spread out to form progressively thicker layers on the mats, whereas ngancau involves only one thin layer of rice. This latter technique is used for drying small quantities of rice, such as glutinous rice, seed rice or, especially, a few days' supply of eating rice.

Whichever technique is used, drying usually is completed in one day. However, a second or even third day customarily is added when drying seed rice (as opposed to eating rice). The Kantu' say that the seed rice must be especially dry before it is stored, or else it will not germinate when it eventually is planted. The Kantu' also add a second or third day when drying eating rice if the first day is overcast or interrupted by rainfall. The Kantu' guard against sudden rainfalls by readying folded kajang 'awnings', made of overlapping pandanus leaves, on the drying platform. In the event of rainfall, each awning is opened into the shape of an inverted 'V' and placed over some of the drying rice.

After the Kantu' have dried their rice, they berayar 'winnow' it. The purpose of this operation is to separate the whole grains from the empty grains and also the bits of stem and stalk that passed through the grill and into the rice box during threshing. This reduces the amount of material that eventually must be carried from the farmhouse back to the longhouse, and it also reduces the amount of material that must be husked preparatory to cooking and consuming the grain.

The Kantu' usually winnow their rice immediately after drying it, while it still is piled on the platform in front of the farmhouse. Therefore, the amount of grain winnowed at one time usually equals the amount that can be dried at one time on the platform, and the number of days of winnowing usually equals the number of days of drying. The Kantu' describe this relationship as, ideally, Sekali' ngelekai, sekali' berayar 'One time of drying, one time of winnowing'. If the drying progresses exceptionally well and quickly on a given day, due to a very hot sun, the winnowing is done that same afternoon. Otherwise winnowing is carried out on the day after drying, either in the early morning or late afternoon. The Kantu' say that it is too vigorous an activity to perform during the midday heat.

Before winnowing begins, the platform is completely covered with plaited mats. In the center of the platform, these mats are overlaid with one or two additional, smaller mats. The worker who pours out the grain stands atop these latter mats, while the worker who fans the grain stands off to one side. The former initiates winnowing by lifting a basketful of threshed and dried grain to his/her head and resting it there. They use either a basket specially designed for this purpose, the timpa', or a larger version of the harvesting basket. Both baskets average about twenty-five liters in volume. Once the basket has been lifted to the head of the pourer, the second worker begins to fan the air with a pendidit 'fan'. This is made of a flat rectangular section of plaited reeds, with a stiff edging of rattan, and has a square area of approximately one-third of a meter. The fan is held near its bottom edges with both hands and is worked (which is called nidit) by being pushed down and pulled up briskly, through an arc of nearly 180 degrees, in front of the fanner. The fanner always stands to one side of the pourer, and whenever the wind changes direction he or she changes position - and the pourer turns, in response - so as always to fan to leeward. When the first worker begins to fan, the second worker tilts forward the basketful of grain (heretofore resting atop his/her head) so that a steady stream of grain pours out. The emptying of an average basketful of grain takes approximately thirty seconds. While emptying the basket, the pourer faces away from the fanner to minimize irritation to his/her eyes from the fine mia' 'hairs' or trichomes mixed with the grain. After the contents of all baskets have been poured out, the mats directly under the pourer are taken up and the grain on them is poured back into the baskets. This is winnowed (viz., poured and fanned, exactly as just described) a second time. After this second winnowing, the central mats are taken up again and the grain on them is carried into the swidden house to be stored or prepared for transportation to the longhouse.

The workers then turn their attention to the grain on the peripheral mats. The grain on these mats was blown a greater distance during winnowing than was the grain on the central mats, so most of it is lighter and more likely to be empty than the latter. Nevertheless, some of the grains on these peripheral mats are quite whole. In order to recover these grains, the peripheral mats are taken up and the grain on them is put into baskets in preparation for a second winnowing, not by the berayar technique just described but by the nampi' technique. This is a one-person technique based on use of the *capan* 'winnowing tray'. This tray has the shape of a 'U' with upturned sides (high at the apex of the 'U', and tapering down to the point where the sides join the open lip of the tray). The tray is held in both hands, with the lip away from the body, and is moved up and down in front of the body through an arc of about forty-five degrees, terminating at waist level. This movement causes the heavy, whole grains in the tray to move to its back (to the apex of the 'U'), while causing the lighter, empty grains and bits of stalk to move to the front of the tray and over its lip. The work is temporarily halted several times, so that the worker can sift the grain by hand. When the winnowing of one tray has been completed, the grain remaining in the tray is removed and put into storage. The material that was winnowed out of the tray, called *ampa' berat* 'heavy empty ears', is gathered off the mats and then winnowed again using the same technique. The material remaining in the tray after this final winnowing is stored for use as chicken feed. The material winnowed out of the tray, called *ampa' rayar* 'shed empty ears', is discarded.

The *nampi'* technique of winnowing enables the workers to recover scarce and lightweight grains from among many empty ones. It achieves finer discriminations than does the *berayar* technique but at a higher cost in labor. Thus *berayar* is used earlier in the winnowing process - when the quantity of grain involved is greater and requires less discrimination in winnowing technique - and *nampi'* is used later on - when the quantity of grain involved is less but requires greater discrimination. In the case of small batches of grain - such as a household's seed rice, a few days' supply of eating rice, or immature rice from the early harvest - the *nampi'* technique may be used exclusively from the beginning of the winnowing process to its end. This technique can be used with no increase in per unit cost on as little as one winnowing tray of grain, whereas the *berayar* technique cannot. In the case of seed rice, however, the use of *nampi'* is dictated not only by economics but also by a ritual proscription against the use of *berayar*. In winnowing as in all else, the critically important seed rice demands the most sensitive and discriminating treatment possible.

RICE YIELDS

There are two important aspects of Kantu' rice yields, namely how large they are, and why they are as large as they are.

Measures of Yields

I measured rice yields in thirty-seven separate sample plots, located in twenty-six separate 1975 and 1976 swiddens. Each plot covered an area of 9.3 square meters. All of the ripened grain in each plot was reaped, threshed on a mat, winnowed by the small-scale technique, and then measured for volume in a graduated cylinder. These measures did not include any grain that was reaped while immature, during the early harvest; nor did they include any grain that was still immature at the time of this measurement (viz., at the time of the main harvest) and that was left for reaping during the late harvest. The measurements in the thirty-seven plots averaged 2.06 liters (standard deviation 1.15 liters) for the area of 9.3 square meters, which is equivalent to a yield of 2215 liters per hectare. I also made a second set of measurements of rice yields, based on measures from entire swiddens, namely thirty-three 1975 swiddens and thirty-two 1976 swiddens. If the harvest of a swidden was transported back to the longhouse, I counted the number of carrying baskets used and then multiplied this sum by the average volume of these baskets. If the harvest was not transported back to the longhouse but instead was stored in the farmhouse, I measured the dimensions of the rice bin and the depth of the grain in it[13]. These calculations produced a mean yield per swidden of 1957 liters (standard deviation 1877 liters) of threshed and winnowed (but unhusked) grain, and a mean yield per hectare of 1497 liters (standard deviation 1197 liters). I have concluded that this last measure of rice yields per hectare, while probably a little conservative, is more representative than the earlier and higher figure based on my measurements in the sample plots.

The Kantu' have their own ways of measuring the yields in their swiddens. They begin with projected measures, based on their reading of omens encountered during the selection of the swidden site. The Kantu' regard the good or ill character of these omens as one of the most important determinants of the eventual success or lack of success of the rice crop. Another attempt to divine the eventual success of the crop is made on the first day of planting, in those swiddens in which stem rice is planted. It is based on the puto' pandak 'short tubes', eight lengths of bamboo that are sunk into the ground in the 'place of the seed rice' early in the morning of this day. In four cases the internode at the bottom end of the bamboo is left intact; in the other four cases the bottoms are cut open; and in all eight cases the internode at the upper end of the bamboo is cut open as well. Each bamboo is filled with rice beer and then covered with a small rice cake. Later in the day, the rice cakes are removed and the level of rice beer in the bamboos is examined. If the level has remained high this augurs well for the coming harvest, but if the level has dropped markedly this augurs ill. The overall interpretation of this omen is based on what happens in the majority of the bamboos. This divination is clearly structured to reassure the farmers, therefore, since the level of rice beer always will remain high in at least the four tubes with sealed bottoms[14]. A second type of divination also may be performed on the first day of planting, based on divination from the liver of a slaughtered pig. The diviners focus their attention on the one lobe of the liver that is by custom associated with one's own longhouse, and also on the attached gall bladder. If both this lobe and the gall bladder are plump and free of blemishes, the coming harvest is divined to be good rather than bad. It is significant that only the wealthier and harder working (or simply luckier) households have pigs available for sacrifice and divination on these occasions, since the wealthier and harder working households also tend to reap better harvests. This circumstance would account for the accuracy of liver divination and thus for the continued belief in it, if the majority of pig livers have an auspicious rather than inauspicious appearance, as seems perhaps to be the case.

After the divinations associated with planting have been carried out, there remains but one attempt to divine the character of the harvest beforehand, based on the growth of singkenyang. Some households plant cuttings of this in 'the place of the seed rice', and if it later flowers (which it rarely does, the Kantu' say), a good harvest is expected. Even if the plant does not flower, the Kantu' say that they can divine the size of the coming harvest from the number of bulbs that it produces. If a plant in the swidden of a given household produces more bulbs than it did the previous year, the harvest is expected to be good; but if it produces fewer bulbs, the harvest is expected to be poor. While there is no further, purposeful divination regarding the coming harvest, the Kantu' are quick to divine from any situation or omen that, while spontaneous, nevertheless is deemed relevant. One such omen is the sighting of a species of hairy caterpillar on a tree in or near a swidden. The Kantu' believe that the color of this caterpillar's hair foretells the health of the rice crop in that swidden. If its hair is white, the rice crop will be light colored and healthy; but if its hair is dark, then the ears of rice also will be dark and sickly.

The Kantu' also attempt to assess the character of coming harvests based on such things as the quality of the burn, the goodness of meteorological conditions during the growth of the rice, the visible extent of pest predation in the swidden, and so on. Late in the growth of the rice plant but well before harvesting begins, the Kantu' make the first direct estimate of the size of the harvest by evaluating two qualities of the plants: the average number of panicles per plant - this characteristic of the crop is termed belayan - and the average number of healthy grains per panicle - this characteristic of the crop is called buah. They say that if both panicles and grains are plentiful the harvest will be good, while if only one is plentiful the harvest will be fair, and if neither is plentiful the harvest will be poor. Once the Kantu' begin to harvest, they make a second estimate of harvest size based on the amount of time consumed in reaping. By this measure, the largest and also rarest harvest is one that cannot be reaped to completion. Liging, the senior male in household #5, recalls the year in which after observing the singular omen of kayu patok Ketupung 'a tree pecked by a Rufous Piculet', he took the largest harvest of his life from a single swidden. Using all the labor in his own household as well as all that he could hire from other households, he reaped 150 carrying baskets of rice from the swidden; and still, he says impressively, he was forced to leave an additional amount of rice to rot on the stem, unharvested.

Once all of the rice in a swidden has been reaped, the Kantu' can make their first reasonably accurate estimate of the harvest, based on the height of the rice (now threshed but still unwinnowed) in the rice box in the farmhouse. This pile of rice will be reduced in volume during winnowing, but since the amount of the reduction should approximate a constant (viz., 25 percent), the size of the pile can be taken as an accurate measure of the harvest. Indeed, when Kantu' wish to know how large a crop another household has taken from a given swidden, they ask what part of one's body (e.g., knee, thigh, waist, etc.) is level with the top of the grain in the rice box. As the rice box is filled during the harvest and as it is extended upward, the threshing grill perforce is moved upwards as well. If the grill eventually is pushed up so high that the threshers must bend over to avoid hitting the roof of the farmhouse, the amount of grain in the rice box is judged to be great indeed. But the ultimate goal of all households is to fill the rice box to the roof itself, to the point at which the offering basket - containing the panicles reaped during the rice-breaking - is hanging. The Kantu' further measure the size of the harvest according to the amount of time needed to dry it. The relation between drying effort and harvest size is evident in this statement by the headman of Tikul Batu, Melayun of household #2: he says that whenever he dreams of climbing a mountain, he will thereafter puas ngelekai 'dry [grain] to the point of exhaustion' - meaning that this particular dream always presages a large harvest.

The best Kantu' measure of the harvest is obtained when, after being dried and winnowed, the grain is stored in boxes, bins or bags in the farmhouse or - and this is most likely - is packed into carrying baskets for transport to the longhouse. Each household keeps close track of the number of baskets used to transport the grain from each of their swiddens, and it is this number that the household gives when asked about the yield of a given swidden. The Kantu' favorably contrast this practise to that of the Islamicized Dayak living along the Kapuas river, who measure not what is carried to the longhouse but what is reaped. The latter always is greater than the former, because while the harvest is being reaped and before it is transported, some portion of it will be consumed by the household and some portion also may be given to other households in payment for harvest labor. The implication is that the Kantu' are more successful swidden farmers than the Islamicized Dayak and can 'afford' to cite the relatively lower carrying figure. The carrying figures that are cited by the Kantu' do not represent all of the harvest that remains to the household even at this point in time. They are measures of eating rice alone and explicitly exclude all seed rice. Also explicitly, they are measures of padi 'threshed and winnowed but unhusked grain', not tangkai 'unthreshed grain', nor renas 'threshed but unwinnowed grain', nor beras 'threshed, winnowed and husked grain'.

After the rice has been carried back to the longhouse, the Kantu' make a further albeit cruder measure of the harvest, based on the number of circular rice bins in which the grain is stored in the loft of the longhouse. The average household requires one to three such bins to store a year's harvest. This means of reckoning the harvest size is illustrated in the statement by Mara', the senior male in household #4, that whenever he dreams of taking enemy heads, his harvest that year is certain to fill a number of rice bins equal to the number of heads taken by him in the dream. If a particular household fills a large number of bins in a given year, it may have to place additional supporting beams in the longhouse gallery below its storage loft - as did household #4 following its 1975 harvest - which is a further measure of a good harvest. The best measure of all is if a household fills so many rice bins that its loft cannot contain them all, and it must borrow space in the loft of another household. Only one household in Tikul Batu has been this fortunate within recent memory, namely household #6, which was storing grain in the loft of household #8 during the years 1974-1976.

Measures of what each household actually harvests aside, the measure of what each household needs to harvest is simple and twofold. Most important is the need for seed rice for the next season's planting. The Kantu' always take care to set aside sufficient seed rice for the next year, before any rice is diverted to consumption. (Their notion of the greatest profligacy is to nutok benih 'pound seed rice' for consumption, thereby depleting the stores of seed for the coming year's swiddens, and severing the vital link from one year's cycle to the next.) A measure of a poor harvest is one that falls short of this seed rice requirement, of which they say Benih pun nadai pulai 'Even the seed did not return'. The percentage of households at Tikul Batu that did not reap sufficient grain for the next year's seed was zero after the very good 1975 harvest, 19 percent after the average 1976 harvest, and 57 percent after the very poor 1974 harvest (the likes of which occur once in ten years, according to the Kantu').

The second need that ideally is filled by the harvest is for enough eating rice to last until the beginning of the following year's harvest. If a harvest satisfies this requirement, the Kantu' say, Ia datai ke taun 'It came to the year'. The harvest marks the end of one agricultural year and the beginning of another, which terminates with the following harvest season. Therefore, when the Kantu' say 'It came to the year', they mean that a harvest satisfied consumption requirements through the end of the following year. The Kantu' also may say of a harvest that lasts until the following harvest,

Bisi' ti ditinggang 'There was that which was covered over'. When a new harvest is carried back to the longhouse, it is poured into the rice bins atop any grain still remaining from the previous year's harvest. Any harvest that is 'covered over' by the succeeding harvest was, therefore, more than sufficient for one year's consumption requirements.

The consumption and seed rice requirements together constitute a minimal measure of the sufficiency of the rice harvest. The Kantu' households attain or surpass this measure about one year out of every three (table 63)[15]. The shortfalls of the other years are partially redressed with grain that is stored during these good years (table 64). Because of the importance of stored

TABLE 63
The 1975 And 1976 Harvests Vis-À-Vis Seed Rice And Consumption Requirements

Year	Household Harvest: < Total sufficiency	Household Harvest: > Total sufficiency	Mean Percentage of Sufficiency	Standard Deviation
1975	8 households	6 households	104%	37
1976	11 households	5 households	85%	56
Total	19 households	11 households	94%	48

NOTE: The 'Total' mean and standard deviation are based on the total sample of thirty households.

TABLE 64
The 1975 And 1976 Harvests Plus Stored Rice Vis-À-Vis Total Requirements

Year	Household Harvest + Stored Grain: < Total sufficiency	Household Harvest + Stored Grain: > Total sufficiency	Mean Percentage of Sufficiency	Standard Deviation
1975	7 households	7 households	117%	55
1976	8 households	8 households	106%	78
Total	15 households	15 households	111%	68

rice in addressing the frequent insufficiencies of the annual harvest, harvest surpluses always are stored as opposed to being sold. The only household in Tikul Batu with a history of selling stored rice is household #6, which has the largest rice stores in the longhouse. Even so, the amount of rice drawn from its stores for sale is relatively very little, numbering only in the dozens of liters per year. While stored rice surpluses are rarely sold, however, they regularly are paid out in exchange for swidden labor. Any household with much rice in storage is obliged to assist any other household in the longhouse that is in need of rice, and this obligation takes the form of exchanging rice for swidden labor - the outright sale of rice being regarded as too calculating a transaction for two households within the same longhouse. These norms relate to a critical aspect of the Kantu' swidden system: while a few households have more than sufficient rice in any given year, and while most have less, the total rice resources of the longhouse usually exceed the total rice requirements of all of its households (table 65). The wage labor system takes advantage of this fact to maximize the utilization of total rice resources.

TABLE 65
Individual Household Rice Requirements Vis-À-Vis Total Longhouse Rice Resources

Year	(A) Total Rice Resources of All the Households in Tikul Batu	(B) Total Rice Requirements of All the Households in Tikul Batu	(A) As a Percentage of (B)
1975	54.0 tons	46.5 tons	116%
1976	49.5 tons	44.5 tons	111%

These two uses of stored rice - to redress harvest shortfalls and to hire wage labor - together mitigate against the buildup of rice stores by any household for more than a few years at a time. The single exception to this in the longhouse Tikul Batu is the previously mentioned household #6. During 1974-1976, household #6 still had rice in storage from harvests ten to fifteen years previously. These stores were sufficient to meet its minimal requirements for over eighteen months, without counting the rice from its 1975 and 1976 harvests. This household's exceptional industry at swidden farming has created and maintained this store, despite recourse to it after the occasional bad harvest and for payment for wage laborers. Part of this store has been untouched for so long that it has rotted away. The members of household #6 say that the grain that is deepest down in their rice bins - and hence oldest - is crumbled and inedible. This damage is not rued by household #6, however, nor does it inspire ill will among the other households in the longhouse. Rather, all households regard the possession of rotting rice stores as in some sense the ultimate measure of successful harvests. Since this measure is so infrequently attained, it is best interpreted as a cultural comment on the wisdom of storing grain, as opposed to empirical evidence of the folly of such storage.

Causes of Variation in Yields

The Kantu' believe that their swiddens are less productive today than they used to be, due to the defilement and damnation of the contemporary world by sexual delicts. They believe that such delicts damage the swidden fortunes not only of the households of the persons involved, but also of other households in the same longhouse, and sometimes the fortunes of other longhouses lying along the same river or even in other river systems. A sexual delict with purportedly very far ranging effects occurred within the memory of living Kantu'. A Kantu' male living near Putussibau on the Kapuas river, distant by some 135 kilometers from Tikul Batu, had sexual relations with a daughter of marriageable age. This act, the news of which was brought to Tikul Batu by travelers, was widely believed to be the cause of an unusually destructive flood that hit all low-lying swiddens in the Empanang valley that year.

There are other types of moral delicts that can bring the wrath of the spirits down upon the swiddens. For example, in 1970 Mandor, the elder male of household #12, had an open argument with the father of his wife (Tikin of household #7) and a sibling of the latter (Gelungan of household #13), in violation of the rules governing interaction with affines. Other Kantu' believe that this violation placed Mandor in a state of ritual disfavor, which has been responsible for the ensuing and continuing poor returns of Mandor's swiddens. A measure of Mandor's relative lack of success is the fact that his 1975 and 1976 harvests ranked eleventh among the fourteen households in the longhouse that farmed both years[16]. A second household that is said to have suffered the wrath of the spirits is household #8. In the 1970s, this household made swiddens several years in a row in or near the sepan 'watering holes' of sambar deer and barking deer. This was seen as an offense to the spirits, who often communicate with humans through the agency of these deer. The other longhouse members associate this offense with several reverses in the fortunes of this household, including the lapse into insanity of one of its members, the partition of the household (into three households, #8, 16 and 17), and the continuing poor yields in its swiddens. This household's 1975 and 1976 harvests ranked last in the longhouse.

Just as the disfavor of the spirits can be incurred, so can their favor be curried, in particular by the strict observance throughout the swidden cycle of all prohibitions pertaining to the swiddens. The Kantu' believe that it is especially important to observe all bird omens and to secure the best omens possible at the beginning of the cycle. For example, many Kantu' attributed the exceptional yield in household #9's 1975 swidden - which was the highest yield in any swidden, of any type, during either 1975 or 1976 - to this household's strict observance of omens during the initial clearing of this swidden. The sacrifice of pigs (and to a lesser extent, chickens) also is believed to win the favor of spirits. When a measure of each household's sacrifices is calculated (averaged over the two years, 1975 and 1976), ranked, and compared with its ranking in terms of harvest success, a rank order correlation coefficient of (.744) is obtained. The strength of this correlation (viz., $P < .01$) suggests that indeed there is an empirical association between the number of animals that a household sacrifices and the size of the harvest that it reaps, just as the Kantu' believe. Whereas the Kantu' draw the causal arrow from sacrifice to harvest, however, the actual causal direction is probably the other way around, as noted in the earlier comment on hieromancy. The households that reap the highest yields tend to acquire - and hence be able to sacrifice - the greatest number of livestock.

Benign spirits are said to sometimes express their favor towards some fortunate Kantu' by giving him or her a charm, in the shape of an unusual stone, an unearthed potshard, a bezoar, or the antlers of the barking deer. The Kantu' call such objects pengaroh padi 'rice charms'. Select households are favored not merely with charms from the spirits, but with the manifest presence of the spirits themselves. It is common knowledge in Tikul Batu that a python inhabits one of the rice bins in household #14's loft, and that a cobra inhabits a bin in the loft of household #4. Both serpents are referred to as antu padi 'rice spirits'. It is believed that these spirits have given the two households good fortune with their swiddens. This good fortune is evident in the fact that the 1975 and 1976 harvests of households #14 and #4 ranked third and fourth within the longhouse. It also is suspected, although no Kantu' is sure, that household #6 has a rice spirit. This suspicion is based on this household's long history of outstanding harvests and rice surpluses.

Belief in the ability of spirits to affect the harvest may be related to the inability of human effort to completely determine the success or failure of the harvest. There are several points during the swidden cycle - namely when trimming felled trees, redressive burning, redressive planting, weeding, and guarding - at which the work can be performed more or less intensely, depending on the industriousness of the households involved. This variation might be expected to have an effect on swidden yields, but in fact it does not (table 66)[17].

TABLE 66
Intensification Of Swidden Cultivation And Variation In Yields

		Extent of intensification in the swidden	
		< Mean	> Mean
Swidden yield	< Mean	4	5
	> Mean	4	6

N = 19 1976 swiddens.
P = .605

This lack of association between intensification of cultivation and higher swidden yields is illustrated by the cases of households #6 and 11. Although they are regarded as respectively the most and least industrious households in the longhouse, the yield per hectare in the 1976 swiddens of the former ranked fourteenth out of sixteen, while that of the latter ranked third. This inability to increase yield per unit of area by increasing labor inputs is characteristic of all extensive systems of agriculture, such as the Kantu' swidden system. Such systems facilitate, rather, increases in the area cultivated as a result of increased labor inputs.

It is this figure that varies according to industriousness and intensification of labor. To return to the 1976 swiddens of households #6 and 11: in terms of total hectares cleared per household that year, the latter ranked only eighth out of sixteen while the former ranked first.

The actual causes of variation in yields are numerous and varied. One important determinant is location in swampland as opposed to dryland: there is a strong tendency for the former to give higher yields per unit of area (X^2_c=21.3, P<.001 for fifty-six 1975 and 1976 swiddens). The higher yields of the swamp swiddens are due to the availability of waterborne nutrients as well as the beneficial action of the water in aerating the soil, decomposing organic material, and retarding weed growth. Among dryland swiddens, the Kantu' say, those located within the floodzone generally give higher yields than those located above it - perhaps due to the alluvial silt that periodically is deposited on floodzone land. On the other hand, floodzone swiddens also are vulnerable to damage from flooding. In 1976 for example, ill-timed floods were blamed for relatively low yields in all of the floodzone, dryland swiddens at Tikul Batu.

A second important determinant of yields is the nature of the burn. If part of a swidden does not burn at all (and if redressive measures also are unsuccessful), it will not be planted [18]. If it burns but not well, it will be planted but - as the Kantu' say - the rice only will flourish k'ujung kayu ngau k'api' pando' ditunu 'at the crowns of the [felled] trees and at the stack burnings'. That is, the rice only will flourish at the concentrations of ash produced by the burning of dense masses of foliage and by the reburning of stacked timber left over from the initial burn. In general, the poorer the burn in a swidden, the poorer its eventual yield (table 67).

TABLE 67
Thoroughness Of The Swidden Burn And Variation In Swidden Yields

		Percentage of swidden burned (and planted)	
		< Mean	> Mean
Swidden yield	< Mean	5	4
	> Mean	1	9

N = 19 1975 and 1976 swiddens.
P = .0495

Another determinant of swidden yield is the timing of planting. As noted earlier, the Kantu' say that they should plant during either the seventh or eighth moons of their lunar calendar. Planting before or after this results in lower yields - as in the 1974 swiddens, which were not planted until the tenth moon. Lower yields also result from planting before or after the normative data for the longhouse as a whole, whenever this happens to be. The greater the divergence between the planting date of a given swidden and the mean date for all the swiddens of the longhouse, the greater the pest burden in the swidden, the greater the losses, and the lower the yield. Among the 1976 swiddens (e.g.), those swiddens with below average yields were planted an average of eleven days before or after the mean longhouse planting date, whereas the swiddens with above average yields were planted an average of just five days before or after the mean date.

Weather conditions, in particular ill-timed droughts, can jeopardize yields just as much as pests. The Kantu' are most concerned about conditions during one particular stage in the growth of the rice plant: this stage begins with the initial swelling of the panicle inside the uppermost leaf sheath, and ends before the panicle has fully emerged from the sheath and before the plant has finished flowering [19]. During this critical stage in the rice plant's growth, they say that the ideal weather conditions are a constant repetition of sunshine and rain (table 68). If instead rain predominates, the maturation of the rice plant will be delayed. On the other hand, if drought predominates the emergence of the panicle from its sheath may be aborted permanently, so that an unusually high percentage of grain will mature as empty kernels [20]. In swampland swiddens, however, the presence of standing water protects the grain against this threat.

A final, major determinant of swidden yield is the length of the interval between the maturation of the rice crop and its harvest. The longer this interval the greater the losses to pests, especially to those pests - in particular the various rats and squirrels - that pose a chronic and omnipresent threat to the swidden crops. Also, the longer this interval the greater the likelihood of losses to lodging and shattering. My observations indicate that serious lodging and shattering do not begin until the average crop has been in the ground for more than approximately twenty-seven weeks. A planting-harvesting interval of twenty-seven weeks was exceeded in only one major dryland swidden in 1976, the Kenua' swidden of household #3. Due to the involvement of this

TABLE 68
Influence Of The Weather On Rice Maturation

Type of Weather	Stage in the Rice Plant's Growth	
	Panicle has not completely emerged from the tiller	Panicle has completely emerged from the tiller
Continuous rain	Slows both the emergence of the panicle and its further maturation.	Slows maturation (with the exception of glutinous rice).
Intermittent rain/sunshine	Quickens maturation.	No effect on maturation.
Continuous sunshine	Halts both the emergence of the panicle and its further maturation.	Quickens maturation.

household in an adat dispute at the commencement of this harvest season, some of the rice in this swidden remained in the ground for as much as 205 days, until 28 March, before being reaped. By 19 March, after 196 days in the ground, they already were saying - facetiously - that they had to use a pemansai 'fishing basket' to gather this grain, because so much of it had lodged and shattered all over the ground. (Even by 4 March, after the grain here had been in the ground for only 181 days, they already were saying that the stalks and panicles - although still intact - were kero' 'brittle'.) This was the only 1976 swidden in which major losses to lodging and shattering were claimed by the owners.

A number of these determinants of interswidden variation in yields also account for variation within individual swiddens - variation that can be equally dramatic (table 69). For example, intraswidden variation in drain-

TABLE 69
Intraswidden Variation In Rice Yields

Swidden #	Yields from all sample plots within each swidden (in kilogram/hectare equivalents)		
	Mean	Standard deviation	Range
41	1178	638	234 - 2,045
44	1097	994	377 - 2,567
67	1277	643	755 - 2,185

age may be associated with variation in yields, as was the case in swiddens #40 and 45 in 1976. The owners of both swiddens claimed that only the swamp rices in these swiddens survived to harvest time, all of the dry rice varieties having been killed by flood waters. Intraswidden variation in the interval from the maturation of the rice to its reaping also is relevant. In the 1976 swiddens, the first rice to be reaped in each swidden was taken on the first day it attained maturity, but the last rice to be reaped was not taken until an average of four weeks after attaining maturity. The latter portion of the crop is much more likely to suffer from predation, lodging and shattering than the former. Crop yields within the swidden also will vary according to variation in the thoroughness with which it was burned. In general the burn tends to be better towards the center of the swidden and worse towards its periphery, and on this basis rice yields also tend to be better towards the center and worse towards the periphery. A related characteristic of the swidden periphery, which the Kantu' frequently cite as a cause of the poor growth of rice there, is the shading of the growing plants by the adjoining forest. The pattern of predation also contributes to this pattern of variation in yields, because it too tends to be lighter towards the center of the swidden and heavier towards its periphery. Finally, when there is no obvious cause for nonetheless striking variation in yields within a swidden, the Kantu' may attribute it to the fact that they heeded ill omens when initially clearing some parts of the swidden, but not others.

LABOR

Several different types of labor are used during the harvest, including household labor, reciprocal labor, wage labor, and guest labor. The rate of labor is more or less the same in each case.

Labor Rate

During measurements in the thirty-four sample plots earlier mentioned, each of which was reaped from start to finish without interruption, it took an average of 18.1 minutes to cut enough panicles to produce one kilogram of threshed and winnowed grain. The range was

5.3 minutes to 63 minutes per kilogram and the standard deviation was 10.3 minutes/kilogram. One cause of this variation was the density of the grain. There was an inverse relation ($r=-.52$, $P<.01$) between the number of minutes taken to reap one kilogram of grain and the number of kilograms of grain in one square meter. The greater the density of the grain, the less worker movement required to reap a given amount, and hence the better (viz., the lower) the reaping rate. The brittleness of the grain also was an important factor in determining the reaping rate. The Kantu' say that greater care and time must be taken to reap overly mature and hence brittle panicles, or else they will fly apart in their hands. As they say, Tangkai ti kero' nadai tau' jamah kaya 'Brittle panicles cannot be gripped hard'. Once panicles have lodged and shattered to the ground, it is very time-consuming to retrieve them. In three sample plots in which the Kantu' recognized a severe problem with brittleness, the reaping rate averaged 21 minutes/kilogram, compared with an average rate of 17.8 minutes/kilogram in the remaining thirty-one plots. Some of the variation in the rate of reaping also relates to variation in the workers' freedom of movement within the swidden. For example, the reaping rate averaged 21.3 minutes/kilogram in eight plots located in swampland, where footing is uncertain, compared with 16.1 minutes/kilogram in fifteen (all secondary forest) plots located on dryland. The presence of more and larger unburned timber in the primary forest swiddens also inhibits worker movement and affects the reaping rate accordingly: it averaged 18.2 minutes/kilogram in eleven dryland plots located in primary forest swiddens, compared with 16.1 minutes/kilogram in the fifteen dryland, secondary forest plots just cited. A final factor responsible for variation in the reaping rate is the gender and age of the worker. In the thirty-four sample plots, the average rate was 15.3 minutes/kilogram for females and 19.5 minutes/kilogram for males. Reapers of both genders over twenty-five years of age worked at an average rate of 15.6 minutes/kilogram, while reapers under twenty-five years worked at a rate of 20 minutes/kilogram. The relationship between age and rate of reaping was least marked among female reapers and most marked among males, due to a difference in the age at which they begin to reap. Females of age twenty-five typically have practised reaping for at least fifteen years, whereas males of the same age typically have practised reaping, to any serious extent, for perhaps half as long.

The rates of reaping during the early and late phases

of the harvest tend to be slower than the rates during the main phase, due to lower densities of reapable grain. When all three rates were averaged together, for a six-swidden sample, an overall rate of 18.7 minutes/kilogram was obtained (table 70). As for work rates when processing the harvest, the rate for threshing atop a grill averages 56 minutes per carrying basket, or 0.63 minute per liter of rice (as measured before threshing). When threshing is performed on a mat instead of a grill, the rate is 1.42 minutes/liter. A minimum of 70 percent is threshed on the grill, producing an overall work rate of 0.87 minute/liter for threshing the harvest [21]. It takes an additional 0.80 minute per liter of grain to carry the threshed rice from the rice box out onto the drying platform, layer it out, and cover it up during rain storms. (No labor is expended in watching out for rainstorms - this is done incidentally by someone who is otherwise engaged in cooking, minding younger children, or threshing.) Winnowing consumes an additional 0.7 minute per liter of grain, for two winnowings by the large-scale technique and two winnowings by the small-scale technique [22]. If only the small-scale technique is employed, as is ritually prescribed for the seed rice,

TABLE 70
Labor Inputs During The Three Harvest Phases

	Swidden #					
	41	42	43	44	67	68
Early harvest workdays	1.2	0.	0.4	0.5	1.9	0.
Main harvest workdays	147.4	19.7	77.1	137.3	73.4	21.4
Late harvest workdays	1.1	0.7	1.1	5.9	1.2	1.0
Total workdays X 350 minutes	150. 52,500	20.4 7,140	78.6 27,510	143.7 50,295	98.9 34,615	
Kilograms in harvest	2,508	380	1,680	2,771	1,801	
Work rate in min/kg	20.9	18.8	16.4	18.2	19.2	

the overall rate rises to 1.0 minute/liter [23]. Given that seed rice comprises 9 percent of the average harvest, the overall work rate for winnowing the average harvest equals 0.73 minute/liter [24]. The overall work rate for the entire sequence of postharvest rice processing is 2.63 minutes/liter of threshed and winnowed grain, or 4.24 minutes/kilogram of threshed and winnowed grain[25]. The overall work rate for both harvesting and harvest processing is 22.9 minutes per kilogram (viz., 18.7 + 4.24 minutes/kilogram). This equals 29.9 days per hectare in primary forest swiddens (given average yields of 457 kilograms/hectare), 49.2 days per hectare in secondary forest swiddens (given average yields of 752 kilograms/hectare), and 86.5 days per hectare in swampland swiddens (given average yields of 1322 kilograms/hectare).

Household Labor

A sizable percentage of all swiddens are harvested solely by the household's own labor force (table 3). This labor force, both ideally and in practise, is composed nearly equally of males and females. During the 1976 harvests of households #3, 4 and 16, each adult male participated in an average of 82 percent of his household's harvest days, while each adult female participated in an average of 84 percent. The participation of juvenile males in the harvest is somewhat less than that of juvenile females (viz., 42 percent versus 54 percent), but it is high when compared with (e.g.) weeding. This is underscored by the participation of juvenile males in harvesting not only their own household's swiddens, but the swiddens of other households as well. In addition to reaping, the household's males are responsible for carrying the grain to the farm house and threshing it. Their greater body mass suits them to these tasks. On the other hand, Kantu' females are better suited - or become better suited - to reaping, as indicated by their superior work rates cited earlier. These gender-based differences are reflected in the fact that adult males stay in the farmhouses during the day more often during the harvest season than do adult females. This allows the household's females to reap and allows its males to thresh (which they do while cooking and minding infants as well).

Reciprocal Labor

The Kantu' use relatively little reciprocal labor

during harvesting. They say that each household ideally harvests kediri' 'alone' rather than bedurok 'reciprocally' in multihousehold work groups. The Kantu' explain this limited use of reciprocal labor by saying that if they used this system during the harvest, the work schedule might work out in such a manner that they would have to work in the swiddens of another household while the grain in their own swiddens becomes overripe, falls to the ground, and rots. Thus, the time pressures of the harvest make the use of the reciprocal labor system unattractive. It may be asked why this is so, since the use of reciprocal labor might be expected to increase the utilization of labor, as in planting, and hence ease these very pressures of time. The answer lies in the fact that while the typical swidden can be planted in one day, it usually cannot be reaped to completion in one day or even two. The rice in a typical swidden rarely ripens in a sufficiently uniform manner to make such reaping possible. Given a lack of uniformity, the harvest is better carried out by a few people - namely, the labor force of the owning household - working every day throughout a lengthy harvest season, than by a large group of people - namely, a longhouse-wide reciprocal labor party - working there just once or twice. In the minority of instances in which reciprocal labor is used, its use conforms to this preferred harvesting pattern (table 71)[26]. Table 71 shows, first, that

TABLE 71
Exchange Of Reciprocal Harvest Labor Between Two Households

1976 Harvest Dates	Contribution Of Reciprocal Labor Workers	
	To the upstream Kenua' swidden of household #2 by household #4	To the Kenua' swidden of household #4 by household #2
7 March	4 workers	
8 ,,	3 ,,	
9 ,,		4 workers
10 ,,	5 ,,	
11 ,,		4 ,,
12 ,,	5 ,,	
13 ,,		8 ,,
14 ,,		5 ,,
15 ,,	5 ,,	
16 ,,	1 ,,	

the reciprocal labor groups are not large. On all of the harvest dates in both swiddens, reciprocal labor was contributed by only a single household (viz., household #2 or 4). Since the number of reciprocal laborers from without the household was so small on any given harvest day, many days of reciprocal labor were needed to complete the harvesting. Table 71 further shows that in neither swidden was all of this reciprocal labor performed on consecutive days. There was an alternation back and forth between the swiddens of the two households, with first one household contributing one or two days of reciprocal labor to the other, and then the second doing the same to the first; with the result that harvesting did not cease for more than one or two days in either swidden during this period. Households #2 and 4 had special motivation to make this exchange of labor possible, because of the close affinal tie between them, a tradition of amicable cooperation, and the proximity of their respective Kenua' swiddens.

The minimal use of reciprocal labor during harvesting is a function of the circumstances that normally prevail during the harvest. When the circumstances change, so may the receptivity to such labor. During the 1972 and 1973 swidden years, any storage of grain in the swiddens (even temporarily) was forbidden by the government, to eliminate possible food sources for guerrillas operating in the region. The normal practise of storing harvested grain in or near each swidden, at least until the completion of the harvest, was no longer possible. In response to this changed set of circumstances, the Kantu' greatly intensified their use of the reciprocal labor system during harvesting. All of the 1972 and 1973 crops were harvested and then carried back to the longhouse at the end of the harvest day by longhouse-wide reciprocal labor parties. After 1973 the government relaxed this prohibition, and the Kantu' promptly reverted to their former practise of harvesting with household labor supplemented mainly by wage labor.

Wage Labor

Some wage labor was used in thirty-five of the sixty-nine 1975 and 1976 swiddens. It accounted for a total of 17 percent of all labor inputs during these thirty-five harvests. As I noted earlier, the Kantu' designate the harvest as one of the two swidden phases (the other being weeding) during which the need for wage labor is greatest. The reliance during the harvest on wage labor as opposed to reciprocal labor is due, in part, to the fact that harvest success varies greatly from one household to another and as this varies so too do demands upon the household work force. Such variation is not conducive to the operation of a reciprocal labor system that - in order to achieve a balance in the exchange of labor - presumes a general equality in the demands upon each household's work force. This variation is more conducive to the operation of a wage labor system. Households with small harvests have more labor than they can use in their own swiddens, while households with large harvests do not have enough. Since the former have both a surplus of labor and a shortage of grain, while the latter have a shortage of labor and a surplus of grain, their exchange of labor for grain is mutually advantageous, but an exchange of labor for labor is not[27].

Two different forms of wage labor are used during the harvest. The form that represents the clearest exchange of goods for labor is that in which workers from one household harvest in the swiddens of another in return for two kolak of threshed and winnowed grain per worker per day, or an equivalent amount of husked rice (two kilograms by weight), currency (300 rupiah), or trade goods (with the value of 300 rupiah), plus one or two meals (table 72). Whatever the medium of payment, the character of this exchange is unambiguous.

TABLE 72
Frequency Of Different Mediums Of Payment For Labor During The 1975 And 1976 Harvests

Medium of Payment	Percentage of Total
Rice	60
(Threshed, unhusked rice)	(42)
(Threshed and husked rice)	(18)
Currency	17
Trade Goods	23
Total	100%

The two kolak of rice or their equivalent are called upah 'wage payment', giving them is called ngupah 'paying of wages', and working for them is called nguli 'working for wages'. In 1975 and 1976, 90 percent of all days of harvest wage labor in the longhouse's swiddens were paid for under this sort of arrangement. Nevertheless, the Kantu' view this as a recent and somewhat unwelcome introduction into their traditional system of harvest wage labor. The Kantu' say that Ngupah dua kolak s'ari 'Payment of two kolak per day' is Cara Singanan, cara Iban 'Way of the Singanan, way of the Iban'. Singanan is a derisive term for downriver Dayak who have become Moslem, while the Iban are the most bitter of the historic headhunting foes of the Kantu'. The Kantu' prefer to think of the calculated payment of wages to neighbors and kinsmen as more in keeping with the character of these disparaged groups than of themselves.

The traditional system of the Kantu', which they admit has been largely but not yet totally supplanted by the system just described, is quite different. Under this system, said to be as old as any remaining or remembered aspect of Kantu' adat 'custom', extrahousehold workers harvest in blocks of three days. On the first two days, the grain they harvest returns to the swidden owners; but on the third day, whatever grain they harvest belongs to them. The privilege of reaping grain for oneself on this third day is, in effect, compensation for reaping grain for the swidden owners on the first two days. This is not, however, spoken of in terms of payment and wage labor. Rather, during the first two days the workers are said to Ngetau ke sida' ti mpu umai 'Harvest for those who own the swidden', while on the third day they are said to Ngami' kediri' 'Take for themselves'. The speed of the workers on the third day, when the fruits of their labor are theirs to keep, is a matter for laughter among the Kantu'. Indeed, the workers commonly fill one carrying basket on the third day.

The wage laborer fares much better under this traditional system. Instead of being paid 5.9 liters of threshed grain for one day, the worker is paid as much as 88.5 liters of unthreshed grain for three days. While the straight wage of 5.9 liters/day is usually paid at the longhouse, however, the wage of 88.5 liters is paid at the swidden, and it is the responsibility of the recipient to transport it back to the longhouse. Since the standard wage for transporting one carrying basket of grain (over an average distance of one hour's travel) is three kolak of threshed rice, the wage of 88.5 liters of unthreshed rice must be lessened by this amount for comparative purposes. Further allowance must be made for the fact that the wage laborer has to thresh, dry, and winnow this grain as well. This work, which produces 49.6 liters of threshed and winnowed grain, consumes an average of 0.26 workday, which should be added to the original three days of wage labor. In summary, the net daily wage under the traditional system is 12.5 liters of threshed and winnowed rice, which is still more than twice as large as the daily wage of 5.9 liters [28]. Use of the traditional system today is limited largely to those households whose kin ties are so close as to make use of the crasser fixed wage both awkward and shameful (table 73).

TABLE 73
Kinship Ties And The Type Of Payment For Harvest Labor

		Two households joined by a direct consanguineal or affinal tie: No	Two households joined by a direct consanguineal or affinal tie: Yes
Type of compensation arranged between two households for harvest labor	Daily wage	29	5
	Traditional 2 days/1 day system	3	4

N = 41 separate pairs of households that participated in a harvest wage labor arrangement in 1975 or 1976.
P = .031

The need to hire wage labor is a function of the ability of the household to reap all of the grain in its swiddens soon enough after maturation to avoid losses to pests, lodging, and shattering. This ability varies according to the ratio of household workers to swidden grain. The higher this ratio, the greater the ability of the household to reap its grain expeditiously; and so too is the reverse true. As this ratio varies, therefore, there is inverse variation in the amount of harvest labor hired by the household (table 74). During the good 1975 harvest, the nontrading households had an average of 1.7 workers per ton of harvest grain, and they hired an average of 14.8 workdays of labor or 4.8 workdays per ton of grain apiece. During the average 1976 harvest, when their ratio of workers per ton of grain climbed to the more favorable figure of 3.3 workers per ton, these households hired an average of just 8.5 workdays or 1.7 workdays per ton apiece. It is notable that even among the seventeen nontrading households

TABLE 74
The Harvesting Burden On The Household Work Force And The Use Of Wage Labor

		Ratio of household workers to tons of grain in the harvest	
		< Median	> Median
Ratio of wage labor days to tons of grain in the harvest	< Median	4	11
	> Median	11	4

N = 14 households during the 1975 harvest and 16 households during the 1976 harvest.
$X_c^2 = 6.53$
$P < .025$

whose 1975 or 1976 harvests fell short of sufficiency, eight took some grain from the harvest to hire harvest labor. Any household whose harvest is fair or better, even if still short of sufficiency, is likely to be approached by kinsmen from other households and longhouses, offering harvest labor in exchange for grain. Such offers are not forthcoming, or if forthcoming can be refused, only when the household's harvest falls considerably and obviously short of sufficiency.

The trading households tend to hire more harvest labor than the nontraders. While the latter hired an average of 3.4 days of labor per ton of grain during the 1975 and 1976 harvests, the former hired an average of 17.9 days per ton. The greater use of wage labor by the traders is due, in part, to their obligation to accept harvest labor to cover trading debts. This is reflected in the fact that in the 1975 harvest, following a year which many households ran up trading debts because of the disastrous 1974 harvest, the trading households each hired an average of 84 days of labor or 135 days per ton of grain in their harvests; whereas during the 1976 harvest, following a year in which fewer debts were run up because of the success of the 1975 harvest, the traders each hired an average of just 38 days of labor or 61 days per ton.

The need to be hired is due to either a present need for grain - to be paid out as the wage - or a past need for grain - advanced earlier in the year and now to be paid back with labor. These needs for grain, in turn, are a function of the size of the household's harvest

- either the present or preceding one - and the amount of grain still in storage at the time it is carried out (tables 75 and 76). The need to be hired, on the part of some longhouse members, is not always in balance with the need to hire on the part of other longhouse members. For example, in 1975 longhouse members performed 107 days of wage labor in swiddens belonging to other longhouse members, and in 1976 they performed 104 days. Thus, the need to be hired was nearly identical during

TABLE 75
The Sufficiency Of The Preceding Harvest Plus Stored Grain And Harvesting For Wages

		Household's preceding harvest plus stored grain	
		< One year's sufficiency	> One year's sufficiency
Household harvested for wages during the present harvest	No	5	11
	Yes	12	2

N = 14 households during the 1975 harvest and 16 households during the 1976 harvest.
$X_c^2 = 11.4$
$P < .001$

TABLE 76
The Sufficiency Of The Present Harvest Plus Stored Grain And Harvesting For Wages

		Household's present harvest plus stored grain	
		< One year's sufficiency	> One year's sufficiency
Household harvested for wages during the present harvest	No	3	13
	Yes	12	2

N = 14 households during the 1975 harvest and 16 households during the 1976 harvest.
$X_c^2 = 16.2$
$P < .001$

the two years (although in 1975 it was related more to the poorness of the preceding harvest, whereas in 1976 it was related more to the poorness of the current harvest). In contrast, the need to hire varied considerably between the two years. In 1975, in the wake of a very poor harvest and in the face of a very good one, the longhouse's households hired a total of 484 days of harvest labor. In 1976 however, in the wake of this good harvest and in the face of just an average-sized harvest, only 240 days of harvest labor were hired by the longhouse's households. Such fluctuations in the need within the longhouse to hire harvest labor, when not matched by similar fluctuation in the need to perform harvest wage labor, are redressed by varying the percentage of labor hired from without the longhouse. Thus in 1975 fully 78 percent of the labor hired during the harvest had to come from other longhouses, whereas in 1976 this figure could be reduced to 57 percent.

The need to perform wage labor and the need to hire wage labor are not always mutually exclusive. Wage labor was both performed and hired by household #8 during the 1975 harvest and by households #7, 11, 16 and 17 during the 1976 harvest. This may occur when a household obtains a large harvest (viz., sufficiently large to require hiring wage labor) but still lacks some other resource (viz., glutinous rice or trade goods) that can be obtained only through wage labor. Alternatively, a household may obtain a large harvest but still have to perform wage labor to pay off a preharvest advance of grain from another household. In such cases, the household may hire wage labor in part just to replace whatever amount of its own labor has to be drawn off to pay this debt. In addition to hiring wage labor to compensate for its own performance of wage labor, a household also may do the opposite, for the reasons given by Liging of household #5: *Udah kitai ngupah urang ngetau d'api' umai kitai, asa kitai ngasai rugi, kitai pun tau' bekuli ngetau d'umai urang* 'After we hire wage laborers to harvest in our swiddens, if we feel the loss [of the wage payments], we can ourselves perform wage labor in the swiddens of others'. To illustrate, household #16 began the 1976 harvest season in debt to household #5 for thirteen days of labor for grain advanced earlier in the year (table 77). Because household #16 accepted this grain in advance of performing the labor, the timing of its repayment was left to the discretion of household #5. Thus, when household #5 asked household #16 to perform the thirteen days of wage labor between 3 and 13 March, household #16 was forced to comply even though its own harvest season was still in progress (it began on 1 January and ended

TABLE 77
The Hiring And Performance Of Harvest Wage Labor
By The Same Household

During household #16's own 1976 harvest season	After household #16's own 1976 harvest season
(1) Household #16 performed obligatory wage labor ↓	
(2) Household #16 hired wage labor →	(3) Household #16 performed nonobligatory wage labor

on 18 March). In partial replacement of this lost labor, household #16 hired seven days of wage labor - performed between 18 February and 11 March - for its own harvest. Subsequently, in order to recoup the wages that it paid out for this labor, household #16 performed an additional nine days of wage labor itself, between 20 March and 9 April, after its own harvesting was finished.

Liging also said that if a household hires harvest labor and then, as is often the case, is unable to recoup the costs through performing wage labor itself, it still is better for it to sustain this *rugi ngupah* 'loss by paying wages' than to sustain a *rugi dipajoh utai* 'loss by predation by pests'. That is, if the household workers cannot reap the matured grain in good time, it is better to lose some of this grain by hiring additional workers than to lose it to pests. Indeed, the cost of completing the harvest one day earlier, through the use of wage labor, may well be less than the loss of ripened grain during one more day of predation[29]. Even when the wages paid out are greater than the losses to pests, these wages are lost to the household that pays them but are gained by the household that earns them; and hence there is no net loss with respect to the longhouse as a whole. On the other hand, grain that is lost to pests is a net loss not just for the household owning the swidden, but for the entire longhouse as well.

Guest Labor

The Kantu' use almost no labor from without the household during the early and late harvests (or during the processing of the harvest). The low return on labor during these phases of the harvest mitigates against the use of wage labor, in particular. This same low return is the basis for the single, exceptional use of extra-household labor during the late harvest, which consists in the owners of a swidden inviting members of other households to join in its harvest. All of the grain reaped by such guest harvesters belongs to them. No grain (or labor) is given in compensation to the household owning the swidden; nor does this latter household give anything to the harvesters, other than the permission to harvest. Nor is any other arrangement - such as wage labor or gift labor - possible during the late harvest. As the Kantu' say, the harvest of late maturing grain must always be *ngega' kediri'* 'harvest for oneself'.

A late harvest was performed in twenty-four of Tikul Batu's thirty-five 1976 swiddens, and the swidden owners permitted other households to join in just eight of these cases. Kantu' say that permission typically is extended only by households that have either more than sufficient supplies of grain without this late maturing grain or less than sufficient supplies of labor to harvest it themselves. Also important is the proximity of kin ties between the household requesting permission and the household from which it is requested (table 78). All this suggests that permission to reap late maturing grain is not a matter of indifference to any household, which also is suggested by the fact that the grantors themselves usually do not abstain from the late harvest.

TABLE 78
Kinship Ties And Participation In Guest Harvesting

		Paired households shared a direct consanguineal or affinal tie	
		No	Yes
One of a pair of households permitted the other to join its late harvest in 1976	No	99	14
	Yes	3	4

N = 120 possible pairings of 16 different households.

$X_c^2 = 7.1$

$P < .01$

Of the seven households that permitted other households to harvest their late maturing grain in 1976, all but one joined in this harvest themselves.

The households that ask for permission to guest harvest invariably are those whose own harvests (plus stored grain) fall short of the coming year's requirements (P=.01 for the sixteen households in 1976). However, major shortfalls cannot be redressed through guest harvesting alone. For example, of the two households that were permitted to participate in the late harvest of household #16's Lubuk Menserai swidden, one reaped a total of only seven kilograms of grain (after threshing and winnowing), and the second reaped a total of only ten kilograms. In both cases, this grain represented less than 1 percent of the visiting household's grain deficiency for the coming year. Thus, the visiting households see this not as an opportunity to eliminate or even significantly reduce their grain deficiencies, but merely as one opportunity - among many that will occur during the year - to redress this deficiency for the space of a few days. This is an attractive opportunity despite its limited scope, because the return on one day's labor may be two to three times as great as the standard daily wage of two <u>kolak</u> (viz., 3.7 kilograms) of rice. For example, the amounts of seven and ten kilograms cited above each were reaped by one worker in the course of one workday or less.

CHRONOLOGY

The main phase of the harvest has its own distinct chronology, as does each of the subsidiary phases (viz., the earlier harvest of immature grain, the subsequent harvest of late maturing grain, and the processing of the harvest).

Main Harvest Phase

The growth of the rice plants is a time of special significance to the Kantu'. Only during this time is a field properly called <u>umai</u> 'swidden'. Before the rice plants have broken ground, the field should be called <u>tegalan</u>; and after all of the rice plants have been harvested, it must be called <u>memudai</u> 'regrowth'. The Kantu' carefully monitor the growth of the rice, dividing it into fifteen distinct, named <u>maia</u> 'stages' (table 79)[30].

TABLE 79
The Stages In The Growth Of The Rice Plant

Stage	English Gloss	Characteristics
(1) *Mulu idung*	'Be a nostril hair'	The seed leaf is just visible.
(2) *Nyarom*	'Pierce with a needle'	One leaf two-fingers' width in height is visible.
(3) *Ngiko' pipit*	'Be like the tail of a rice sparrow'	One leaf three-fingers' width in height is visible.
(4) *Musa'*	'Become concentrated'	The shortest leaves die, leaving three leaves of equal height.
(5) *Masah kain Singanan*	'Wet the dress of the Malays [a sarong]'	The plant is below the knee cap in height.
(6) *Pala' patong*	'Knee cap'	The plant is as high as the knee cap.
(7) *Darah*	'Blood [ref. to hymen]'	The plant is as high as the thigh.
(8) *Tumok ngelasok*	'Thrust back'	The tillers branch.
(9) *Kanong laki*	'Male pregnancy'	The panicle begins to swell inside the tiller.
(10) *Kanong ino'*	'Female pregnancy'	The panicle further swells the tiller.
(11) *Terbit*	'Break forth'	The panicle completely emerges from the tiller.
(12) *Berat buah*	'The grain is heavy'	The tiller bends over from the weight of the grain.
(13) *Mansau mulut*	'The lips are ripe'	A little of the panicle is golden in color.
(14) *Netak tangkai*	'Cut the panicle'	One-half of the panicle is golden.

Table 78 - Continued

Stage	English Gloss	Characteristics
(15a) Mansau rupus	'Ripe all over'	All of the panicle is golden.
(15b) Suma' garih	'The boundary to reaping by the suma' technique'	The grain is ripe enough to be reaped by the main technique ngetau, rather than by the early technique suma'.

These stages are differentiated according to the height, number, and kind of the plant's leaves or tillers (important in differentiating the early stages), and the presence, weight, and color of the grain (important in differentiating the later stages). The interval from the first stage to the last (viz., from planting to harvesting) consumes a minimum of 157-158 days. Neither the testimony of the Kantu' nor my own data suggest that this interval markedly varies between one rice variety and another, with one exception. The swamp rices usually take longer to mature than the dry rices. This is due to the fact that most swamp rice plants are transplanted, which necessitates trimming both their leaves and roots, inevitably setting back their growth.

The period of maturation does not vary according to the date of planting, but the date of maturation does. The Kantu' say that if they plant their rice during the eighth moon, when Orion is visible in the center of the sky, then it will be ready to be reaped when Orion again is visible in the center of the sky at dusk. (When Orion is observed at this latter date in this position, they call it muai ruman 'throwing away the rice stalks', ruman referring to the stalk of the rice panicle after it has been separated from its grain during threshing.) On the other hand, if they plant their rice in the seventh moon, then they say that it will be ready for reaping when Orion is visible in the eastern sky at dusk. Thus, the date of maturation varies - from the date on which Orion is in the eastern sky to the date on which it is in the center of the sky - according to the date of planting. The rices are period fixed as opposed to date fixed. Consequently, the first swiddens to be planted within the longhouse tend to be the first to be reaped. If the longhouse's 1976 swiddens are ranked in terms of both planting dates and harvesting dates, the Spearman rank-order correlation coefficient shows a significant association ($r_s=.57$, $P<.01$). Similarly, in a year in which the longhouse as a whole plants relatively early, the harvest as a whole will be relatively early as well. In 1974-1975 the mean planting date for the longhouse was 30 August and the mean harvesting date was 22 January; which compare with mean dates of 12 September and 18 February, respectively, in 1975-1976. Since the interval from planting to harvesting is something of a constant, whereas the interval from the initial selecting to planting is not, the overall length of the swidden cycle may vary from one year to the next. As the Kantu' say, if they burn and plant their swiddens earlier than usual (viz., sooner after slashing and felling), the swidden year (viz., measured from site selection to the end of harvesting) will be shorter than usual - what they call pandak taun 'short year'.

Once the rice plants have attained maturity, they must be reaped quickly in order to minimize losses to predation, lodging and shattering. The Kantu' speak explicitly of the threat of such losses and of the consequent time constraints upon the harvest. They say that the harvest is one of the two swidden phases (weeding being the other) that must be completed within the shortest possible period of time, or else Padi nadai temu 'No rice will be found'. Accordingly, households intensify the use of their labor during this phase of work. In 1976 the average household maintained a work rate throughout its own harvest season (in the harvesting of its own swiddens) of 14.1 kilograms of grain per worker per day[31]. Since the average worker can reap no more than 19.3 kilograms in a single day (given a 350-minute workday and a labor rate of 18.1 minutes per kilogram), this means that the average household worked to an impressive 73 percent of capacity throughout the 1976 harvest season. (The work rates of individual households varied in inverse association with their worker/grain ratios: P=.03 for twelve households in 1976.) This high intensity of labor is achieved, in part, by minimizing the number of ritual work proscriptions during the harvest season. The Kantu' are not any less disposed to shorten or abort a harvest workday upon observing an ill omen, nor are they less disposed to honor work proscriptions arising from rituals pertaining to illness, death, the swidden cycle, or the phases of the moon. However, they do restrict the rituals themselves: the frequency with which rituals are staged and work proscriptions are incurred during the harvest season is not at all comparable with the rest of the year (table 80). This norm reflects a cultural recognition of the pressures on labor during the harvest season.

TABLE 80
Ritual Work Proscriptions During The Harvest Season

		Time of Year	
		Harvest season (26 January - 10 April 1976)	Nonharvest season
Work in the swiddens was ritually proscribed for the entire longhouse	No	74	209
	Yes	2	81

N = 366 days (1 May 1975 - 30 April 1976).
$X^2 = 21.9$
$P < .001$

Subsidiary Harvest Phases

While the main harvest of mature grain can begin, at the soonest, 157-158 days after planting, the harvest of immature grain can begin up to one week earlier. Actual starting dates vary according to the household's need for the grain. The Kantu' say that households suffering lapar 'famine' will begin the early harvest when only one-third of the grain in a given rice stand is ready to be reaped, whereas everyone else will wait until at least two-thirds of the grain is ready. The Kantu' say that it is preferable to postpone the early harvest until a large percentage of the grain can be reaped, because otherwise it is too difficult to select the harvestable grain from the unharvestable grain.

The timing of the early harvest also is influenced by ritual considerations: the rice-breaking ceremony can be performed, and the early harvest can begin, only during bulan mansang 'waxing moons'. Neither should be carried out during bulan surut 'waning moons' (or during the full moon or the period when there is no moon in the sky). They say that if this prohibition is violated, the harvest will be poor - which relates to the symbolism of the waning moon. Surut means not only 'wane', but also 'ebb', 'dwindle', 'become shorter or smaller'. All of these denotations are contrary to what is desired during both the growth of the rice (viz., upward growing, tall plants) and its harvest (viz., rice boxes and bins being filled to the rim with the harvest). The lunar prohibition often is honored in its violation. For example, at the end of January 1976 the members of household #16, eager to begin the first phase of the harvest in their Lubuk Menserai swidden, declared that they would not wait to begin until the waxing of the next moon on 11 Feburary - which was the earliest possible beginning date, under the prohibition - but would wait only for the beginning of the next bulan perintah 'government moon' (viz., the month of February) on 1 February. The household justified this violation of the prohibition with an ad hoc attack on its logical basis. They said that since the Kantu' do not count the moons at this time of year (cf. table 49), they need not ngau bulan 'use the moons' in a matter such as this harvest prohibition. This household's attack on the prohibition was prompted by the exceptionally ill-timed maturation of the grain in its swidden. This grain attained sufficient maturity for harvesting to commence just as the moon became full, on 27 January. In order to honor the lunar prohibition, therefore, the household would have had to postpone its harvest by the maximum interval of sixteen days (viz., the interval between the last day of one moon's waxing, on 26 January, and the first day of the next moon's waxing, on 11 February). In the event, not only did household #16 not wait for the next moon to begin waxing, they did not even wait for the beginning of the next government moon, but performed their rice breaking on 28 January. Such violations of the lunar prohibition are not uncommon (table 81).

TABLE 81
The Commencement Of The 1976 Harvest And The Lunar Calendar

1976 Lunar Calendar	Early Harvest Begun in
Waxing moon .... 12/1 - 26/1	......1 swidden
Full moon & waning moon 27/1 - 7/2	......8 swiddens
No moon .... 8/2 - 10/2	......1 swidden
Waxing moon .... 11/2 - 25/2	.....13 swiddens
Full moon & waning moon 26/2 - 8/3	..... 7 swiddens
No moon .... 9/3 - 11/3	..... 1 swidden
Waxing moon .... 12/3 - 27/3	..... 2 swiddens

N = 33 1976 swiddens.

The timing of the late harvest is a function of the timing of the main harvest. In thirteen stands of rice in four 1976 swiddens, the length of the interval between the main harvest and late harvest varied inversely with the length of the interval from planting to the beginning of the main harvest ($P=.001$)[32]. That is, as the interval from planting to the main harvest decreased or increased, according to whether the harvest of a given rice stand was begun as soon as it reached maturity or long afterwards, the interval from the main harvest to the late harvest correspondingly increased or decreased, respectively. This suggests that the timing of the late harvest ultimately is a function of the timing of planting. This also is suggested by the fact that the interval from planting to the late harvest is relatively fixed in length, in comparison to the interval from the main harvest to the late harvest (their coefficients of variation in the thirteen-stand sample were 3.2 percent and 32.8 percent, respectively). This further suggests that the grain reaped during the late harvest is largely padi matak 'late maturing rice' from the original planting, and not semeli' 'new rice' produced by a second, postharvest flowering of the rice plants. If it were mostly the latter, the interval from the main harvest to the late harvest should be more fixed in length and the interval from planting to the late harvest should be less so.

The timing of harvest processing is related directly to the timing of the harvest itself. On average, when harvested grain has accumulated for just two days it is threshed. This puts threshing in potential conflict with harvesting for available labor: 95 percent of the days on which threshing was performed in 1976, by households #3, 4 and 16, were also harvest days. The Kantu' deal with this conflict, in part, by threshing at night. In 1976, for example, household #3 performed fully 63 percent of its threshing at night. In addition, some threshing is done during the day by workers who are not available for harvesting. For example, if it is necessary for someone to stay in the swidden house during part or all of the day, to mind infants and/or prepare the household's meals, that person also can thresh. In contrast, large-scale winnowing cannot be carried out by a single worker who otherwise is required to remain in the farmhouse during the day. Nor can either drying or winnowing be performed at night after the day's harvesting is done. Hence the Kantu' perforce dry and winnow their grain after the main phase of the harvest has been completed or nearly completed.

Chapter Nine
Carrying

The final stage in the linear sequence of nine swidden stages consists in ngangkut 'carrying' the rice harvest from the fieldhouse back to the longhouse and storing it in the loft there[1]. The purpose of this is, first, to ensure the safety of the harvest. There is a greater chance of losing the harvest to theft, predation or other misfortune when it is stored untended, in a distant swidden, than when it is stored in the household's own loft at the longhouse. Storage at the longhouse also is more convenient. Assuming that the household members are going to spend little or no time during the following year in or near a just harvested swidden, the continued storage of grain there makes access costly in travel time. For these reasons, the harvest usually is carried back to the longhouse for storage. This was the case in fifty-two of the sixty-nine 1975 and 1976 swiddens.

In a minority of cases the need for safety and convenience can be served by continuing to store the harvest in the swidden house. This is possible when the household lives in or near its just harvested swidden during a significant portion of the following year, as when the household makes a swidden in close proximity to the past year's swidden. For example, households #1, 6, 8, 11 and 15 stored the harvests from their 1975 Kenua' swiddens at their swidden houses, and each cited plans to make a 1976 swidden nearby in explanation (cf. figure 10). The association between these two decisions was statistically significant (table 82). While the proximity of the coming year's swidden is a necessary element in the decision to store a harvest in the farmhouse, it is not always the sole element involved. Members of household #11 maintained that one reason for not carrying either their 1975 or their 1976 harvests back to the longhouse was the lack of space in their longhouse quarters (which, at twenty-three square meters, was the least of any household, and well below the mean of seventy-eight square meters). In the case of households #16 and 17 in 1976, neither of whom had as yet constructed

TABLE 82
The Storage Of The 1975 Harvest And The Location Of The 1976 Swiddens

		Household made a 1976 swidden in the vicinity of its 1975 swidden	
		No	Yes
Household stored the 1975 harvest in a swidden house	No	16	13
	Yes	0	5

N = 34 1975 swiddens.
P = .03

NOTE: One swidden was considered to be in the 'vicinity' of another if it was adjacent or else was separated by no more than one or two other swiddens or sections of forest.

their own living quarters at the longhouse, there was little alternative to storing their 1976 harvests in the swidden house. Subtler factors also can be involved. The members of household #6 make a practice of storing their harvests in their swiddens (as they did in both 1975 and 1976), and of living there themselves, due at least in part to the misanthropic nature of the household head. A final factor is the size of the harvest. The two reasons for storing the harvest in the longhouse - protection and ease of access - become proportionately less compelling as the size of the harvest shrinks. At the same time the desire to keep the size of one's harvest secret becomes more compelling. A household that has reaped a poor harvest is <u>malu</u> 'ashamed' to exhibit this fact by carrying it back to the longhouse, under the gaze of the whole community.

TECHNOLOGY

Before carrying properly commences, the Kantu' direct some preparatory attention to the footpaths leading from the swiddens back to the longhouse and in particular to the <u>panti'</u>, the simple log bridges spanning the innumerable watercourses of the Kantu' territory. This is the one time each year that they receive any attention. Rotting logs are replaced and handrails are added to particularly treacherous crossings. This work is done by whichever households own the swiddens serviced by particular footpaths and bridges. Other preparatory activities focus on the refurbishing, or more rarely new construction, of the lanyi', the baskets used to carry the rice. These baskets, the largest made by the Kantu', are stoutly made of rattan and are fitted with a backstrap for each shoulder, as well as a tumpline to pass across the forehead or be held in one hand. They also are fitted with four rudimentary feet so that they can stand upright when loaded. Goods other than rice usually are not transported in these baskets. The Kantu' say that they never carry cucurbits or tubers (e.g.) in them because the weight would be too great. When filled with rice these baskets weigh approximately fifty-five kilograms, a load that approaches the limits of strength and endurance of the mature, healthy, Kantu' male. The Kantu' say that they can lift their own body weight but no more; and indeed fifty-five kilograms approximates the average weight of adult, male Kantu'. Of course, there is considerable variation in the individual capacities of both men and baskets. The Kantu' employ a rule of thumb to determine whether a particular carrying basket, filled with grain, can be carried by a particular person. They say Alah dikatong, alah d'angkut 'If it can be picked up [by a man], it can be carried [by him]'.

There is one more preparatory task, which consists in gathering and filling the baskets. A household planning to use a multihousehold work group, as opposed to just its own work force, usually must borrow baskets from other households to supplement its own. These are brought to its farmhouse several days before the date set for carrying. The household members then spend a day or two filling the baskets with their threshed and winnowed grain. They pack the grain tight using either their feet or a hardwood pestle. When each basket is filled, a layer of rice chaff (set aside from the threshing) is tied on top to prevent any spillage of grain on the trail.

On the morning of the day set for carrying in the harvest from a given swidden, again assuming use of a multihousehold work group, the longhouse males gather at the farmhouse where the harvest has been stored. The day begins relatively late, because everyone eats before leaving the longhouse (which is not the case in every swidden phase). At the swidden house the men stop again to drink coffee and rice beer or arak. As the men drink each one picks out, from among the filled and waiting baskets, the one he wants to carry. The younger males amiably bully one another into taking on a larger basket than desired. Upon completion of these preliminaries the men leave the farmhouse in single file, each carrying one basket, and head towards the longhouse.

Unless the distance from the swidden to the longhouse is relatively short, the men do not carry the baskets in without a break. Typically they carry them to a prearranged alai mentu 'rest place', where the carrying baskets are dropped to the ground and leaned against previously prepared gangah (horizontal bars erected approximately one-half meter off the ground), and the men are given food and drink by waiting women. These rest places are spaced at regular intervals along the main trails and tend to be reused one year after another. The distance between successive rest places approximates a constant, as reflected in the Kantu' custom of measuring some distances as Pemayoh kali' mentu ngangkut 'The number of rest stops when carrying a harvest'. This constant is about one statute kilometer. Thus, the men of Tikul Batu usually made two rest stops when carrying in the harvests from their 1975 and 1976 swiddens, which were located approximately three kilometers up the Kenua' stream.

After a break at the initial rest place, the carriers usually do not proceed on in the direction of the longhouse. Rather, they return to the farmhouse, take one more carrying basket apiece, and bring them on to this rest place as well. Only then, after another rest, do they proceed on down the trail carrying the first lot of baskets. This sequence is repeated until the longhouse is reached. In this typical case, therefore, the men have in transport twice as many baskets as they can carry at any one time. This system intensifies the use of labor during carrying and so brings it to an end quicker. Any further intensification is perhaps too demanding physically, as I have no record of any group transporting three or four times (e.g.) as many baskets as can be carried at any one time (viz., three or four times as many baskets as men). If the number of carrying baskets exceeds the number of carriers by a factor of more than two, the group will make a second trip late in the afternoon of the same day. Since the carriers will be fatigued from the first trip, the intensity of labor is kept lower during second trips. For example, when carrying in the harvest from swidden #41 in 1976, eleven men transported twenty-two baskets during one trip early in the day, and then they carried in an additional eleven baskets during a second trip later that afternoon. The maximum number of baskets that can be transported in a single day is equal to about three times the number of workers involved, as in the case just cited. When this ratio is exceeded, the carrying is spread over two days. This happens in 20-25 percent of the cases in which a swidden harvest is transported by a multihousehold work group.

Whenever possible the Kantu' transport their harvests on water by perau 'canoe' instead of by foot on land. These canoes, averaging three meters in length, ideally are made of iron wood. They consist of a one-piece hull, to which low gunwales are sown with rattan. The Kantu' have been known to make them but they usually obtain them through trade with the Iban. Within the territory of Tikul Batu, only the Empanang and Kantu' rivers are navigable by canoe (although the lower reaches of the Kenua' stream sometimes are passable by canoe after heavy flooding). Among the sixty-nine 1975 and 1976 swiddens, twenty-three were located close enough to the Empanang or Kantu' for their harvests to be brought by canoe to the longhouse, which is located only seventy-five meters from the banks of the former and only a little further from the banks of the latter (figure 33). (Daily travel between most of these swiddens and the longhouse also is by water, the Kantu' preferring under almost any circumstances to paddle rather than walk.) These twenty-three swiddens represent almost one-half of the 1975 and 1976 swiddens from which the harvest was transported back to the longhouse. The ability to thus reduce the onerous burden of overland transportation by foot, in one out of two swiddens, is important to the Kantu'. Some have said, for example, that they could not stand to live in the hills to the southwest (where a related ethnic group lives), because there is no water transportation at all there.

The next phase of this operation commences when the carriers reach the longhouse. If any ill omens were heard on the trail, the baskets are gathered at the longhouse steps where the household, whose harvest it is, must sacrifice a pig to propitiate the spirits. Then the harvest is brought into the longhouse. The carrying baskets initially may be stacked in the gallery outside the apartment of its owners, while a pegela' 'offering' is prepared within[2]. This offering is made only when a sizable harvest from a stem rice swidden is carried back to the longhouse by a large, interhousehold work group. It is not obligatory, except for the first harvest to be carried into the longhouse each year. In 1975, Tikul Batu went so far as to summon the

FIGURE 33
Transport Of The 1975 And 1976 Swidden Harvests By Canoe

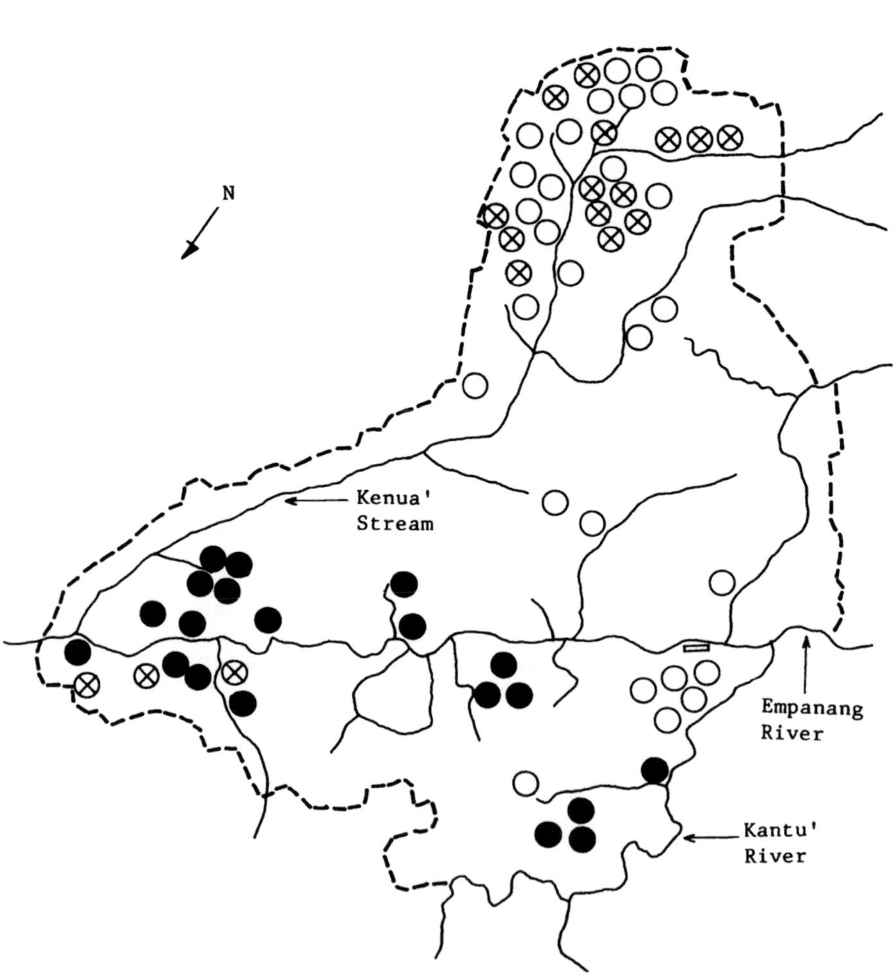

SCALE: 1:18,000 (swiddens not drawn to scale).

KEY: ● signifies that the swidden's harvest was transported to the longhouse by canoe.

○ signifies that the swidden's harvest was transported to the longhouse by land.

⊗ signifies that the swidden's harvest was not transported to the longhouse at all, but was stored in the swidden.

area's Panglima Adat 'Law chief' from a downriver Iban settlement to officiate at the carrying in of the first harvest that year, one of the rare times during the swidden cycle in which the Kantu' defer to an extralonghouse authority.

While the offering is being prepared, the carriers gather in the apartment of the host household to eat, drink rice beer and arak, and sing. When all is ready, the offering is carried up into the household's sadau 'storage loft' (figure 3), part is hung in a basket over the bark rice bin in which the harvest is to be stored, and part is placed in the bin itself. Also placed in the bin are the first panicles of rice reaped during the harvest (during the ritual rice-breaking) as well as the last (from the vicinity of the offering post in the 'place of the seed rice'). Then a small quantity of rice beer is poured onto the base of the bin and also onto its vertical seam, to the accompaniment of the mantra, Awak ke kenyang, awak ke melepong; awak ke jimat, awak ke nadai tau' amih 'So that it will be full, so that it will rise; so that it will be frugal, so that it cannot be finished'[3]. This is followed by sprinkling some popped rice into the bin, the symbolism of which (e.g., expansion and fullness) is in keeping with the preceding mantra. Finally, seven handfuls of rice are counted out of the first carrying basket and cast into the bin to the accompaniment of further ritual invocations[4].

With the completion of these preparations, some men begin to carry the baskets from the gallery up a ladder to the loft, where others wait to tip them into the bins. This is called nyimpan 'storing'. (It usually is done on the same day as the carrying but may be postponed until the following day if the carrying finishes very late.) One man, usually a member of the host household, stands in the bin and levels off and packs down the rice while it is being poured in, using his feet. The rice is poured on top of any grain remaining from previous harvests. There should always be a little old grain in the bottom of the larger rice bins, because of a proscription against completely emptying them: Padi nadai tau' amih dalam gentung besai 'Rice must not disapear from the large rice bins', even during a period of famine.

The gentung padi 'rice bins' are made of bark, which cannot be chewed through by rodents. (In contrast, all plaited containers, whether of bamboo, rattan or reed, are vulnerable to rodents.) The base of the bin consists of a flat sheet of bark, and the sides consist of a single circular sheet with overlapping ends. A second sheet may be positioned atop this one to increase the height of the bin. To obtain these bark sheets, the Kantu' first fell a suitable tree (not all species provide suitable bark), and then slice away the outer bark with their brush swords. Then they beat the inner bark with wooden clubs until it begins to separate from the trunk and can be pried away with levers. Finally they heat this inner bark above a fire and manipulate it to increase its natural curvature.

The different rice varieties are poured into the bins in a careful order. Since the first rice put into the bin will be the last taken out and hence the longest in storage, the swamp rices are put in first because they store best over time. Not only does swamp rice not deteriorate as quickly but, from one perspective, it actually improves with age. The Kantu' say that if swamp rice is stored for a long time, it will swell greatly when it is cooked. This is desirable because, the Kantu' say, they need not cook as much rice as they otherwise would. After the swamp rices have been stored, storage of the dry rices begins. The large-grained and small-grained dry rices are kept separate by means of laying plaited mats into the bin. The Kantu' say that if they mix a large-grained variety with a small-grained one, they will not be able to husk the latter properly. After the last of the rice has been stored in a given bin, the household places its pengaroh padi 'rice charm' into the bin. They believe that this charm, consisting of the special pigs' tusks formerly carried into battle, will keep the grain safe.

Not all of the household's harvest is carried in and stored in this manner. Seed rice is brought back to the longhouse separately, several days after the main carrying operation (if it is not brought back before), using strictly intrahousehold labor. Handling of the seed rice by other than household members is thought to show a lack of respect for it. The carrying of the seed rice is distinguished terminologically from the carrying of the rest of the harvest. The latter is called ngangkut, meaning 'transport', but the former is called ngema', which specifically means 'to carry on one's back', a denotation more suited to the sensitivities of the seed rice. The seed rice also is treated distinctively in storage, as are the glutinous rices. To avoid mixing them with other grains, both are stored not in the rice bins but in carrying baskets, old rice sacks, or any of several plaited containers (e.g., belansai or kelekit) designed especially for the storage of small quantities of grain.

LABOR

I said earlier that a single worker can carry in to the longhouse, from swiddens at the farthest reaches of the longhouse territory, up to three baskets in one day (assuming two round trips between swidden and longhouse). The norm is somewhat lower, however. During 1975 and 1976 the average work rate for multihousehold work groups engaged in transporting the harvest was two carrying baskets per day, which was equal to 0.21 workday per basket per statute kilometer. The rate was faster for work groups using river transportation, averaging 0.15 workday per basket per kilometer. Work on the footpaths and bridges consumed another one man day of labor per kilometer, and the filling of baskets consumed a final 0.1 workday per basket. Overall work rates varied among swidden types according to systematic variation in the size of the swidden, the size of the harvest, the tendency to store it in the swidden as opposed to the longhouse, the ability to transport it to the longhouse by water instead of land, and the distance to the longhouse. In 1975 and 1976 the overall work rates were 3.8 days per hectare in primary forest swiddens, 4.9 days per hectare in secondary forest swiddens, and 9.6 days per hectare in swampland swiddens[5].

The Kantu' speak of carrying in the harvest as one of their most physically demanding tasks. They say, Semua urang nadau tau' ngangkut padi 'Not everyone can carry rice'. Young, old and infirm males are de facto excluded from this work. The Kantu' also exclude females, at least conceptually, saying that carrying the harvest is men's work. They add that a minority of women are capable of doing it, but in practise participation by even the strongest and healthiest women is quite limited. Rarely if ever does a woman participate in the carrying of a harvest belonging to a household other than her own. Even in the case of her own household's harvest, a woman is at most likely to assist in carrying the baskets from the near river bank into the longhouse. Full female participation in the long, overland carries is nonexistent. In any case, the Kantu' women have their own distinctive role in this stage of the swidden cycle. They precede the carriers along the trails and provide them with food and drink at each resting place. In addition, they prepare the feast at the longhouse with which the carriers are greeted when they finally arrive. Also important in explaining the minimal participation of women in the actual carrying itself, is the near absence of time constraints on this swidden stage: the labor of women is not really needed.

In twenty-eight of the fifty-two 1975 and 1976 swiddens whose harvests were carried back to the longhouse, all of the carrying was done by the household's own work force, which is called ngangkut kediri' 'carried alone'. In the remaining twenty-four cases the carrying was done by a multihousehold work force, which is called berangkut 'carrying together'. The labor arrangement whereby the latter is accomplished is called berimpoh 'cooperative labor'. As I noted in the introductory chapter, this differs in important respects from the more common bedurok 'reciprocal labor' arrangements. Whereas bedurok is based on the strict calculation and reciprocation of man-days, berimpoh is not. As the Kantu' say of this stage, Asa berangkut kami nadai ngerikin tuboh-ari 'When carrying we do not reckon body-days'. What they do reckon, under berimpoh arrangements, is the participation of entire households throughout the duration of a specific task, in this case carrying in the harvests from one of each household's swiddens. It is assumed that each household will participate to the fullest extent of its work force throughout the duration of this task. In practise, in 1975 and 1976, an average of 82 percent of the able-bodied carriers in each participating household joined in each day of carrying under berimpoh arrangements. (This can be compared with an average of just 56 percent for the households that participated in carrying on a gift labor or wage labor basis.) Since there is considerable variation from one household to another both in the number of carriers that they can contribute to other households and - because of variation in harvest size - the number of carriers that they require themselves, most households participating in a berimpoh carrying arrangement do not attain an exact balance between labor expended and labor received in any given carrying season. Among the fourteen households at Tikul Batu in 1975 and the sixteen present in 1976, only six struck an exact balance between labor inputs and outputs in carrying, while seventeen realized negative balances and seven realized positive ones. This is permissible within the cooperative labor system, based on the principle that each year's imbalances will even out in the long run. The same households do not continually achieve negative or positive balances. For example, when each household's net input or output of labor in 1975 was compared to the same for 1976 a negligible correlation coefficient ($r=.06$, $P>.05$) was obtained.

Berimpoh 'cooperative labor' also differs from bedurok 'reciprocal labor' in its scope. The Kantu' characterize the former as sekayu rumah 'longhouse wide', meaning that all households in the longhouse ideally are involved; as opposed to reciprocal labor arrangements, which they characterize as pilih pintu 'selecting by door', meaning that each household chooses with whom it wants to work. All households in the longhouse are expected to participate in cooperative arrangements for carrying in the year's harvests, even including - although to a lesser extent - those households not planning to carry in the harvests from their own swiddens. Among the fourteen households that farmed in 1975 and the sixteen that farmed in 1976, a total of fourteen did not carry their own harvests in to the longhouse; but twelve of these latter still participated in carrying in the other households' harvests. Their participation was called not berimpoh 'cooperative labor' but rather betolong 'gift labor' or, if they were paid a wage, bekuli 'wage labor'.

The standard wage for carrying grain back to the longhouse, overland, is three kolak of threshed, unhusked grain per basket (assuming that the harvest is being carried in from one of the more distant swiddens, averaging at least 2.5 kilometers from the longhouse.) Since the average work group carries in two baskets of grain per worker per day, the total wage for one day of this work averages six kolak. This remarkably high level of payment, three times as high as in most other stages of the swidden cycle, is explained both by the difficulty of the work and by the fact that it immediately follows the harvest. This timing exacerbates the difficulties of attracting wage labor, because even the worst off households are likely to have some grain of their own at this time and so can refuse this arduous labor unless the wage is made sufficiently attractive. The food and drink provided to the workers during this stage are unusually attractive as well. In addition to cooked rice and assorted vegetables, the carriers typically are provided with glutinous rice cakes, noodles, chicken, often the flesh of the highly esteemed river turtle, and occasionally pork as well; and sweetened coffee, rice beer and/or arak to drink. (The value of such a feast is equalled or exceeded during only one other stage of the swidden cycle, namely planting.) In 1976 for example, the equivalent of 39,103 rupiah were devoted to feasting the 114 workers who carried in Tikul Batu's harvests. This worked out to an average value of 343 rupiah per worker, or 86 percent of the basic daily wage of 400 rupiah. This was clearly an added inducement to the participation of not only the wage laborers but also the cooperative laborers and, especially, the 'gift' laborers for whom this food and drink was the only direct compensation or reciprocation that they would receive.

In part because of all the costs and obligations associated with the use of cooperative labor groups, they are not used in more than one-half of the cases in which a swidden harvest is carried back to the longhouse, as I noted earlier. One determinant of whether or not cooperative labor is used in each case is swidden size (table 83). Particularly in the tiny swamp

TABLE 83
Swidden Size And The Use Of Cooperative Labor To Carry In The Harvest

		Swidden size	
		< 1.7 ha	> 1.7 ha
The harvest was carried back to the longhouse by	The household work force alone	24	4
	A cooperative work group	5	19

N = 52 1975 and 1976 swiddens whose harvests were brought back to the longhouse for storage.
1.7 = mean size of the swiddens in this sample.
$X^2 = 22.15$
$P < 0.0001$

swiddens, measuring just a fraction of a hectare, there is little need for a large carrying party. Household members often carry rice from these swiddens back to the longhouse on the same day it is harvested, without even threshing it first. This is made possible by the small amounts involved, and it is made necessary by the typical absence of a swidden house in or near such swiddens, in which the harvest can be accumulated and processed. A second determinant of the use of cooperative labor is the accessibility of the swidden to transportation by canoe (table 84). The Kantu' are less anxious to enlist extrahousehold labor when transporting grain by canoe than by foot, because of the comparative ease of the former: they do not want to receive labor on water that they will have to pay back on land.

TABLE 84
Access To Water Transport And The Use Of Cooperative Labor To Carry In The Harvest

		The swidden was accessible to water transportation	
		No	Yes
The harvest was carried back to the longhouse by	The household work force alone	7	21
	A cooperative work group	17	7

N = 52 1975 and 1976 swiddens whose harvests were brought back to the longhouse for storage.
$X^2 = 10.5$
$P < 0.005$

CHRONOLOGY

Transportation of the harvest back to the longhouse must follow the completion of harvesting (and also processing the harvest), so the chronology of the former is to some extent a function of the chronology of the latter[6]. The interval between harvesting and carrying must be sufficiently long to allow time for refurbishing bridges and trails, carrying the empty baskets out to the swidden, and packing them with grain. In 1975 and 1976 this interval averaged thirty-two days, but it ranged from as little as two weeks to as long as ten weeks. The first swiddens in which harvesting and processing are completed are the first to be scheduled by the longhouse for carrying. They try not to schedule these cooperative carries on consecutive days because, the Kantu' say, the work is so onerous. Thus, in 1975 they spread twelve carries over fifty-seven days, and in 1976 they spread ten carries over nineteen days. The intensity of labor use during these two carrying seasons equalled 13 percent and 24 percent, respectively, which is as low or lower than any other stage in the swidden cycle[7]. These low rates of labor usage are due not only to the onerousness of the work, but also to the absence of any particular time constraints on the execution of this swidden phase, as I noted earlier. Households that do not use cooperative labor follow no set schedule in carrying in their harvest, but simply bring it in a few baskets at a time, whenever they have the time.

Chapter Ten
Harvesting Nonrice Cultigens

The harvest of the ngkayu 'nonrice cultigens' or 'relishes' is distinct from the rice harvest. The Kantu' do not refer to harvesting relishes as ngetau. Nor do they regard this harvest as anywhere near as important within the swidden cycle as the harvest of rice. This is reflected in their calling a field an umai 'swidden' until the last of the mature rice has been harvested, whereupon they begin to call it memudai 'secondary forest', even though the harvesting of most relish types has not been completed by that date (if even begun). The distinctness of the relish harvest is due to the difference in maturation periods between the relishes and rice and to consequent differences in harvesting patterns.

MATURATION

The relishes belong to the same swidden - meaning the same act of cultivation - as does the rice. The same initial clearing and burning of the forest that enables the rice to be cultivated also enables the relishes to be cultivated. No special cultivation efforts are directed towards the relishes alone, with but one or two exceptions, the most important of which involves post-rice harvest weeding. Within one or two months of the end of the rice harvest, most households nyiang 'clear away', using the brush sword, all weeds in the vicinity of the relishes growing near the swidden house. These efforts usually focus on just two or three of the most important relish types, meaning sugarcane, bananas and the various tubers. If extensive clearing is done in a large stand of cassava or sweet potatoes, the resulting stand is called a kebun 'garden'. A relatively large, cleared stand of tubers covers no more than 5-10 percent of the average-sized swidden, and only that limited, cleared area is called a 'garden'. The presence of such a garden does not alter the field's overall status as *memudai* 'secondary forest', which it acquired with the completion of the rice harvest.

There is one exceptional case in which the relishes are not grown within the same cycle of cultivation as the rice. If a household's annual rice crop and/or relish crops fail, they may decide to carry out a second, mid-year planting of relishes. This is done just after the completion of the rice harvest. First, the household slashes the *jerami' padi* 'rice stubble' and any regrowth that has sprouted up since the swidden was burned or weeded. As in the making of gardens, this is done in only a small portion of an average-sized swidden. After slashing and after a brief drying period, the slashed material is burned. After burning, the household plants a few of the major relishes, usually cassava, sweet potatoes and maize. The area thus cleared, burned and planted is called a *tempalai*. As in the case of gardens, the presence of a *tempalai* in a newly harvested field does not alter the *field's* overall status as secondary forest. The Kantu' distinguish the *tempalai* from the aforementioned *kebun* on the basis that the former represents a new and distinct cycle of cultivation while the latter does not. The Kantu' distinguish the *tempalai* from the *umai* on the basis that rice never is planted in the former and always is planted in the latter. A *tempalai* cannot be called an *umai* nor, in making a *tempalai*, can one say *bumai tempalai* 'make a tempalai using swidden techniques'. Instead, according to the Kantu', one must say *nanam tempalai* 'plant a tempalai'. *Tempalai* thus may be glossed as 'relish swidden'. They are made with infrequency in the swiddens of Tikul Batu and are made with even greater infrequency in true secondary forest (viz., as opposed to just-fallowed swiddens), although this is not unknown.

With the exception of this 'relish swidden', the relishes are planted early in the swidden cycle. within one month (before or after) of the rice planting. They grow side by side with the rice and mature before or after it, according to their individual characteristics. There is great variation among the types of relishes in this respect, their maturation periods ranging from a low of about one month to a high of eighteen months or more. Overall, the total number of relish types harvestable at any given time is a function of the amount of time that has passed since the swidden was planted. According to table 85, the number of harvestable relish types is zero in the first month after planting, rises to a high in the seventh to ninth months after planting, and then again declines to zero after the twenty-second month[1]. This does not mean that there are months in which

TABLE 85
The Chronology Of Maturation Of Relishes Within A Single Swidden

Number of Months After the Swidden Was Planted	Number of Relish Types Harvestable (out of a total of 25)
0 - 1	0
1 - 2	2
2 - 3	3
3 - 4	7
4 - 5	6
5 - 6	10
6 - 7	14
7 - 8	15
8 - 9	15
9 - 10	13
10 - 11	12
11 - 12	9
12 - 13	8
13 - 14	8
14 - 15	7
15 - 16	5
16 - 17	3
17 - 18	2
18 - 19	3
19 - 20	2
20 - 21	2
21 - 22	1
22 - 23	0

no relishes at all are available to the Kantu'. The harvest period in a given swidden extends over twenty-one months (excluding the first month in which no relishes are harvestable), but after the twelfth month a new set of swiddens is planted as well. Consequently, the availability of relishes during most months is a function of the harvests in the swiddens of not one year but two years[2]. Table 86 shows that there is no point in the year at which no relishes, from either the past or the present swiddens, are available to the average household. Indeed, during no month of the year does the number of relish types harvestable fall below approximately one-third of the total number of types planted (viz., eight types if the maximum number of twenty-five types is planted), and in some months it rises to nearly two-thirds of the total (viz., sixteen types). There is a near perfect intermesh between the two harvest periods, such that after a given type of relish ceases to be productive in the old swidden it soon becomes productive in the new one. For example, sweet potatoes become productive in the new swidden within only one to three months after they cease

TABLE 86
The Chronology Of Maturation Of Relishes In Two Succeeding Years' Swiddens

Current Swidden		Previous Swidden		Both Swiddens
Number of months after planting	Number of relish types harvestable	Number of months after planting	Number of relish types harvestable	Total number of relish types simultaneously harvestable
0 - 1	0	12 - 13	8	8
1 - 2	2	13 - 14	8	10
2 - 3	3	14 - 15	7	10
3 - 4	7	15 - 16	5	12(-1)
4 - 5	6	16 - 17	3	9(-1)
5 - 6	10	17 - 18	2	12(-1)
6 - 7	14	18 - 19	3	17(-1)
7 - 8	15	19 - 20	2	17(-1)
8 - 9	15	20 - 21	2	17(-1)
9 - 10	13	21 - 22	1	14
10 - 11	12	22 - 23	0	12
11 - 12	9	23 - 24	0	9

to be productive in the old one. The single exception to this is cassava, which is simultaneously productive in both years' swiddens for six months.

HARVESTING

The Kantu' most frequently refer to the harvest of relishes as ngega' ngkayu 'look for relishes', but they also may say ngami'/mai' ngkayu 'get/bring back relishes'. When specific relish types are involved, they may refer to the particular technique or tool used in their harvest. Thus, the Kantu' say nebang pisang 'fell bananas' [with a brush sword]', matah tebu 'cut sugarcane [with a brush sword]', and ngali ubi 'dig up cassava [with a brush sword]'. They invariably use such terms to refer to relish harvests that are relatively brief and intense. Thus, the Kantu' say mantap/marang lingkau 'cut/chop maize [with a brush sword]', ngetau/mwar jawa' 'reap/pluck millet [with a finger knife]', and mwar lingkau lesit 'pluck job's tears [with a finger knife]'. In all cases the harvesting is done by household members working singly or in pairs. When the members are living in or near one of their swiddens, they will harvest relishes just before midday, in preparation for the midday meal. When they are living at the longhouse, one or two of them periodically will go to the swidden, spend perhaps one hour harvesting the relishes, and then return with them to the longhouse. In either case, this harvesting is done more by adults than juveniles and, especially, it is done more by females than males[3].

Condiments and snack foods are harvested on an approximately equal number of days, but the overwhelming number of harvest days are devoted to those relishes eaten as side dishes (table 87)[4]. Among these latter the cucurbits are the most important, followed closely by leafy vegetables and tubers, and then by legumes and nonrice grains. This comparison of harvest days reflects the comparative importance of each relish type in terms of weight or volume harvested and consumed with one exception, the nonrice grains. As previously noted, the nonrice grains in each swidden are harvested over just a few days, and a large amount is taken on each day; whereas all of the other relishes are harvested over a period of months, and comparatively little is taken on any given harvest day. For this reason, the number of harvest days is not a good indicator of the importance of these grains. In terms of volume actually

TABLE 87
The Relative Importance Of Each Relish Type

Relish Type	Harvesting Days Per Year Per Household
I. Main Side Dishes	89 days
1. Leafy vegetables (Spinach, sweet potato leaf, taro leaf, cassava leaf, cucumber leaf, pumpkin leaf, others)	(29 days)
2. Cucurbits (Wax gourd, pumpkin, cucumber, eggplant, bitter gourd, dish-cloth gourd, others)	(32 days)
3. Tubers (Sweet potato, taro, cassava)	(22 days)
4. Grains (Millet, maize, job's tears, grain amaranth)	(6 days)
5. Legumes	(9 days)
II. Condiments (Chives, sesame, ginger, lemon grass, mustard leaf, chili pepper, others)	9 days
III. Snack Foods (Bananas, sugarcane)	10 days
IV. Other	3 days

NOTE: The harvesting of one type of relish on one day is counted as 'one harvesting day'.

harvested, the nonrice grains (in particular maize), far from ranking last among the side dish relishes as their harvest day totals suggest, might well rank first.

There is an association between brief harvest periods

and good qualities of preservation on the one hand, and long harvest periods and poor qualities of preservation on the other. None of the nonrice grains, with their brief harvest periods, store well on the stalk. All are highly susceptible to damage from weather and pests. In constrast, these grains can be stored for months after harvesting. In addition to storing well as grain, maize and millet can be made into fermented mash and stored in that form as well. In contrast, few of the nongrain relishes, with their long harvest periods, store well after being harvested. Nor are any of the nongrain relishes processed into storable forms. Cassava sometimes is processed into dried and storable chips, but this is done only infrequently and on a small scale. While the nongrain relishes do not store well after harvesting, they do keep well on the stalk. This especially is true of all leafy types, cucurbits, and tubers.

The chronology of harvesting the relishes also varies according to their accessibility. A household is more likely to harvest relishes if its members are living in a swidden house in or near the swidden in which the relishes are growing, than if they are living in the longhouse at some distance from this swidden. The frequency of relish harvesting varies, therefore, according to residence in the swidden houses as opposed to the longhouse, which in turn varies according to the demands of the swidden cycle. In particular, there is a marked increase in the frequency of relish harvesting in each swidden coincident with its rice harvest (table 88)[5]. This increase cannot be attributed to coincidental

TABLE 88
Seasonal Variation In The Incidence Of Relish Harvesting

Number of Relish Harvesting Days Per Month Per Swidden During the		
Last month preceding the rice harvest	One or two months of the rice harvest	First month following the rice harvest
9 days	18 days	11½ days

increases in the number of harvestable relishes in the swidden. This number peaks in May and then gradually begins to decline in June (nine months or more after the swidden was planted), whereas table 88 shows a precipitous decline in relish harvesting immediately after the rice harvest is finished, in March or April (just six to seven months after the swidden was planted).

This decline, in the face of increasing numbers of harvestable relish types, reflects the fact that the incidence of relish harvesting is a function not just of the availability of the relishes, but also their accessibility, in this case due to the presence of workers in the swidden for the purpose of the rice harvest. This variation in the incidence of relish harvesting, according to the presence or absence of people in the swiddens, is only partially counterbalanced by variation in the quantities harvested (table 89)[6]. Even when the Kantu' can harvest only infrequently, that is, they do not harvest large amounts at one time, owing mostly to problems of storage. As a result, the total amount of relishes harvested per month rises during the rice harvest but drops before and after it.

TABLE 89
Seasonal Variation In The Incidence And Quantity Of Relish Harvesting

	During the last month before or the first month after the rice harvest, compared with the month(s) of the rice harvest	
	Lower	Higher
Number of days per month on which relishes are harvested in the swidden (=A)	82% of all cases	18% of all cases
Amount of relishes harvested each day in the swidden (=B)	32% of all cases	68% of all cases
Total amount of relishes harvested per month in the swidden (=AxB)	70% of all cases	30% of all cases

ALTERNATIVE RELISH SOURCES

There are two sources that the household can tap when its access to relishes decreases but its demand for them does not: one is relishes growing in swiddens belonging to other households. The transfer of relishes between households is subject to several constraints. Relishes can be taken from a swidden (or former swidden) by anyone, without the permission of its owners, if they are eaten raw and on the spot. This would include, for example, a stalk of sugarcane or a bunch of ripe bananas. The Kantu' say that since such relishes are taken only to satisfy an immediate hunger or need, their harvest by passersby can be neither prohibited nor protested by the swidden's owners. This freedom does not extend, however, to relishes that cannot be eaten in their raw state (e.g., most tubers, cucurbits, legumes, leafy plants, etc.). Because such relishes must be taken back to the farmhouse or longhouse to be cooked before being consumed, their consumption is a matter not of a single person's immediate need for a snack, but of an entire household's need for a meal, and for this the permission of the swidden owners is necessary. (In practise edible greens such as the leaves of the cassava plant usually are exempted from this proscription, although the cassava tuber itself is not.)

Persons seeking permission to harvest the relishes of another household usually say that they wish to meli 'buy' relishes. Typically, the latter household then tells the former one to harvest what they need, with no compensation being necessary. The Kantu' regard relishes not as a cash crop but as a subsistence crop. Even rice, which is not regarded as a cash crop either, is more capable of being borrowed, traded or bought than are relishes. The sale of relishes would be regarded as both a crass upgrading of their value and a crass downgrading of the value of the relations between the two Kantu' households involved. (When it is a case of relations between Kantu' and non-Kantu', however, relishes are traded and sold freely.) In most cases, requests to harvest relishes are honored, even when this means that the donor household itself will run short in the near future. The incidence of requesting and giving permission to harvest relishes reaches a peak in January (table 90), which is the last and hence worst month of the famine period preceding the rice harvest in February and March[7]. Cassava, the traditionally preferred starch staple of the famine period, when rice either is unavailable or in short supply, was involved in six of the eight January requests in table 90. Some of these requests were prompted by a simple lack of relishes in the recipient household's own swiddens, but others were prompted by a lack of, in particular, accessible relishes.

Accessibility clearly was a factor in the cases in which the same household both gave and took in the same month. This was the case with household #3 in October 1975, when it both gave and requested permission to harvest cassava. The former 1975 Kenua' swidden in which they gave permission to harvest cassava was distant

TABLE 90
Seasonal Variation In Interhousehold Requests To Harvest Relishes

		Household						
		#3		#4		#16		
Month	Year	Request	Give	Request	Give	Request	Give	Total Requests
May	1975	1	2	0	0	1	0	4
June		0	2	0	1	0	0	3
July		0	3	0	0	0	0	3
August		0	1	0	2	1	0	4
September		0	0	0	0	1	0	1
October		1	1	0	0	2	1	5
November		1	1	0	0	1	0	3
December		0	1	0	0	0	0	1
January	1976	0	3	0	1	3	1	8
February		1	0	0	3	0	1	5
March		0	0	0	0	0	2	2
April		3	1	0	0	1	0	5

from the longhouse (in which the members of household #3 were then living) by one hour; whereas the former 1975 Kantu' swidden of household #4, in which #3 requested permission to harvest cassava, was distant from the longhouse by less than one-quarter of an hour. There was great variation among households #3, 4 and 16 in the degree of accessibility of relishes in the former 1975 swiddens to household members living in the 1976 swiddens. Household #4 was best off because its 1975 Kenua' upriver swidden adjoined its 1976 Kenua' swidden, and its 1975 Kenua' downriver swidden was within fifteen minutes of this latter swidden as well. Its 1975 Kantu' swidden, while distant from its 1976 Serangir swidden by twenty minutes, still was more accessible to this latter swidden than was any swidden belonging to another household. In the case of household #3, its 1975 Kenua' swidden adjoined its 1976 Kenua' swidden, but no 1975 swidden (of this same household) lay near its 1976 Lubuk Kepayang swidden. Finally, in the case of household #16, its 1975 Sebeliya' swidden was distant from its 1976 Lubuk

Menserai swidden by fifteen minutes and, in any case, was almost barren of relishes. Its 1975 Kenua' swidden, while richer in relishes, was distant from its Lubuk Menserai swidden by at least ninety minutes. Therefore, in terms of the relative accessibility of the relishes in their 1975 swiddens while living in their 1976 swiddens, household #16 had the greatest need to obtain relishes from the more accessible swiddens of other households, while household #3 had a lesser need, and household #4 had the least need of all. These differences are reflected in the ratios of taking to giving in table 89: 2/1, 1/2, and 0/7, respectively, for households #16, 3, and 4.

In addition to obtaining relishes from the swiddens of other households, there is a second way in which a household can fill a gap between its demand for, and supply of, harvestable and accessible relishes. This involves substituting uncultivated relishes for cultivated relishes as side dishes in the daily meals. These noncultigens are collectively called utai ti disumai 'things that are cooked', and the gathering of them is called ngega' 'look for'. The most important of these wild relishes are tubo' 'edible bamboo shoots', pako' 'edible ferns', and kulat 'edible fungi'[8]. The incidence of harvesting these wild relishes varies inversely with variation in the incidence of harvesting the cultivated relishes, especially at the time of the rice harvest (table 91)[9]. Thus, when the rice harvest begins the inci-

TABLE 91
Seasonal Variation In The Harvesting Of Cultivated Versus Noncultivated Relishes

Category of Relish	Number of Harvesting Days Per Month Per Household During		
	The last month preceding the rice harvest	The one or two months of the rice harvest	The first month following the rice harvest
Cultivated	16	26	16
Noncultivated	8	5	12

dence of harvesting cultivated relishes increases and the incidence of gathering wild relishes decreases; and when the rice harvest ends, the reverse occurs. The Kantu' prefer their cultivated relishes to the wild relishes. However, the cultivated relishes are accessible only when the Kantu' are living in or near their swiddens and not when they are living in the longhouse. In contrast, the noncultivated relishes are accessible when the Kantu' are living in the longhouse but not when they are living in the swiddens. Most of these noncultigens do not grow well in the mixed primary/secondary forest associations that surround the swiddens, whereas they flourish in the rubber groves that cover much of the land within one-half hour of the longhouse (figure 5). In the course of one year the average household gathers wild relishes on slightly more than one-half as many days as it harvests cultivated relishes, which suggests that the former are approximately one-half as important to household consumption as are the latter[10].

Chapter Eleven
House Making

During part of the year most Kantu' households experience certain needs for shelter, in particular the need for shelter in or close to their swiddens, that cannot be met by the longhouse alone[1]. This need for shelter, and the method of satisfying it, varies from one part of the swidden cycle to another.

SLASHING AND FELLING

The first need for shelter in or near the swidden is experienced during the slashing and felling stages. The men engaging in this work minimally require shelter in which they can hone their tools and smoke before beginning work in the morning, in which they can eat and again hone their tools during a midday break, and to which they can return in the event of a rainstorm. It is rare for a shelter to be built specifically for this purpose. Typically the workers make use of an already standing langkau 'farmhouse' in the vicinity of the swidden being cleared. A shelter need not be located in the swidden to satisfy these particular needs, and houses that are distant by five or even ten minutes are acceptable for this purpose. Workers even may use a nearby farmhouse belonging to another household, if the latter grants its permission.

Sometimes a household makes a swidden in an area in which no already standing house (whether its own or another household's) is available for its use. In such cases the household may construct a daytime shelter specifically for use during slashing and felling. These shelters may be called langkau tanah 'ground houses' - because they may contain no elevated floor but merely a plaited mat placed on the ground - langkau kajang 'mat houses' - because they are roofed not with hardwood shingles or even bark sheeting, but merely a coarse mat - or simply langkau penebas 'slashing houses'. They are distinguished not only by the construction of their roofs and floors, but also by their small size (viz., as small as three or four square meters) and by the frequent absence of any siding (figure 34). All of these characteristics suit the slashing houses to little more than shielding their occupants from daytime sun or rain. The Kantu' say that these are not the kind of shelters in which they can cook or sleep. The labor needed to construct such houses is correspondingly minimal. Even the better constructed examples of this type do not require more than 1.5 days of labor.

FIGURE 34
<u>Langkau Penebas</u> 'Slashing House'

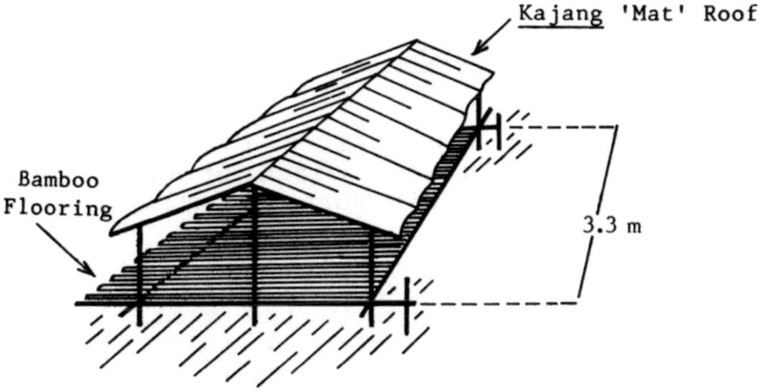

If a swidden is distant from the longhouse by more than about thirty minutes, the need for shelter becomes greater. At this distance a nighttime shelter is necessary to avoid spending an inordinate amount of time traveling back and forth between the swiddens and the longhouse. In the typical case, with a round trip between longhouse and swidden of ninety minutes, use of a nighttime shelter can save up to 102 hours of travel time during the slashing and felling of primary forest swiddens and as much as 51 hours during the slashing and felling of secondary forest swiddens[2]. The nighttime shelters differ from the daytime shelters in their greater size, better construction, and higher cost in labor. Nighttime shelters made at this stage are not as large as those made later in the swidden cycle, however. The Kantu' are reluctant to devote too much labor to house construction before a swidden has been burned. If the house is constructed within the boundaries of the swidden itself, it will be destroyed during (or require dismantling before) the burn. If it is constructed outside of the swidden, it will not completely satisfy the postburn requirements for shelter. In either case, therefore, there is some waste of labor, and with large nighttime houses the waste may exceed the amount of travel time saved by their construction. Therefore, whenever a household is compelled to construct a nighttime shelter during slashing and felling, it will make it smaller than normal.

A somewhat different solution to the need for shelter during slashing and felling involves use of the <u>dampa'</u> 'subsidiary longhouse'. A <u>dampa'</u> resembles the <u>rumah</u> 'longhouse' in design and differs from it only in the poorer quality of materials used, in its lesser dimensions (typically one-half those of a normal longhouse), and in its lesser number of <u>pintu</u> 'doors' or constituent households (figure 35)[3]. Until World War II, <u>dampa'</u> and not <u>rumah</u> were the principal dwellings of the Kantu'. They ascribe this to the fact that chronic warfare often forced them to abandon their dwellings, which made them reluctant to invest labor in the more substantial longhouses. The less substantial <u>dampa'</u> were used for a few years and then abandoned, whether necessitated by warfare or not, at which point the inhabitants made a new <u>dampa'</u> in another area. This settlement pattern maximized the defensive capabilities of the group, because it enabled the constituent households of the <u>dampa'</u> to make each year's swiddens close both to it and to one another as well. The Kantu' regarded proximity of workers and hence of swiddens as critically important during this historic period of warfare. This proximity could not have been attained by the households of a <u>rumah</u>, the greater permanency of which necessitates a larger territory and a greater scattering of people and swiddens[4].

After World War II the incidence of warfare declined and so too did the defensive advantages of the <u>dampa'</u>, as a result of which the Kantu' began to make <u>rumah</u> instead as their principal residences. <u>Dampa'</u> are still made upon occasion, however, but not for all of the same reasons as formerly. The present use of subsidiary longhouses resembles the Iban system described by Freeman[5]. A <u>rumah</u> is used as the principal, central dwelling of the group, and <u>dampa'</u> are used as subsidiary dwellings close to each year's swiddens. The Kantu' say that they

FIGURE 35
<u>Dampa</u>' 'Subsidiary Longhouse'

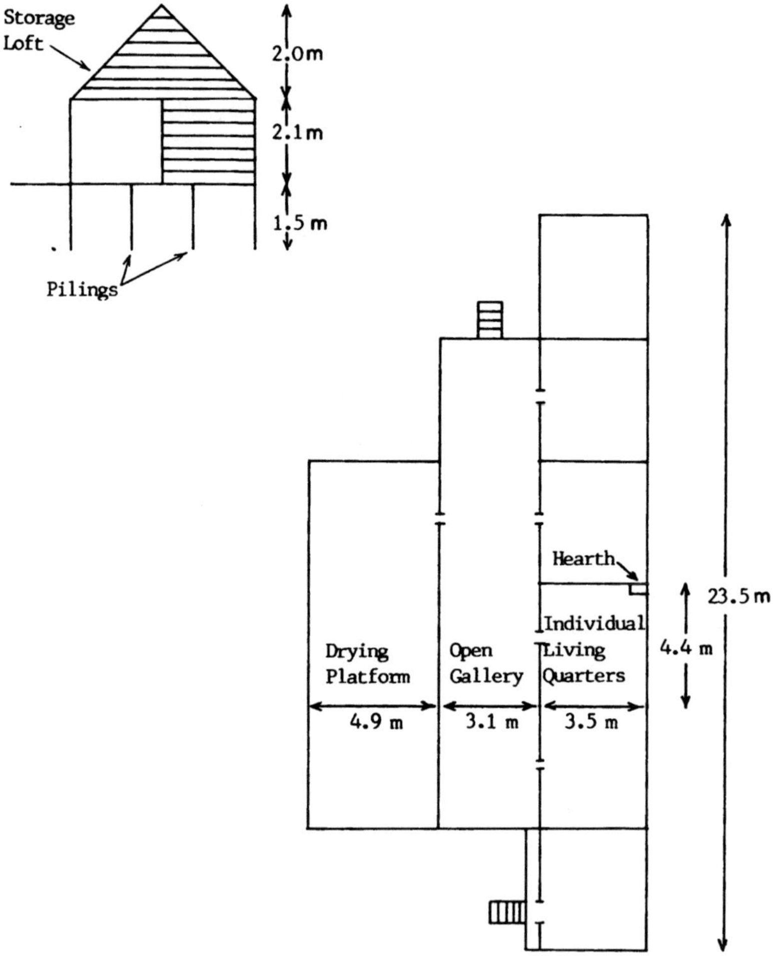

only construct a subsidiary longhouse when they make their swiddens in forest that has never been farmed and that is far from the main longhouse. The subsidiary longhouse always is made before the clearing of this forest begins, so that it can be used throughout slashing and felling. The last subsidiary longhouse made by the Kantu' of Tikul Batu is typical. This was made in 1971 and occupied for three farming years. It was located in primary forest near the headwaters of the Kenua' stream, at a distance of approximately one hour from the longhouse Tikul Batu. Like all _dampa'_ it was relatively small. At that time there were fourteen households in Tikul Batu, and only six (viz., households #2, 3, 4, 5, 6 and 8) participated in its construction and use.

The construction of a subsidiary longhouse under circumstances such as these serves several functions. First, the Kantu' say that it heightens the security and well-being of the women and children. This is a particular concern during slashing and felling because these are largely male activities and separate the men from the women and children. This situation is improved by the construction of a subsidiary longhouse, which can be located close to where the men are working and also affords the women and children of any given household the mutual protection of women and children in a number of other households. The year 1971, in which the Kantu' of Tikul Batu made their last subsidiary longhouse, was a time of general insecurity because of insurgent activity in this area. To some extent, therefore, the former defensive purpose of the subsidiary longhouse still is relevant to its construction.

The Kantu' also say that use of a subsidiary longhouse obviates the necessity of carrying in the harvest. Households that build a subsidiary longhouse usually store the year's harvest there, instead of carrying it back to the main longhouse. This is possible for two reasons. First, the plurality of households increases the likelihood that someone always will be present, which makes the storage of rice there a safer prospect. Second, again because of the plurality of households, the subsidiary longhouse is more acceptable as a residence for extended periods of time. The presence of other households means that the social life of the _dampa'_ approximates that of the _rumah_, as a result of which households living in the former feel less inclined to return to the latter during slack periods in the swidden work. The subsidiary longhouse may be contrasted to the single household swidden house, which provides less safety to rice stores and less social life to the household, and which (for these reasons among others) typically is not used for storage of the harvest.

A probable third benefit of making subsidiary longhouses involves savings in time spent traveling back and forth between the swiddens and the longhouse. When clearing forest for the first time in an area - which is the normal circumstance in which subsidiary longhouses are made - travel time potentially is burdensome. This is so both because more workdays are needed to clear primary than secondary forest, and also because there are no old swidden houses in such areas in which the workers can spend their nights. For both reasons the construction of a dampa' is desirable before beginning to slash and fell the forest in a virgin area. Because subsidiary longhouses are located between as opposed to within the swiddens of its constituent households, there is no waste of labor as in the preburn construction of single-family, single-swidden houses.

A final advantage of making dampa' is their greater use life: the typical subsidiary longhouse is used for three years, whereas the typical, nighttime single-family farmhouse is used for just one or two years. As a result, although two-thirds more labor is required to construct one household's section of a subsidiary longhouse than to construct a single farmhouse, the average yearly cost of the former is almost one-quarter less than the average yearly cost of the latter[6].

BURNING AND PLANTING

There is a need for shelter during both redressive burning and then planting. Because these operations follow so closely upon burning, there is sufficient time only to build a day house, similar to the ones sometimes built during slashing and felling (none of which would survive the burn, of course.) The day houses built for burning and planting tend to be somewhat larger, however, ranging in floor area from six to as much as twelve square meters. This greater size is particularly necessary in swiddens planted by large, multihousehold work groups, since the day house will be used for the storage, preparation and perhaps consumption of the day's food and drink. This greater size also reflects the greater proportion of the household's members involved in swidden labor after the burn, compared with the small proportion involved before it. Both sexes participate in redressing burning and in planting. Also, as the household's women come to the swidden so too do its children, and a place is needed in which to shelter them during the day while the able-bodied members of the household work in the swidden. The Kantu' try to minimize the exposure of their children to the elements, believing that wind and rain can make a child sick and the sun's rays can make a child dark (of skin) and unattractive. These several purposes are best served by a house located right in the swidden, and such a house usually is built, even if there is an old house outside of but proximate to the swidden in question.

WEEDING

Weeding presents similar needs for shelter, needs that are heightened by the relatively large number of workdays involved in this swidden stage. The Kantu' can rely solely on the longhouse to meet these needs only in the case of swiddens that are both very close to the longhouse (viz., within a thirty-minute walk) and very limited in size (viz., considerably less than one hectare in area). This applied to fewer than 20 percent of the 1975 and 1976 swiddens in which weeding was performed. In the remaining cases some type of swidden shelter was needed, the particular type - daytime or nighttime - depending, again, upon swidden size and, especially, the distance from the swidden to the longhouse (table 92)[7]. In the more distant swiddens there

TABLE 92
Swidden-Longhouse Distance And The Use Of Daytime Versus Nighttime Shelter During Weeding

		Swidden-Longhouse Distance	
		< 30 minutes	> 30 minutes
Shelter for swidden furnished by	Day house alone	7	0
	Night house (instead of or in addition to a day house)	8	24

N = 39 1975 and 1976 secondary forest swiddens that had access to a farmhouse.
P = .0004

is a particularly great need for nighttime shelter during this stage of the swidden cycle. Weeding is carried out in the open in direct exposure to the sun, the discomfiture of which is readily acknowledged by the Kantu' themselves. Accordingly, the first and last parts of the day, when the sun's rays are most oblique and weak, are preferred for weeding. The mean temperatures just after sunrise (6:00-8:00) and just at sunset (17:00-19:00) are lower by 9.2° C and 2.8° C, respectively, than the temperatures during the hottest part of the day, the early afternoon (table 9). If the household members can spend the night in or near their swiddens, they can work for part of this time just after sunrise and just before sunset, thereby taking advantage of the coolest of the daylight hours. If they have to spend the night in the longhouse, however, most of these early and late hours will be spent in preparing for, undertaking, and recuperating from the trip between the longhouse and the swidden.

When nighttime shelter is needed in a swidden, the owners may either construct a new langkau umai 'swidden house' within the swidden or use a nearby langkau mudai 'secondary forest house' standing in the regrowth on a previous year's swidden site. Langkau mudai, which can be glossed as 'old swidden houses', comprised ten of the thirty-two cases in 1975 and 1976 in which nighttime swidden houses were used during weeding. In four of these cases, a single old house provided shelter for not one but two separate swiddens belonging to the same household. The Kantu' are anxious to use old night houses whenever possible because this obviates the need to expend labor on the construction of a new house. This savings in labor is important enough to make the presence of old houses a factor in the selection of new swidden sites, as noted in chapter one. Whether or not an old house can be used for a given swidden is a function of the distance involved. If an old house is located within six to eight minutes' walk of a secondary forest swidden, it probably will be used. If the distance is greater than this, it probably will not be used. The reuse of an old house also is a function of its age. After the first year in which a night house is made and used, it is further usable for a second and at most third year. During the second and third years of use, an additional three to four workdays each year is needed to repair weather damage to the bark siding and the bamboo drying platform. After the third year of use, however, deterioration will have proceeded too far for restoration to be practical. A four-year old house is never reused, no matter how favorable its location.

In the event that all standing houses are too old or too distant to be utilizable, the household needing nighttime shelter must construct a new house. The first step is to select a location within the swidden. The primary consideration in this decision is proximity to a water source (typically a small brook), from which the household members can take water for consumption, and in which they can bathe, launder their clothing, and soak rushes and hardwoods to be used in plaiting and carving. A second important consideration is proximity to the trail that leads from the edge of the swidden to the principal trails leading to the longhouse. The importance of proximity in both of these cases is to minimize the walking time over the most often traveled trails. These considerations result in about one-half of all houses being located right at the swidden's edge, while the other half are located somewhere within the swidden. Rarely is a house located beyond the swidden's edge, in or under the eaves of the adjoining forest. While this would have the advantage of affording some protection from sun and rain, of more importance to the Kantu' is their belief that such a location is conducive to illness. In addition, such a location makes it more difficult to protect the swidden against crop pests.

After selecting a location, the next step consists in marking out the dimensions of the house, using long poles laid on the ground. The typical night house is roughly square in configuration, with a floor area of thirteen to fourteen square meters. There is some variation above and below this figure, due to variation in the size of the household and the planned number of years of use. Some variation in size also is due to variation in the household's supply of roof shingles. After the dimensions of the house have been marked out, construction commences with the frame. This is made with logs, trimmed but otherwise unfinished, selected from amongst the half-burned or unburned timber lying on the surface of the swidden. The uprights are secured in the ground by repeatedly being lifted up and thrust forcefully down, and then twisted back and forth in the holes thus made. Horizontal poles are lashed to these uprights with lengths of vine. Natural forks in the uprights are used for the cross poles whenever possible. The flooring of the house is located at one <u>depa</u>' 'fathom' (the arms' span of a man) above the ground. It is made of freshly cut bamboos, split to produce relatively flat slats. The slats are laid down in parallel rows and lashed into large sheets, each of which then is tied down to cross poles.

The sides of the house extend upwards from the flooring for one fathom. The siding is made from bark sheets that are holed and then lashed to cross poles. The siding is broken by only a single doorway, which is covered when necessary with a sheet of bark. There is a gap of twenty to twenty-five centimeters between the top of the bark siding and the roof. This admits light and air into the house, while the long overhang of the roof (extending beyond the house sides by another twenty to twenty-five centimeters) keeps the rain out. The roof of the house slopes down on both sides of a horizontal ridge pole that is located directly above a line bisecting the house. The ridge pole is positioned one fathom above the house's sides (and hence two fathoms above the floor and three fathoms above the ground). This height produces a pitch to the roof of about forty degrees (departing from the horizontal) in the typical house. The Kantu' use this steep pitch, despite the consequently high ratio of roof area to floor area, in part because a shallower pitch is less able to withstand the force of the torrential rains. In addition, the high ceiling that is produced by this steep pitch makes it cooler inside the house and provides needed room for threshing. The roof is made from wood shingles. Each shingle is one-half to one centimeter in thickness, as wide as a man's hand, and four times that in length. Each shingle is holed, laid down (rather than across) the slope of the roof, and then lashed individually with rattan to cross poles. The shingles are put on in parallel rows running across the slope of the roof, with each higher row partially overlapping the row below it. The roof extends down to and beyond two sides of the house. On the other two sides (those perpendicular to the ridge pole), there is a triangular gap between the roof and the siding. In some houses, these gaps or eaves also are covered with bark sheets. Ideally, however, they are covered with overlapping rows of large leaves, each leaf holed and lashed to cross poles. The use of leaves rather than bark sheets is preferred because of their superior ventilation properties.

The frame, flooring, siding and roofing constitute the major structural features of the house. In addition, a storage loft or platform may be constructed, running between two sides of the house at the point where these meet the roof. The loft may cover as much as one-half of the house's floor space, and always is located against the side opposite the doorway. A hearth is the other principal feature of the interior. It consists of a box constructed of poles on the floor in one corner of the house. It is filled to a depth of five or six centimeters with earth, atop which the cooking fires are made. Half-way up the house's side, above the hearth, a rack for drying and smoking foodstuffs and other materials is made. To one side of the hearth another rack is constructed, to serve as a storage place for the household's cooking and eating utensils. Finally, a large log is cut to length, notched for steps, and laid against the doorway to provide access to and from the ground. The log is dropped to the ground when all household members are away, to inhibit access by either domestic or feral creatures.

This construction, which is carried out by the adult and juvenile males of the household, consumes an average of sixteen workdays and is not undertaken lightly[8]. This is reflected in the rarity with which any given household constructs two such houses in a single year. Among the fourteen households that farmed in 1975 and the sixteen that farmed in 1976, twenty-two made new night houses; and in only four of these cases did one household make as many as two houses in a single year (and in no case did one make three). Moreover, in four cases a household made a single night house fulfill its need for shelter in two separate swiddens (figures 11, 12). The swiddens in these cases were separated by an average of ten to fifteen minutes, and the night house always was located in the one in which the need for shelter was greatest (viz., the larger swidden and/or the one cut from secondary forest - where weeding is required - as opposed to primary forest - where little or no weeding is done).

The construction of night houses is costly not only in terms of labor, but also in terms of materials, referring in particular to the bark sheets used for siding and the wood shingles used for roofing. Both items are scarce goods today. This scarcity is not a function merely of the considerable labor that is required to fashion siding and shingles from the raw materials (e.g., three workdays to fashion the shingles for a night house of average size). Rather, the raw materials themselves are scarce. It now requires a special trip of several days duration, either downriver or into the hills, to find trees with the particular bark or wood needed to make siding or shingles. The high cost of finding and then fashioning these two items is lessened by their reuse, however. After a given lot of siding and shingles have been used for one, two or even three years on one house, and that house finally is abandoned, they are salvaged and used in other houses for an additional number of years.

GUARDING

The need to guard swiddens against the rice sparrow, in particular, creates special needs for shelter. In a swidden in which this pest poses a real problem, even if a night house has already been built, the household is likely to build one or more langkau penginang pipit 'houses for guarding against rice sparrows'. Members of the household sit in these shelters and work the 'chase lines' used to scare off the sparrows. With a roof made of matting, no siding, and a floor area of only one or two square meters, their construction is a matter of just one-half workday apiece.

HARVESTING

The harvesting of the rice crop presents the greatest need of all for shelter in the swiddens. At this stage there is a need not only for a place in which the household members can take shelter, cook and sleep, but also for a place in which they can store and process their rice crop before carrying it back to the longhouse. In the secondary forest swiddens these needs are met with houses that were constructed earlier in the swidden cycle (or during some previous cycle). But if no house was constructed earlier in the cycle, none is constructed now. This is not true in the primary forest swiddens, in which the harvest represents the first stage in the cycle in which there is both a need for a full-fledged house and an opportunity to construct one.

Of twenty-two primary forest swiddens made in 1975 and 1976, all but one had access to a nighttime house during their harvests. In thirteen of these cases an old house was reused. The Kantu' are more likely to use old houses, in particular more distant old houses, for primary forest swiddens than secondary forest swiddens. This is due to the lesser number of work days that are spent in the former during weeding, guarding, and harvesting. (However, when using a relatively distant old house for a given primary forest swidden, especially if it is a large swidden, the Kantu' may construct a day house in the swidden as well.) In the remaining eight cases, a night house was constructed in the swidden at the time of the harvest. Completion dates ranged from ten days prior to the start of the harvest to several days afterwards. The lateness of this construction and the relative brevity of the houses' use (viz., during the harvest and harvest processing alone, compared with use during weeding and guarding as well in secondary forest swiddens) is counterbalanced by the greater swidden area served by them (viz., primary forest swiddens average one-third larger than secondary forest swiddens).

In both secondary and primary forest swiddens, the open platform for drying and winnowing grain is added on to night houses just prior to, during, or just after the harvest. (If an old house is being reused, its platform is restored at this time.) This platform is built at the level of the farmhouse's floor, along the side of the house containing the doorway. It is constructed of bamboos lashed down in rows parallel to the side of the house (which enhances the footing of the people crossing it on their way to or from the doorway). The platform typically covers an area equal to approximately two-thirds of the floor area of the house itself, and it requires approximately 3.5 workdays to construct.

CARRYING

There is no need for shelter during or following the last phase of the swidden cycle, carrying, unless the harvest is stored in the swidden. Usually only night houses are used for such storage. A day house is rarely used for such prolonged storage other than of seed rice, when it is called a langkau benih 'seed house' (e.g. the day house in swidden #27.) In any case, no swidden house is ever constructed for the purpose of storage alone[9]. However, the anticipated need for a place in the swidden in which to store rice may be one important factor in the initial determination, earlier in the swidden cycle, to build a house or not.

From selection through carrying, the average overall labor costs of shelter in each type of swidden are 3.2 days per hectare, 4.0 days per hectare and 13.0 days per hectare in primary forest, secondary forest, and swamp swiddens, respectively[10].

PERENNIAL FACTORS

There are additional considerations in the construction and usage of swidden houses that are not specific to any particular swidden stage. One such consideration is intrahousehold amity in households containing not one but two married couples (viz., typically one couple belonging to a junior generation and another belonging to the senior, ascending generation). The tensions that sometimes exist between two such couples are mitigated when each couple lives and works in a separate swidden with its own night house; and indeed, there appears to be some association (although not established at the .05 level) between the presence of two couples in a single household and the use of two night houses in a single year (table 93)[11]. The state of relations not

TABLE 93
The Number Of Couples And The Number Of Swidden Houses Per Household

		Number of couples in the household	
		One	Two
Number of night houses used by the household in a given year	One	12	5
	Two	4	9

N = 14 households in 1975 plus 16 households in 1976.

$X_c^2 = 3.23$

$P < .10$

within but among households also is a consideration in the construction and use of swidden houses. Certain households, the members of which are chronically unable to get along with some or all of the other members of the longhouse, try to spend no more time in the longhouse than is absolutely necessary (although continuing to maintain their official residence in the longhouse). To this end, these households construct and use night houses in some circumstances in which other households would live in the longhouse. Household #6 exemplifies this misanthropic type of household. Bungin (its senior male) characterized himself as Urang ti diau di langkau umai 'One who dwells in swidden houses', as compared with Urang ti diau di rumah 'One who dwells in the longhouse'. Household #6 is atypical in the extent to which it has renounced life in the longhouse in favor of life in the swidden house, but all households utilize their swidden houses to some extent as a refuge from the problems of longhouse life.

One of the foremost of these problems is the disputed ownership and outright theft of chickens. The fowl belonging to one household often cannot be distinguished from those belonging to another (with the exception of the large and distinctive fighting cocks). In addition, the Kantu' rarely cage their fowl but allow them to range free so that they can seek some of their own food. Given these circumstances, it frequently happens that one household will take a fowl actually belonging to another household and sacrifice it in a ritual offering. The threat of this causes each household to keep its domestic fowl at its swidden house for as much of the year as possible[12]. There are other reasons as well why the Kantu' prefer the environment of their swiddens for raising fowl. The Kantu' say that the vicinity of their longhouse Tikul Batu, which lies only eighty-five meters from the Empanang river, is so muddy as to jeopardize the survival of the young chicks. This is not true of the swiddens, most of which are located on higher and drier ground. In addition, the swiddens provide the free ranging fowl with a better food supply. If there are grain bearing rice plants in the swidden, this grain is available to the fowl for consumption. Of greater importance, any insect pests in the swidden also are available for their consumption. The Kantu' say that they doubly benefit from this situation, in that their fowl are fed and the pest populations are reduced. They also say that fowl eat certain young grasses (especially lalang 'Imperata cylindrica [L.] Beauv.) that otherwise would require weeding.

The presence of fowl in the swiddens necessitates the presence of the swidden owners as well. While predators rarely trouble fowl in the vicinity of the longhouse, due to their timidity near this center of human activity, this is not the case in the swiddens. Fowl commonly are taken in the swiddens by lang 'Brahminy Kites', iba' 'serpents' munsang 'civets', menturun 'bearcats', and ngkulau 'clouded leopards'. Rare is the household that loses no fowls to these predators during a given year. Nonetheless, the Kantu' say that their 'hearts' feel better if they keep their fowl in the swiddens and lose them to predators, than if they keep them at the longhouse and lose them to thieves. Notwithstanding their willingness to accept these losses, they still do everything possible to guard their fowl, and this has a marked impact on the extent to which they occupy their swidden houses. When fowl are being kept in a swidden, one or more members of the household usually will stay there as well, even in the absence of any swidden work, simply to guard them. The Kantu' commonly explain their presence at a swidden house between stages of swidden work, by saying Kami nginang manok 'We are guarding the chickens'.

A final consideration of residence in the swidden houses is the freedom thereby afforded from the ritual proscriptions attendant upon residence in the longhouse. Whenever a given household sponsors a curing ceremony or makes a major placatory offering of any kind, all of the other households in the longhouse must observe (to some extent) the attendant pantang 'proscriptions'. Typically, this involves a proscription against any work in the swiddens on the following day. All households must observe similar proscriptions for major rituals staged in neighboring Kantu' longhouses as well. In all cases, however, observance of a proscription is predicated upon knowledge of the ritual giving rise to it. Observance of proscriptions for a ritual that is staged at the longhouse or that is made known to the inhabitants of the longhouse is expected only of those people sleeping in the longhouse at that time, not of those sleeping out in the swiddens. The Kantu' say, Laban sida' ti bemalam d'umai nadai nemu kerja kitai dirumah, sida' nadai masuk pantang 'Because those who are spending the night in the swiddens do not know of our [ritual] activities at the longhouse, they are not included in the proscriptions'. In practise this rule is observed more in the letter than the spirit. If one household plans to commence a curing ceremony in the longhouse in the evening of a given day, for example, any other household that leaves the longhouse before nightfall to sleep in its swidden house is exempted from the following day's proscription: and this often appears to be less an effect than a cause of departure from the longhouse. The more misanthropic households are particularly noted for such timely departures.

At times in the past the swidden houses have served the Kantu' as more than a base for farm work and a refuge from social obligations. During the past century the Kantu' have had to abandon longhouses on several occasions (due to ill omens, illness, warfare, or other factors), and on each occasion several years passed before construction of a new longhouse could be completed. During these intervals between longhouses, each household lived the year-round in a swidden house. This demonstrates, therefore, that exclusive reliance upon the swidden house for all shelter needs is possible. It also demonstrates that the Kantu' live in longhouses, whenever possible, because of more complex social and economic needs.

Chapter Twelve
Tool Making

The various tools used by the Kantu' in their swiddens can be grouped into three basic categories, based on the method of manufacture: tools that are plaited, tools that are forged, and tools that are hewn.

PLAITED TOOLS

The Kantu' make a wide variety of articles by plaiting the stalks and leaves of local plants. These products are collectively called utai ti' d'anyam 'things that are plaited'. Most of them are used extensively within the swidden cycle, in particular in the processing, carrying and storing of rice (table 94). Among the eighteen distinct tools listed in table 94, eight are used exclusively for these swidden-related tasks and nine are used predominantly for them (the single exception

TABLE 94
Plaited Swidden Tools And Their Uses

Rice Processing	Rice Carrying	Rice Storage	Other
Baka' 'Basket'	Bangkat 'Basket'	Belangsai 'Bag'	Tongui 'Sun Hat'
Capan 'Winnowing Tray'	Ladung 'Backpack'	Kelekit 'Basket'	
Kajang 'Roofing'	Lanyi 'Basket'		
Kelabingkai 'Winnowing Fan'	Ligit 'Basket'		
Kelaya' 'Mat'	Raga' 'Basket'		
Pelangkus 'Threshing Grill'	Sedung 'Basket'		
Samir 'Awning'	Takin 'Basket'		
Timpa' 'Basket'			

being the kelaya' 'mats', most varieties of which are used for nonswidden purposes). The manufacture of these tools, therefore, is in effect an additional aspect of the swidden cycle.

The Kantu' at Tikul Batu acquire some plaited goods through trade, chiefly the ladung 'backpack', pelangkus 'threshing grill', and some types of kelaya' 'mats'. These usually are obtained from the Sekapat and Remai, Kantu' subgroups that live in the hills forming the watershed to the southwest of the Empanang river valley. Prices (or the equivalent in trade goods) vary from 225 to 600 rupiah for mats, 500-1000 rupiah for a backpack, and 1050-1500 rupiah for a threshing grill. Most Kantu' at Tikul Batu know how to make all of these articles. There are several different reasons why they may obtain them through trade instead. First, in the hills where the Sekapat and Remai live, there is a greater abundance of the varieties of rattan preferred for plaiting than there is in the territory of Tikul Batu. Second, the inhabitants of the hills also have more time available for this work. The Kantu' at Tikul Batu often say that they purchased some plaited article from the hills only because they had no time to make it themselves. These Kantu' have access to more profitable means of earning cash - in particular by tapping rubber or by engaging in trade - than do the hill peoples - who have less rubber and who are more removed from the main trade routes to Sarawak.

The Kantu' at Tikul Batu trade only one article of their own manufacture, namely the tongui 'sun hat'. The Kantu' plait the base of their sun hats from pandanus leaves, and then they sew (using thread made from the leaf of the pineapple plant) broader leaves from a certain palm over the topmost side, thus ensuring that the hat keeps off rain as well as sun. The Iban groups living between Tikul Batu and the Indonesian-Malaysian border know how to plait the base of the hat, but not how to sew on the overlapping leaves, and so their hats are not rainproof. These groups always are willing to buy sun hats from any Kantu' traveling through their territory; but this trade is intermittent and incidental, not matching in importance the trade between the Kantu' and the Sekapat and Remai. Even this latter trade does not match in importance the manufacture of woven goods within the household for the use of the household.

Materials from a number of different plants, both cultivated and noncultivated, are used in plaiting swidden tools (table 95). The noncultigens used in plaiting can be taken freely by anyone from wherever they are

TABLE 95
The Plants Used In Plaiting

Cultigens	Plaited Tools In Which They Are Used
Beman 'a vine'	bangkat, capan, ligit, raga', takin
Buloh 'bamboos'	raga', takin
Perupok 'pandanus'	belansai, kajang, kelaya', lanyi', tongui
Wi 'rattans'	ladung, lanyi', takin, timpa' (plus minor use in the rim of most tools)
Noncultigens	**Plaited Tools In Which They Are Used**
Buloh 'bamboos'	raga', takin
Gerenih 'a palm'	takin, tongui
Purun 'a sedge'	belansai, tongui
Singan (unidentified)	baka', capan, kelabingkai, kelaya', kelekit, raga', takin, timpa', tongui
Tando' 'a rattan'	lanyi', takin
Wi 'rattans'	capan, ladung, lanyi', pelangkus, sedung, takin, timpa' (plus minor use in the rim of most or all tools)
(Various trees)	Bark is used for cord, straps, and siding; wood is used for struts or siding.

found growing, even from secondary forest or rubber groves belonging to another household or another longhouse. However, cultigens planted by a household in its swiddens, secondary forest, or rubber groves cannot be harvested by another household without prior permission. As a rule, only one of the four cultigens, pandanus, is sufficiently scarce to justify seeking such permission. Pandanus is scarce in part because its planting is ritually proscribed for anyone other than a bujang 'bachelor' or ino' darah 'blood woman' or maiden. Many Kantu' marry without having planted pandanus and, consequently, must ask other households for permission to cut pandanus when some is needed. A request for pandanus or for any of the other cultigens used in plaiting cannot be refused if both households belong to the same longhouse. Nor is any compensation ever made to the donor household.

After a given plant has been cut and gathered, it is subjected to a series of preparatory operations. The series for one important plant, singan, is as follows: (1) ngancau dua ari 'dry for two days' in the sun, (2) ngerining 'slice off' all ancillary growth, leaving just the single main trunk of the plant, (3) mai' kerumah 'carry to the longhouse', (4) ngupak 'peel' each leaf off of the main stalk, (5) ngulong 'roll' the two ends of each leaf together so that it forms a circle, with the curve against the natural curve of the leaf, and then tie it, (6) nusok 'thread' lengths of rattan through numbers of these circles so as to tie them into bunches, (7) ngancau tiga ari panas terit 'dry for three continuously hot days', (8) ngemai 'expand' by soaking in the river, (9) ngempat pala' 'cut the head' or tied ends away from each leaf, and (10) mila' split' each leaf, the number of divisions being determined by the fineness of plaiting needed for a particular piece of work. After this the singan is ready to be plaited.

Plaiting itself consists of three named stages, namely (1) nentan 'start' in which the first steps towards making a piece are taken, establishing the basic outline of the work to be done, (2) nganyam 'plait' in which the main body of the piece is plaited, and (3) nemo' 'finish' in which the plaiting is completed and the edges are finished off. For most plaited goods nemo' is followed by nelas 'lash', an operation that is distinguished from nganyam 'plait' as narrowly defined. Nelas consists in affixing strips of rattan or bamboo to the rims and edges of plaited goods to strengthen them. Niang 'make a post' is a second supplementary operation, which consists in making and affixing vertical posts to the longer carrying baskets (sedung and lanyi') to give them greater strength. Another distinct operation is nusok 'thread', used in cases for which nganyam is not suited, as in making awnings (viz., kajang or samir). It consists in threading together long pandanus leaves with a length of rattan. Ngelalin 'interlace' is another alternative to nganyam, performed on the sides and bottoms of certain carrying baskets (e.g., bangkat, ladung and sedung). The equivalent parts on all other baskets are made by the nganyam 'plaiting' technique. Interlacing differs from plaiting in that it produces a much coarser plaiting, with gaps of one to two centimeters. Finally, there is a distinct operation for mending plaited tools, called nyilup 'insert', which consists in plaiting new plant material into worn areas on an article.

The finished, plaited article may be used jointly by the entire household or it may be used customarily by just one of its members. The latter usually is the case with tongui 'sun hats', ladung 'backpacks', raga' pemenih 'sowing baskets', and takin pengetau 'harvesting baskets'. For example, if there are four persons in a given household who can harvest, the household will own four different harvesting baskets, each one customarily used by just one harvester. The household member who normally uses a given plaited article is said to mpu 'possess' it. Another household member has the right to use that article only when it is not being used by this customary user. The ownership of plaited goods ultimately resides with the household as a whole, however, and not with the individual household members. Thus, household members who marry out of the household are not allowed to take with them any of the plaited goods that they customarily used. Out-marrying household members may receive gifts of various kinds upon departing, but only barang beteras 'things that have heartwood', meaning things made of iron, bronze, silver or gold, not things plaited from rushes, bamboos or rattan. Only in household partition can the members of each of the resulting households take with them the particular plaited articles that they have been accustomed to use.

When the use life of plaited goods is finished, after just one or two years in most cases, they are discarded. Until recently, however, there was a ritual proscription against discarding such goods without first making an offering of thanks and propitiation. The Kantu' based this proscription on a story of a couple who threw out all of their plaited goods to make room in their swidden house for a bumper crop of rice. The gods saw this as rua' 'wasteful' and caused the crop to disappear in retribution.

The specific operation of nganyam 'plaiting' always is performed by women, as is mending; while interlacing, lashing, post making and threading all are performed by men. The men also may assist in gathering and transporting the various plants used in plaiting, but they will not assist in preparing them for use. The role of gender in this division of labor is especially clear in the case of lashing versus plaiting. When both operations are performed on the same tool, the Kantu' explicitly speak of lashing as the proper male contribution to its construction and plaiting as the proper female contribution. Overall, the proportion of work that is male specific is quite small, with the result that the total contribution of men to plaiting (broadly defined) is much less than that of women. During the 1976 swidden year, in households #3, 4, and 16, females contributed an average of 82 percent of all plaiting labor, while males contributed an average of just 18 percent. To a lesser extent, there was variation by age as well. Among males, the senior married men were more likely to find time to work on plaited goods then were the junior married men. The reverse was the case among females: the younger married women tended to contribute the most labor to plaiting.

A girl first learns the rudiments of plaiting from her mother or mother-in-law. She may later ask for instruction from more learned women in other households. In any case, her instruction must proceed by ringkat 'stages'. This especially applies to learning to plait kelaya' 'mats'. The different types of mats are ranked according to the level of skill needed to plait them, the more difficult mats being those that are larger or contain designs. Women are proscribed from attempting to plait a mat of a given stage without having first learned how to make the mats of all lower stages, in order. Further, if a woman has a mimpi jai' 'bad dream' while in the midst of plaiting a particular type of mat for the first time, she must pull apart what she has done and never again attempt to plait that type of mat. The basis for these proscriptions is an association of personal danger - to the worker and perhaps to her household as well - with the more difficult plaiting techniques[1]. Because of this association, there are some techniques that no woman at Tikul Batu, even the oldest and most skilled, will attempt. For example, there are no tongui lelambak (sun hats covered with designs) in Tikul Batu today, although the skill to make them is possessed by a number of women[2]. The making of this type of hat is proscribed for anyone other than a tungul pemandai 'leader of skills', a status of which no contemporary woman is sufficiently confident to attempt to plait this hat.

Women (as well as men) tend to concentrate their plaiting labor during certain times of the year (table 96)[3]. This seasonal variation is determined by two factors,

TABLE 96
Seasonal Variation In The Labor Devoted To Plaiting

Percentage of Year's Plaiting Labor Performed Each Month												
Jan.	Feb.	Mar.	Apr.	May	June	July	Aug.	Sep.	Oct.	Nov.	Dec.	Total
25	12	8	15	10	1	4	8	3.5	6	2.5	5	100%

one of which is the varying ability of the household during the year to devote its labor resources to making plaited goods. This ability largely varies according to the demands of other activities upon the labor of the household's women. The greatest seasonal demands upon female labor occur during the planting of the new swiddens in September; their weeding during part of October, November, and part of December; the harvest during part of February, March, and part of April; and the preparation and staging of the postharvest feast during part of June and part of July. There is little or no demand for female labor during the months of January, May and August, and table 96 shows that fully 43 percent of the year's plaiting labor is carried out during these three months.

The second factor accounting for seasonal variation in plaiting labor is the varying need for plaited goods during the year. The greatest need for them during the entire swidden cycle occurs during the harvest and harvest processing. As table 96 shows, this need is responsible not only for the annual peak of plaiting labor just before the harvest season begins, but also for the continued devotion of relatively large amounts of labor to plaiting throughout this season, in spite of the conflicting demand for female labor in the harvest itself. This conflict is resolved by scheduling much of this plaiting labor at night (table 97)[4]. Table 97

TABLE 97
Seasonal Variation In Plaiting During The Day Versus Night

Percentage of Each Month's Plaiting Labor Performed at Night											
Jan.	Feb.	Mar.	Apr.	May	June	July	Aug.	Sept.	Oct.	Nov.	Dec.
33	56	72	40	45	9	16	19	49	59	83	50

shows that the percentage of plaiting labor done at night rises to high points not only during the principal harvest month (viz., March), but during the principal weeding month as well (viz., November). This percentage drops to low points during January and August, when female labor is not being used during the day in the swiddens and consequently can be turned to plaiting. The lows during June and July are explained by the fact that much of the preparation and staging of the postharvest feast takes place at night and hence conflicts directly with any other nighttime usage of female labor.

The total amount of labor devoted to plaiting varies from year to year and from household to household according to the size and type of the household's swiddens. These two variables are determinant because they are broadly correlated with harvest size, and a large proportion of these tools are used in processing, carrying and storing the harvest. I estimate that 13.0 workdays are devoted to plaiting for each hectare of primary forest under cultivation, 21.4 days for each hectare of secondary forest, and 37.7 for each hectare of swampland[5].

FORGED TOOLS

The second major category of swidden tools is called utai ti ditempa' 'things that are forged'. These comprise the duko' penebas 'brush sword', beliong 'adze', bingkong 'weeding sickle', lungai 'plaiting knife' and senapang 'gun'. The craft of hand forging tools is long established among the Kantu'. Indeed, they often are cast in the role of ironsmiths to neighboring groups. They sell firearms to Iban living upstream from them (often accepting roof shingles or canoes in trade), and they sell firearms and occasional brush swords to Malays and Islamicized Dayak downstream (often accepting sawn timber in trade). Neither of these latter groups possesses any knowledge of hand forging iron. The origins of their metallurgy are now beyond the memory or oral history of living Kantu', with one exception. They say that they first learned how to make firearms by copying the matchlocks brought into their area by Dutch troops during the colonial era. In the mid 1960s they began making breechloading shotguns and sold most of their matchlocks to neighboring groups. They acquired this new technology from a traveling gunsmith from a related group in the Ketungau river valley.

The Kantu' forge their tools from scrap iron (often old automobile springs) obtained from the border bazaars of Sarawak or, less frequently, from the distant provincial capital, Pontianak. This besi 'iron' is worked in a charcoal fire, the fuel for which is made by burning stacked tiers of ironwood. The fire is fanned with a peputan, a bamboo piston bellows. In a typical sequence the iron is heated, then ditempa' 'beaten' into shape with hammer and anvil, dikikir 'filed' to give it an edge, d'ansah 'honed', and then reheated and disepoh 'tempered' in water. The finished brush sword, weeding sickle and plaiting knife all are hafted to ulu 'hilts' of wood, using heated resin to cement the joint. (The Kantu' have experimented with scraps of sheet plastic as a substitute for resin, but they find its bonding qualities inferior.) The hilts usually are further tightened with a fine binding of rattan. The adze is affixed to its peredah 'handle' with not resin but only a complex lashing of rattan, from which it is easily removed when traveling. Brush swords are provided with a sarong 'scabbard', fashioned from two wooden slats lashed together with rattan. Each of these tools also requires some periodic maintenance. Edges are rehoned at least once a day when a tool is in use, using water and a hard smooth rock. After being honed many times, the thin, sharp edge of a blade becomes too thick to hold an edge for any length of time. When this happens, it is ditempait 'given a new edge', in a repetition of the original forging sequence.

The average household devotes eleven workdays per year to the manufacture and maintenance of its forged tools. This assumes a typical household inventory of three brush swords, two adzes, three weeding sickles, two plaiting knives and one shotgun. The iron invested in this inventory has a cash value of approximately 4800 rupiah. Given an average use-life of three years for the cutting tools and six years for the shotgun, the yearly cost of iron is approximately 1300 rupiah, or 3.2 workdays at standard wage labor rates. Thus, combined labor and capital expenditures on forged tools equal 14.2 workdays per household per year. The major part of this expenditure is related to swidden cultivation. Brush swords find their primary use in slashing the forest and in the construction of swidden houses, adzes are used almost exclusively in felling the forest, and weeding sickles are used exclusively in the swiddens (unless a pepper garden also is being cultivated). The goods made with the aid of the plaiting knife also are used primarily for swidden related needs. Finally, most of the game taken with the shotguns are taken in or near swiddens with standing crops, so the resources devoted to firearms directly contribute to the defense of swiddens from pests. On average, nonswidden activities account for at most 20 percent of the usage of these five tools. If the preceding figure of 14.2 workdays is discounted by this amount, a final figure of 11.4 days per household per year is obtained. Since the average household cultivates 4.6 hectares per year, this represents an expenditure of 2.5 workdays per hectare.

HEWN TOOLS

The third and final category of swidden tools is termed utai ti digawai 'things that are made' or, since the material of manufacture in this category is wood, 'things that are hewn'. Of principal concern here are the perau 'canoe' and pengayoh 'paddle'. The Kantu' make their own paddles but they usually obtain canoes in trade from the Iban. The average canoe is 2.5 depa' 'fathoms' in length (approximately four meters), costs 2500 rupiah (the standard being one thousand rupiah per fathom), and has a use-life of six years. The yearly cost for the average household, owning 1.7 canoes, is therefore 708 rupiah or 1.8 workdays at standard wage labor rates[6]. Periodic maintenance, chiefly makal 're-caulking' the canoe's seams, replacement of the lantai 'floor' of bamboo slatting, and the manufacture of paddles, double this figure to about 3.6 workdays. The canoes are used widely for all sorts of purposes, and only about one-half of this expenditure (viz., 1.8 days) contributes directly to swidden cultivation. This refers to the cultivation of secondary forest and swampland swiddens alone: the contemporary primary forest swiddens are not accessible by canoe. Since the average household cultivates 2.6 hectares of secondary forest and swampland swidden per year, the cost of this category of swidden tools amounts to 0.7 workday per hectare in these swiddens.

LABOR SUMMARY

The total cost of all swidden tools averages 15.5, 24.6, and 40.9 workdays per hectare in primary forest, secondary forest, and swampland swiddens, respectively (table 98).

TABLE 98
Labor Devoted To Tool Manufacture Per Swidden Hectare

Tool Type	Swidden Type		
	Primary forest	Secondary forest	Swampland
Plaited	13.0	21.4	37.7
Forged	2.5	2.5	2.5
Hewn	0	0.7	0.7
Total	15.5 days/ha	24.6 days/ha	40.9 days/ha

Conclusion

I will devote this concluding chapter to a brief summary of the roles of labor, time and space in the Kantu' swidden system. I will finish with a comment on the overall significance of the study, in particular for development planning.

LABOR

The chief resource devoted by the Kantu' to their swidden system is their own labor. Total labor expenditures, over the course of an entire swidden cycle, average 88, 168, and 275 workdays in primary forest, secondary forest and swampland swiddens, respectively (table 99)[1]. This places the Kantu' system into the category

TABLE 99
Overall Labor Inputs Per Hectare Throughout The Swidden Cycle

Swidden Stage	Swidden Type		
	Primary forest	Secondary forest	Swampland
# 1 Selecting	0.6	0.8	0.
2 Slashing	8.0	9.2	11.5
3 Felling	12.0	4.0	0.
4 Burning	1.4	3.6	10.5
5 Planting	12.9	14.7	91.9 (5 and 6 combined)
6 Weeding	0.	48.8	
7 Guarding	1.0	4.0	11.0
8 Harvesting	29.9	49.2	86.5
9 Carrying	3.8	4.9	9.6
10 Harvesting of relishes	(not included)	(not included)	(not included)
11 House making	3.2	4.0	13.0
12 Tool making	15.5	24.6	40.9
Totals	88.3 days/ha	167.8 days/ha	274.9 days/ha

TABLE 100
Comparative Production Data From Other Swidden Groups

	Kantu'	Iban	Hanunóo	Lamet	Raiapu Enga	Lua'/ Karen
Labor Inputs/Ha Primary forest..	88 days	138-175 days	545 days	NA	NA	NA
Secondary for-est/grassland	168-275 days	123-163 days	501 days	NA	1108 days	NA
Area Cultivated Per worker/year	0.97 ha	0.53 ha	1.28 ha	0.46 ha	NA	NA
Per consumer/yr	0.59 ha	0.32 ha	0.61 ha	0.31 ha	NA	NA
Rice Yields Per hectare	457 - 1322 kg	810 kg	2410 kg	1335 kg	NA	NA
Per worker	4.5 - 5.2 kg	5.4 kg	4.5 kg	NA	NA	NA
As a % of requirements..	30-40%	32%	NA	NA	NA	29%/ 59%

NOTE: All workdays have been recalculated based on the 350-minute workday of the Kantu'. All rice is measured in a threshed and winnowed but unhusked state. NA = Data not available.

exemplified by the rice swiddens of the Sarawak Iban (table 100)[2]: labor expenditures are relatively low. In the mixed grain and tuber swiddens of the Hanunóo of Mindoro, the labor inputs range several times as high. In the nonrice swiddens of the Raiapu Enga of central New Guinea, labor inputs range yet higher. As this comparison suggests, the Kantu' swidden system is exceptionally land extensive as opposed to labor intensive. This also is reflected in the measures of land farmed. Each worker cultivates an average of 0.97 hectare per year, which amounts to an average of 0.59 hectare per consumer (cf. table 100).

Yields among the Kantu' average 457, 752, and 1322 kilograms of unhusked rice per hectare in primary forest, secondary forest and swamp swiddens, respectively. This seems to place the Kantu' at the lower end of the scale of swidden groups in terms of swidden productivity (table 100). Their system is seen to be more typical when a measure of productivity more appropriate to an extensive agricultural system is used, namely yields per unit not of land but of labor. By this measure yields average 5.2, 4.5 and 4.8 kilograms per 350-minute workday in primary forest, secondary forest, and swampland swiddens respectively (cf. table 100). The fact that there is relatively little variation among the different swidden types in returns to labor also commends the use of this measure, since it suggests that what is being measured is a basic and pervasive aspect of this agricultural economy. If the labor that is incidentally devoted to the cultivation of the nonrice crops is factored out, or if their harvests are added in, then the return on labor is still higher, increasing by about 50 percent in each case[3]. In general, this level of returns meets or exceeds subsistence requirements in 30-40 percent of the households each year, which is not atypical for Southeast Asian swidden systems (table 100).

TIME

One of the most interesting features of the Kantu' agricultural calendar is the amount of disjunction among swidden types (table 101). The work seasons in one swidden type often coincide with the slack seasons in one or both of the other types. As a result, when a household is experiencing a slack period in one of its swiddens, it often can employ this labor in another of its swiddens of a different type. In this way the household can eliminate, at least in part, the slack periods otherwise attendant upon swidden cultivation and thereby intensify its overall utilization of labor. Just as the underutilization of labor is a problem during some stages of the swidden cycle, so is the overutilization of labor a problem during others, notably during planting, weeding and harvesting. These three stages alone account for 48 percent, 67 percent and 65 percent of the total labor inputs in primary forest, secondary forest and swampland swiddens, respectively; yet all three perforce are carried out in a total of just eight weeks in the primary forest swiddens, eighteen weeks in the secondary forest swiddens, and eleven weeks in the swampland swiddens. Because of the need to invest such large amounts of labor in the swiddens in such short periods of time, the utilization of labor is very high during these stages. It averages 73 percent (of the maximum possible) during harvesting and 45 percent during both planting and weeding. These measures of labor intensity identify these three stages as the true constraining stages of the

TABLE 101
The Calendars Of Work In The Major Swidden Types During The 1975–1976 Farm Year

Month	Year	Swidden Type: Primary Forest	Secondary Forest	Swampland
February	1975	Selecting		
March		Slashing		
April				
May		Felling		
June			Selecting	
July			Slashing Felling	
August		Burning	Burning	Slashing
September		Planting	Planting	Burning Planting
October				Transplanting and Weeding
November			Weeding	
December				
January	1976			
February				
March		Harvesting	Harvesting	Harvesting
April		Carrying	Carrying	Carrying

swidden cycle. That is, the average Kantu' household only can cultivate as much land as it can plant, weed and harvest during these fixed periods of time.

SPACE

The household's ability to exploit the differing chronologies of different swidden types depends upon the cultivation of two or more separate swiddens each year. The ability to minimize risk and at the same time maximize the exploitation of different microenvironments within the longhouse territory also depends upon the cultivation of multiple swiddens. The overall importance of multiple swiddens in the agricultural strategy of the average Kantu' household is seen in the fact that each year each household makes an average of 2.3 new, separate swiddens. This strategy, based on a plurality of fields per household, is not well described in the literature on swidden agriculture. In most case studies the number of swiddens cultivated per household per year is not reported, but the implication is that each household cultivates just one. In his study of the Iban, for example, Freeman implies throughout that each household cultivated just one swidden per year[4]. On the basis of Conklin's data, it can be calculated that the Hanunóo cultivate 1.8 swiddens per household per year; but Conklin does not go into the significance (if any) of multiple swiddens within the Hanunóo system[5].

The household's ability to cultivate two or three separate and sizable swiddens each year depends upon a relatively low population/land ratio. With a population at Tikul Batu of 115 persons in September 1974, and a longhouse territory measuring approximately ten square kilometers in area, the population land/ratio then equalled 11.5 persons per square kilometer. In the Iban and Hanunóo systems with which the Kantu' have been compared, the comparable figures are 3.5 persons/square kilometer and 24.2 persons/square kilometer, respectively[6]. Conklin maintains that the Hanunóo territory could accomodate a 60 percent increase in population, raising the population/land ratio to 38.7 persons/square kilometer. In the Kantu' case a population increase of just 38 percent, to 159 persons or 15.9 persons/square kilometer, theoretically could be accomodated under the extant pattern of land use. This assumes that land continues to be cultivated at the rate of 0.59 hectare/person/year, it assumes an average, minimal fallow period of seven years, and it assumes that approximately 75 percent of the land in the territory is arable[7]. Any further increase in population should entail some intensification of the agricultural system (if degradation of the environment is to be avoided), such as already has occurred among the Hanunóo.

SWIDDEN AGRICULTURE AND DEVELOPMENT

The primary challenge of Kantu' life is to secure their livelihood, by means of swidden agriculture, in an environment characterized by great diversity and uncertainty. This study has shown that the Kantu' exploit this diversity and cope with the uncertainty by means of a subsistence system that maximizes diversity both within and among household swidden strategies, at the same time as it maximizes interhousehold interdependence in the exchange of both labor and the products of labor. This system is not static. To the extent to which historical data have been available for analysis, they have shown an ongoing process of adjustment and readjustemt in the subsistence system, as the Kantu' respond to changes in their social and political as well as physical environment.

The Kantu' adaptation to their environment involves not merely their subsistence but, in more general terms, their entire society. The local environment and the nature of their adaptation to it is the dominant, independent variable in Kantu' life[8]. Many aspects of behavior that would not otherwise be glossed as 'economic' are determined by the economic and environmental imperatives previously discussed. This characterization of Kantu' reality is reflected in the cognitive schema of the Kantu' themselves. For example, there is no distinction within this schema corresponding to that between the analytic categories of 'production' and 'ideology'. Most of the Kantu' belief system is bound up with swidden related ritual, and all such ritual is glossed by the Kantu' as integral steps in <u>bumai</u> 'making swiddens', with no ontological differentiation from slashing, felling, burning, and so on.

By the time of this study (1974-1976), the Kantu' were beginning to feel the effects of various governmental programs designed to develop Indonesia's isolated, tribal swidden agriculturalists. What the Kantu' may expect from these programs can be predicted from the problematic experiences of other swidden groups in Kalimantan, as well as elsewhere in the country[9]. In addition to a desire to change the pattern of tribal life to something more familiar to the central government, the government is concerned to interdict the swidden cultivation of the country's tropical forests. In the face of this concern, the abundant evidence in this study that the Kantu' are rational and sophisticated users of their natural environment is important, since this belies any imperative for change and development based on purported ignorance and misuse of this environment - which is one of the most frequently cited justifications for government intervention.

The need for more intensive exploitation of the land also is cited as a justification for intervention. And indeed, in a country like Indonesia with a large, growing and unequally distributed population, the intensification of cultivation in swidden areas arguably is desirable. Care must be taken, however, to distinguish between intensification that achieves higher yields per unit of area and intensification that achieves higher yields per unit of labor. While the former tends to be most valued in intensive agricultural systems, in which land is scarce, wage labor is abundant, and capitalization is advanced, the latter invariably is more important in extensive systems in which it is land that is abundant and labor that is scarce. Because the latter systems conserve labor while the former conserve land, the latter tend to achieve higher returns on labor despite much lower returns on land. Thus, the typical Indonesian swidden may well yield more kilograms of rice per workday than the typical Indonesian wet rice field. As a result, the average swidden farmer may be proceeding with the best of economic motives, and with the most correct of calculations, when he rejects the government's invitation to develop his extensive swidden system into an intensive irrigated system[10].

A related problem in measuring and evaluating agricultural intensification involves the distinction between increasing outputs and increasing the size of the population supported. It is widely assumed that the latter automatically follows from the former, but in fact it does not. For example, the output from one hectare of forest in one year may be many times higher under a system of industrial timber extraction than under a system of swidden farming, but the absolute size of the population supported actually may be lower, due to the fact that loggers typically have a much higher standard of living than do swidden farmers[11]. This one example points to the need for care in 'costing' a change from swidden agriculture to some other system of land use[12].

All of these points, involving measures of input and output as well as overall impact on the environment, are relevant to the decision as to whether a given agricultural system is in successful adaptation to its environment. By most of these measures, the Kantu' system can be so defined. Realistically however, it is likely that the future of the swidden system of the Kantu' and similar groups will be decided not on the basis of such fine definitions and evaluations, but on the basis of political considerations. If politics decides the fate of these societies, however, pragmatics should dictate that change, if it must come, will be based on extant practises and knowledge. Peoples such as the Kantu' have unique knowledge of their environment and of how to exploit it. Successful development, however defined, is most likely to be achieved if this traditional knowledge is both appreciated and utilized.

Notes

Introduction

1. Netting (1974:21).
2. Condominas (1957/1977), Conklin (1957), and Freeman (1955/1970). See also Frake (1955), Izikowitz (1951), and Geddes (1954). The African study by DeSchlippe (1956) and the general survey by Gourou (1953) also were influential.
3. See (e.g.) Miles (1972), Netting (1965, 1968), Sahlins (1957), and Worsley (1956) on social structure; Carneiro (1961), Harris (1972), and Wittfogel (1957) on political structure; and Harris (1966) and Rappaport (1968) on religion.
4. See (e.g.) Clarke (1966), Spooner (1972), Waddell (1972), and Dove (forthcoming b) on Boserup; Durrenberger (1984) and Sahlins (1971, 1972) on Chayanov; and Barlett (1980), Cancian (1972), and Johnson (1971) on risk taking and decision-making.
5. See the reviews by Barlett (1980a), Cancian (1977), and Hoben (1982).
6. Useful surveys of anthropological research on swidden agriculture have been provided by Bennett (1973) and Netting (1974: 24-28). More general bibliographies of this now voluminous literature are contained in Conklin (1963) and Watters (1971).
7. See (e.g.) Conklin (1954) on ethnobotany; Allan (1965), Carneiro (1960), and Street (1969) on carrying capacity; Padoch (1978) and Pelzer (1948) on migration; Harner (1970), Rappaport (1968), and Vayda (1961) on warfare; Bennett (1973) on environmental degradation in general, and Conklin (1959) and Dove (1981c, 1983a) on grassland succession in particular.
8. See (e.g.) Rappaport (1968) and Rambo (1981) on energy tranfers; Vayda and Jessup (1980) on patch theory; and Beckerman (1983) and others in the same volume on forest mimicry.
9. See (e.g.) Burling (1965), Geertz (1963), Wallace (1970), and Dove (forthcoming b).
10. See (e.g.) Clarke (1966), Hanks (1972), Netting (1965, 1968), and Schlegel (1979).
11. See (e.g.) Kunstadter et al. (1978) on Thailand and Grijpstra (1976) on Sarawak.
12. See (e.g.) Friedman (1975) and Dove (1983a, forthcoming b).
13. Although this is the first study to focus on swidden cultivation in Kalimantan, it quickly was followed by others. In 1978-1979

Allen Drake (of Michigan State University) studied the economy of the Mualang, a swidden group in West Kalimantan (see Drake 1983). Beginning in 1979, Andrew Vayda, Christine Padoch, Timothy Jessup, Nancy Peluso, and Carol Colfer (in the context of UNESCO'S Man and Biosphere program) began research in East Kalimantan, focusing on the human ecology of swidden cultivation among other topics (see Vayda 1981; Padoch 1983; Jessup 1981; Colfer 1981, 1983). In addition, Paul Bucher (an American agricultural extension worker) carried out work in 1978-1979 on the swidden system of a group of Kantu' related to those whom the author studied (see Bucher 1980).

14. See the efforts by Howard (1963) and Moerman (1968) to prepare 'grammars' of farming.
15. Cf. Barlett (1980b:547).
16. Cf. Barlett (1980b:547), Cook (1973), Gudeman (1978), and Orlove (1977).
17. Hempel (1958), Nagel (1961).
18. Harris (1979), Ross (1980).
19. See Dove (1980b, 1981b, 1982b, 1983b, 1984). Cf. Miles (1972, 1979) and Boserup (1965).
20. Cf. Rappaport (1979:145-172) and Vayda and McCay (1975).
21. Cf. Hoffman (1983).
22. Rappaport (1979:97).
23. See Dove (1977a) for further details of the methodology employed.
24. Cense and Uhlenbeck (1958:10-13).
25. Hudson (1970:304-306).
26. The major Dutch sources are Bouman (1924), Dunselman (1955), Enthoven (1903), Kielstra (1893), and van Naerssen (1951-1952). The relevant writings are by King (1972, 1973, 1976b and 1979) and by Bucher (1980).
27. The common origin of the Iban with the Kantu' and related groups has been examined briefly in historical works on the Sarawak Iban by Pringle (1970) and Sandin (1967).
28. Enthoven (1903 I:70-71).
29. The separate households of Tikul Batu are numbered here just as they are numbered by the Kantu' of Tikul Batu. Bilek 'household' #10 was uninhabited upon my arrival at Tikul Batu, and then was inhabited by me until the conclusion of the study. Two new households (#16 and #17) were created in 1975, adding to the fourteen that had existed in the fall of 1974 when my study was begun. Neither of them had constructed their own living quarters by the time the study finished.
30. Compare Freeman (1981) on authority in Iban society.
31. See Freeman (1957, 1960a, 1961, and 1970) for relevant, pioneering analyses of Iban social structure. See Appell (1976a) and King (1978) for comparative analyses of social structure in a number of different Bornean societies.
32. The strength of the rule, 'One swidden, one household', is illustrated by the circumstances that attended the creation of household #11 during the partition of household #1. Household

#1 had made just one large swidden that year; and when the partition took place, the rice in this swidden had been planted but was not yet harvestable. As a result, when the household was partitioned the swidden also was partitioned by laying tree trunks end to end through its center. Each of the households resulting from the partition, #1 and 11, cared for just one of the halves of the swidden, as thus divided, through the end of that year's cycle.

33. The partition of household #2, in which household #14 was created, originated in a quarrel between two sisters-in-law over the distribution of fish roe among their respective children.
34. One of the few exceptions to this rule involves the longhouse-wide division of most wild pig and deer taken by longhouse members (Dove 1981d).
35. The July 1975 partition of household #8 created two new households, #16 and #17. Household #16 moved out of the living quarters of #8 and temporarily began sharing those of household #6; but household #17 remained with #8, saying that it was sharing quarters temporarily with the latter. There arose some controversy within the longhouse as to whether household #17 still in fact belonged to household #8. In trying to settle this point, the headman of the longhouse weighed the cooking arrangements (viz., together versus separate) of the parties involved as critical evidence. In this instance, indeed, shared consumption was regarded as a more important definition of the household than was shared residence.
36. See Dove (1980d) for an analysis of the role of women in the Kantu' work force.
37. See Dove (1980a, 1982a, 1982b, 1983a, 1983b) for further discussion of Kantu' property rights, especially rights to forest. On Iban land rights see Freeman (1970:chapter 3). For a more general discussion of land tenure in Borneo, see Appell (n.d., 1971), Dixon (1974), King (1975a), and Weinstock (1979). Major works on adat land law in Indonesia include those by ter Haar (1948), Hooker (1978), and Vollenhoven (see Holleman 1981).
38. This relative lack of mobility can be contrasted to the situation in pioneer Iban communities, as reported by Freeman (1970: 76-77); but see Padoch (1978, 1982) for another view of Iban mobility.
39. See Dove (1982b) for a futher explanation.
40. The 'Big Kenua'' referred to here is not shown in figure 6, which maps the streams and rivers within the territory of Tikul Batu. The 'Little Kenua'' referred to here is shown in figure 6, but not under the same name. In the context of intralonghouse as opposed to interlonghouse geography, it is renamed the 'Big Kenua'' and one of its tributaries is named the 'Little Kenua''. The watershed between these two, last named streams is not, therefore, the watershed mentioned in the text as the boundary between the territories of the Kantu' longhouse Tikul Batu and the Iban longhouse Empakan.

41. The use of interhousehold labor arrangements in Dayak agriculture is well known. It was mentioned in many of the early ethnographies, and Provinse (1937) devoted an article to its occurrence among the Siang. Crain (1976) has written an interesting account of the social and ritual aspects of a multihousehold work party among the Lun Dayeh.
42. For futher discussion of interhousehold labor arrangements among the Kantu', see Dove (1979, 1981b, 1982b, 1983a, 1984, forthcoming a).
43. These are, in increasing order of importance, complexity, and cost: (1) belian berencah 'changing ceremony', belian berunda' 'suspending ceremony', (3) belian betanam pentik 'ceremony of planting the offering post', (4) belian ke tanyo' 'ceremony on the drying platform', (5) belian ke perabong 'ceremony on the ridgepole of the longhouse', and (6) belian ke tanah 'ceremony on the ground'. See Freeman (1967), Jensen (1974:141-150), and Sutlive (1976) for analyses of Iban shamanism.
44. See Dove (1979) for an analysis of Kantu' feasting and Dove (n.d.). for an analysis of ceremonial sacrifice. See Jensen (1974) for a relevant, comprehensive discussion of the Iban belief system.
45. These same parts of the longhouse territory have been planted with a wide variety of fruit trees as well.
46. Much of this technology is pan-Malayan and is well described in Burdon and Parry (1954).
47. These households came from the Jelemuk villages on the Sekapat, a tributary of the Ketunggau river.
48. See Maxwell (1979) for a similar analysis of farm taxonomy among the Kedayan of Brunei, and see Moerman (1968) on a Thai system.
49. A single swidden cannot be called both paya' 'swamp' and pun 'stem', because only swamp rice varieties can thrive in a swamp swidden and all stem rices are dryland varieties. Further, because the absence of stem rice is assumed for any swampland swidden, the use of the term dijap 'opposed' or 'nonstem' for such swiddens is superfluous. Neither can a single swidden be called both paya' 'swamp' and lempa' 'floodzone', because the latter term denotes land that is covered with water when the rivers flood but that is otherwise dry. Nor is a single swidden ever called both paya' 'swamp' and bukit 'mountain', because the latter denotes land with a steep gradient and hence excellent drainage and an absence of any swampy areas. Finally, a single swidden never is called both pun 'stem' and mpukap 'early', because the Kantu' never plant the all important stem rice in an early or midyear swidden.
50. See Maxwell (1978) for a more general discussion of Bornean toponymy.
51. There is some overlap between the territory of Tikul Batu and the territories of the surrounding Kantu' longhouses, due to the fact that most of these Kantu' have common ancestors who

lived in the same longhouse in this same area earlier this century. In calculating the size of the territory of Tikul Batu, however, these overlapping areas were estimated to roughly balance out.

52. Following Conklin (1957:20-28). For a more general introduction to the place setting, see the excellent bibliography of West Kalimantan by Avé et al. (1983). For a useful bibliography of the non-Indonesian provinces of Borneo, see Cotter and Saito (1965). Research on the island as a whole is reported in two scientific journals, the Sarawak Museum Journal (see Loh 1980) and the Borneo Research Bulletin.
53. Robequain (1954:219).
54. In contrast, the Hanunóo of Mindoro in the Philippines, which is subject to a southwest monsoon, southeast typhoon, and north and northwest trade winds, recognize six different wind directions (Conklin 1957:35).
55. See Whitmore (1975:129-138).
56. For the definitive studies of soil under swidden agriculture, see Nye and Greenland (1960) and Sanchez (1976:chapter 10). For Kalimantan in particular, there is a much cited study of the impact of swidden agriculture on podzolic soils, carried out by Driessen et al. (1976).
57. Other Kantu' regard these plates and bowls not as the detritus of human occupation, but as the gift of antu' 'spirits' placed in the earth for discovery by worthy humans.
58. In contrast, Nicholson (1965:table 12) suggests that trees in true primeval forest attain ages of 110-720 years.
59. See Dove (1981c, 1983a) and Dove and Sumitro (in press) for further analysis of the genesis of grasslands in Kalimantan. See Seavoy (1975) for an opposing viewpoint. Richards (1952) and Whitmore (1975) provide excellent general discussions of the tropical forest, including much material that is specific to Borneo. Burkill (1935) provides an equally valuable inventory of the economically useful products of Malayan forests and fields. For studies of cultivated plants in this region see also Ochse (1981).
60. See Medway (1977:139). In this as in other matters, Medway provides the authoritative work on Bornean mammals. Smythies (1981) provides the authoritative work on Bornean birds. Shelford (1917) and Hose (1929) have written general studies of Bornean natural history. An outstanding, relevant work on Malayan primates is edited by Chivers (1980).
61. Authoritative works on crop pests in Indonesia have been written by Dammerman (1929) and Kalshoven (1981).

Chapter One - Selecting

1. Cf. Ellen's analysis of farm site selection among the Nualu of Seram in eastern Indonesia (1973: chapter 6).
2. Permanent exchanges of forest sections typically are made by

two households in different longhouse that each - perhaps as the result of former household moves - own sections that are nearer to the other's longhouse than to their own.

3. During litigation of a land dispute, the elders hearing the case may have to rely on the memories of the interested parties as well as other people in the longhouse. When memories conflict, the one held by the greatest number of people is given the greatest weight.
4. This calculation assumes an average size of 2.54 hectares for a secondary forest swidden; it assumes an average longhouse-swidden distance of forty-five minutes' walk (one-way); and it assumes that labor inputs approximate the means in table 99 for selecting, slashing, felling, burning, and planting secondary forest swiddens.
5. 'Reuse' here refers to the use of an old swidden house throughout the course of an entire swidden cycle in a new swidden. This does not refer to the common reuse of an old swidden house just through the first part of a new cycle (e.g., through planting), at which point a new house is constructed in the swidden and the old house is abandoned.
6. Scott (1976) has carried out one of the most convincing studies of the importance of risk avoidance as opposed to production maximization in the subsistence strategies of Southeast Asian peasants.
7. For 1975 (14-1) x (14÷2) = 91 pairs; and for 1976 (16-1) x (16÷2) = 120 pairs; hence there are 211 total pairs.
8. There used to be an additional advantage to the proximity of different households' swiddens. When headhunting still was endemic in this area, all of the households in a longhouse would make their swiddens in a cluster, for mutual protection. The Kantu' say that no household would have dared then to make an isolated swidden off by itself in some corner of the territory, as sometimes is done today.
9. Assume that a given swidden A is adjoined by another swidden B, if the latter is located in any of the four forest sections adjoining the four sides of swidden A (assuming that the swidden is square shaped, as is usually the case). The probability that swidden B will adjoin swidden A is equal, therefore, to the probability that it will be located in one of these four forest sections as opposed to one of the remaining forest sections in the territory. This probability is equal to 4 out of 279, the latter representing the total number of forest sections in the territory minus one, which represents the swidden A in question. In any given year the probability that a swidden A will be adjoined on at least one side by another swidden is equal to the product of the total number of swiddens made that year, minus one (representing swidden A), and 4/279. The total numbers of swiddens made in 1975 and 1976 were thirty-four and thirty-five, respectively. However, two swiddens were made outside of the longhouse territory in 1975, as were an

additional three swiddens in 1976, so the 1975 and 1976 swidden totals must be reduced futher by two and three, respectively, before taking the product of these totals and 4/279. One further correction is necessary. In 1975 five swiddens adjoined the boundary of the longhouse territory and in 1976 six swiddens did so. These swiddens could only be adjoined on three sides and not four by other swiddens lying within the longhouse territory. As a result, in the above two yearly calculations, the two products (viz., 31x4 and 31x4) must be reduced by five and six, respectively, before dividing by 279. The results of these final calculations equal the probability that any given 1975 or 1976 swidden was adjoined by at least one other swidden due to chance alone.

10. The primary forest section aborted by household #12 in 1976 was farmed by it in 1977. The sections aborted by households #13 and 14 in 1976 were the same sections aborted by them in 1977. The section aborted by household #8 in 1977 had been aborted by it once previously, in 1974, the reason then being the failure of any other household to make a swidden in the same area.
11. The calculations in table 17 are based on estimates that the proportion of floodzone land to highlands in the longhouse territory is approximately .14/.86.
12. The calculations in table 18 are based on estimates that the proportion of swampland to dryland in the longhouse territory is approximately .05/.95.
13. See Dove (1983b).
14. The calculations in table 19 are based on estimates that the proportion of primary to secondary forest in the longhouse territory is .11/.89.
15. The desire to clear additional primary forest does not vary according to the amount of (secondary) forest already owned. There was no association between the clearing (or not) of primary forest by each household during 1975 or 1976 and the number of forest sections already owned (P=.45 for 30 households). The desire to establish rights to forest irrespective of rights already held is due to relatively recent developments that have given land a value (viz., in sale or in rubber growing) beyond its value in subsistence agriculture alone (see Dove 1980a).
16. Seavoy (1973) completely omits this question of restoration of biomass and hence of available nutrients in his study of the fallow period in West Kalimantan. Instead he focuses only on the suppression of herbaceous growth, which also is important but clearly less so
17. Flooding poses the greatest danger to dry rice plants. Swamp rice plants are more likely to be flooded (given the low lying location of most swampland), but they are less likely to be damaged because, according to the Kantu', they have urat dalam 'deep roots'.

18. This refers to successive croppings in rice. Some of the nonrice cultigens that also are planted in the dryland swiddens are cultivated into the second year of the swidden's cultivation.
19. Freeman (1970:276-281).
20. There may be a second explanation as well. The primary forest that the Kantu' are farming today is not true primeval forest, but probably dates just from the latter half of the nineteenth century. In contrast, during some parts of the nineteenth century and at earlier periods in Kantu' history, they were almost certainly farming true primeval forest. It is possible that second-year cropping is agronomically sounder in the latter than in the former and that this explains why the Kantu' once practiced it but do not do so today. However, no current research, as on the accumulation of forest biomass, provides any strong reason to consider this likely.
21. Two caveats must be appended to this description of the initial omen taking in a swidden. First, this is a description of a common, perhaps ideal augural technique, but actual techniques vary widely from one augurer and one household to another. Second, the description given here in any case is quite abbreviated. See also the accounts by Jensen (1974:134-138, 159-168) and Sandin (1980).
22. Work is proscribed during the succeeding phases of the swidden cycle as a result of omens not only from the seven omen birds, but also from a variety of omen animals (viz., sambar deer, barking deer, mouse-deer, sun bears, tarsiers, and red-headed kraits) as well as certain natural phenomena (e.g., falling trees, rainbows, and eclipses).
23. There are exceptions, however. Thus, the augurer may make a pegela' 'offering' to the spirits on his first day at the swidden site, and this - or perhaps more accurately, the spirits' reaction to this - becomes the primary omen, and it is called pun pegela' 'offering stem'. Similarly, the Kantu' may perform an act of divination called betenong kempang 'divining from the kempang tree' for this primary omen, which then is called pun kempang 'kempang tree stem'. Kempang is a species of tree, the shrinking or lengthening of a stick of which is the critical element in the divination.
24. Freeman (1960b). For other studies see Harrisson (1960), Jensen (1974), King (1977), Metcalf (1976), and Sandin (1980).
25. Delacour (1947:175, 234, 347), Glenister (1951:192), Smythies (1981:201, 233, 244, 301-302, 426).
26. See (e.g.) Harrisson (1960), Roth (1896 I:195), and the data in Freeman's own (1960b) study.
27. The author currently is working on an ecological analysis of the Kantu' augural system, taking as examples the pioneering analyses in this field by Harris (1966), Moore (1957), and Rappaport (1968).
28. Cf. Freeman (1970:247-248).
29. In making these calculations, any small bit of adjoining primary

forest that was included in a 1975 or 1976 secondary forest swidden was deducted from the swidden's size, whereas any bit of adjoining secondary forest included in a 1975 or 1976 primary forest swidden was added to the swidden size. The total size of contemporary predominantly primary forest swiddens reflects the desired size, whereas the desired size of historic primary forest swiddens is reflected in contemporary secondary forest swiddens only after deducting any later primary forest additions. In making these calculations the twenty-six 1975 and 1976 secondary forest swiddens located predominantly in swampland were excluded from consideration, because swampland swiddens typically are cut from only a fraction of a forest section. Finally, swiddens composed of approximately fifty percent primary forest and fifty percent secondary forest (or fifty percent dryland and fifty percent swampland) were treated as ½/½ in these and all succeeding calculations. In two-by-two tables in which Fisher's exact test is employed, such fractional figures are rounded off to even integers in some cases and dropped from the sample in others.

30. See Dove (1977b, 1983b) for a discussion of the impact of headhunting on swidden labor.
31. Dove (1984), Boserup (1965), Miles (1972, 1979).
32. Dove (1981b, 1984). See also Chayanov (1966) and Sahlins (1971, 1972).
33. In each of these four exceptional cases, moreover, most or all of the different microenvironments still were included in the household's swiddens. Thus in 1976 household #5 made two swamp swiddens, but one of them (swidden #45) was composed of only two-thirds swampland and fully one-third dryland. Similarly, household #7 made two secondary forest swiddens in 1976, but whereas one (swidden #50) was located almost entirely on dryland, the other (swidden #51) was divided equally between swampland and dryland. In the third instance of repetition, household #2 made two swampland swiddens in 1975 (swiddens #4 and 5). However, the household also made a large dryland swidden that year (swidden #3), and it was divided almost equally between primary and secondary forest. In the fourth and final instance, household #4 made two separate primary forest swiddens in 1975 (swiddens #9 and 10), but it also made additional, separate, secondary forest and swampland swiddens that year (swiddens #11 and 12, respectively).
34. There is some disagreement among the Kantu' as to whether these traditional prescriptions and proscriptions regarding swidden shape are still in force or not.

Chapter Two - Slashing

1. Use of the brush sword is a male characteristic in Kantu' society. However, there are a number of occasions when women must use one as well, as when cutting firewood, gathering

fruit, or assisting in the slashing of brush in the swiddens. On such occasions the sexual loading of the sword is maintained by the use of the scabbard. Men always carry their swords in scabbards, worn on their left hips. In contrast, women never use scabbards. They carry their brush swords in their raised right hands, with the heel of the blade resting on the shoulder and the cutting edge pointing upwards (cf. Lumholtz 1920 I:183). This method of carrying is more dangerous physically, but it also is more benign symbolically.

2. Among trees not slashed but left for felling with the adze, I found the smallest 10 percent to have a mean diameter of eight centimeters.
3. The largest tree that I measured that was cleared during slashing had a diameter of 12½ centimeters, and the smallest tree measured that was left for felling had a diameter of just 5½ centimeters.
4. Assume that a given household contains one worker and that he makes a single swidden requiring 163 man hours to slash. If the worker from this household carries out the slashing by himself, and works an average of 350 minutes each day, he will finish in a minimum of twenty-eight days ([163x60]÷350). Assume instead that the slashing is carried out by a reciprocal work group, consisting of this one worker plus four workers from other households. In that case, the slashing can be completed in a minimum of four days of 495 minutes each ([163x60]÷[495x4]). The swidden owner then will have to slash in the swiddens of the other households in his work group, in order to pay back the labor that they gave him; but he will have to work only sixteen days to pay off this debt (viz., four days times four workers). Added to the four days in which he worked with the others in his own swidden, his total participation in slashing amounts to only twenty days, compared with the twenty-eight days it would have taken him to slash his own swidden on his own. Thus, participation in multiple household work groups typically results in the faster completion of a given work phase, as the Kantu' themselves believe. In the example just cited, the work actually is not completed faster but is just compressed into fewer, longer days. However, since the atmosphere of genial competition in a multiple household work group stimulates a higher than normal intensity of labor, participation in such a group usually does result in the work being completed not just in fewer days, but after the expenditure of fewer total man hours.
5. This figure is probably conservative, given a likelihood of underreporting in my data.
6. It is during the first stage of the swidden cycle, in which the swidden sites are selected, that I think omen taking serves an ecological function, as discussed on pages 86-91.
7. (11.5 years + 7 years) ÷ 2 = 9.25 years.
8. N equals 211 possible, different pairs of households (90 in

1975 and 121 in 1976).

9. In both cases N equals thirty households (fourteen in 1975 and sixteen in 1976), and the 'slashing burden' is reckoned as the average number of slashing workdays per adult male.
10. See Schieffelin (1975) for an anlysis of a reversed, Papuan system in which planting precedes clearing.
11. These intervals are calculated from the midpoints of the swidden stages in question.
12. As in the case of the number of days cut short by omens, this figure is probably conservative due to the likelihood of under-reporting.
13. N equals thirteen households, 'duration' is measured in days and 'onerousness' is measured as the average number of slashing workdays per adult male in the household.

Chapter Three - Felling

1. There are contemporary swidden systems elsewhere in the world in which lopping is still prominent (e.g., Rosaldo 1980:104-106).
2. This tree, probably the waringin, also has special status among Sarawak Iban (Jensen 1974:85n).
3. Approximately one-half of all secondary forest swiddens are cut from very young forest (or grassland) that can be cleared completely with the brush sword alone, with no felling with the adze at all being necessary. The average labor input in felling secondary forest swiddens, therefore, is just four days ([8 days+0 days]÷ 2).
4. These measurements were made at cutting height on trees that were not cleared during the preceding, slashing stage.
5. Two or three of the longhouse's adult women were said to have joined in felling in the past, however.

Chapter Four - Burning

1. Under Kantu' property adat 'law', it is possible for one household to discover, mark, and claim - for their value as timber - one or more individual trees in a forest section that already is claimed - for the purpose of making swiddens - by some other household.
2. There is one major but infrequent exception to the chronology of the burn as outlined here. If the Kantu' encounter exceptionally thick brush of the type called madang when clearing forest (it typically is found in secondary forest), they may use fire to kill it. They first slash as much of the brush and surrounding vegetation as possible, then they dry it and finally burn it, hoping thereby to kill whatever madang they could not kill by slashing. Only then do they proceed to the felling stage, and then to the burn of the swidden as a whole.
3. This calculation is based on the ratio of surface area to volume of a cylindrical log of constant girth. The surface of the

area in cross section of the two ends of the log is not included in this calculation of surface area.

4. The effects of decomposition are more problematic with respect to the burn. The Kantu' say that if a cleared primary forest swidden is set aside for one year, it acquires the potential to burn well, because the felled trees will completely die in the interim; but this potential only can be realized if the burn is preceded by a lengthy drought. The Kantu' cite two reasons for this. First, in the course of decomposition during this one-year interval, the felled timber will acquire moisture. Second, during the course of this decomposition the timber is likely to be bored into by sampok 'termites'. The Kantu' say that the termites carry earth into their holes and that this inhibits the timber's combustibility.
5. This was the season during which all dryland swiddens were burned. Eight grassland-swampland swiddens were burned after these dates. I did not define the season under analysis so as to include these later burnings, because burning swampland often is problematic regardless of prior rainfall patterns and hence often is not influenced by such patterns - whereas it is precisely this influence that is being studied here.
6. Within the memory of Kantu' alive today, the burning-planting interval in primary forest swiddens was much longer. The Kantu' say that formerly the nonrice cultigens (e.g., maize) that they sowed in their primary forest swiddens immediately after the burn were ready for reaping and consumption by the time of the rice planting. This suggests that the interval from burning to rice planting formerly averaged three to four months.
7. There is no such observed association in the secondary forest swiddens, probably because the Kantu' always can achieve the relatively short drying periods that are required for a good burn in these swiddens.
8. In fact, all edges of the swidden tend to burn poorly, due not only to blockage of wind and sunlight by the adjoining forest, but also to greater regrowth. Any regrowth that takes place after the swidden is cleared and before it is burned tends to inhibit the burn, and such regrowth always is likeliest to take place along the swidden's edges, near parent plant material in the adjoining forest. The overall impact on the burn along swidden edges is reflected in this statistic: among twenty swiddens for which data are available, unburned patches along the swidden edges (defined as being within fifteen meters of the edge) comprised over 7 percent of each swidden's total area, whereas unburned patches further away from the swidden's edges comprised just 3 percent of each swidden's total area.

Chapter Five - Planting

1. This is not always the case among swidden societies in this region. Among the Hanunóo of the Philippines, for example,

people ask one another not 'What is your relish?', but 'What is your main dish?' This can be rice but it also can be maize, sweet potatoes, or yams (Conklin 1954:187).

2. It is noteworthy that the Kantu' believe that the food of their ancestors was charcoal, given that the ash that today is produced by the burning of their swiddens becomes the food of the rice plants, which in turn becomes the food of the Kantu'. The consciousness of this connection (whether couched in mythical terms or not) is not limited to the Kantu'. Among the swidden cultivating Bau of Batu Lantai subdistrict in western Sumbawa, the cultural idiom for describing their source of livelihood is 'We eat charcoal'. For further analysis of the myth of Bui Nasi, see Dove (1983c).
3. Some Kantu' maintain that the swamp rice varieties similarly derive from the spirits, but others attribute their origin to experimental breeding and planting - by the Kantu' - around the beginning of the twentieth century. The comparative lack of ritual in swamp rice cultivation strongly suggests that this latter story is more accurate - more accurate in the sense that the swamp rices represent a relatively recent element in the Kantu' swidden system.
4. In table 39 the division of each level IV category into two named varieties at level V and four named subvarieties at level VI is for the purpose of illustration alone. The actual number of named varieties and subvarieties varies from one household to another.
5. Viz., swiddens #41, 42, 43, 44, 67 and 68.
6. The Kantu' also say that one or two rice varieties grow 'too well' in primary forest swiddens, by which they mean that their excessively robust growth in these swiddens makes them especially attractive to pests.
7. There probably is an additional etic explanation for this proscription, which is closer in spirit to its emic justification. Two (or more) swiddens belonging to the same household usually are located in different microenvironments. As a result, the characteristics of a given rice variety that might suit it to one swidden necessarily will differ from those that would suit it to the other. As these characteristics vary, so will vary the characteristics of the rice (always the healthiest looking rice) selected for seed in each swidden. In the course of time, therefore, a single variety that is planted annually in two or more separate swiddens will be transformed into a mixture of subvarieties selected for different reasons in different microenvironments. This would be undesirable from the standpoint of the Kantu', because it would make it difficult for them to predict the performance of this variety in a given environment; and such predictions are critical to each household's decisions, each year, as to which of its rice varieties should be planted in which of its swiddens.
8. Kantu' adat 'law' stipulates that in household partition, the

couple that is leaving to establish its own household need not be given a share of any grain that was harvested before its marriage. It is rare for a departing couple to be denied grain on this basis, however, because it is rare for the Kantu' to be able to store grain for more than a year or two.

9. Table 42 is based on the acquisition of forty separate varieties of rice by households #3, 4,and 16.
10. The longhouse comprised fourteen separate households that year but one, household #12, did not participate fully in that year's swidden cycle. Its members made only one, small swampland swidden, due to the absence of its sole adult male on a trading expedition.
11. It is true among the Kantu' today, and it was all the more true in their recent past, that a piece of iron from which a tool can be made is a scarce, valued good. Thus, new households that are short of both tools and scrap iron with which to make them are called miskin besi 'iron poor'.
12. N equals 120 different paired households (=[16-1]x[16+2]).
13. According to their Kantu' names (for which Linnaean designations and English glosses are provided in the 'Glossary'), these are as follows: Arom, Belaut, Cangkok, Gemalai, Genok, Jawa', Jejamu, Jejebung, Kawa, Keladi, Kenas, Kucai, Lenga', Lengkuas, Lepang, Lia', Limih, Lingkau, Nsabi, Ntimun, Pari', Pisang, Pusut, Rangki, Retak, Semengkok, Tebu, Tepus, Terong, Ubi.
14. Unlike the rice varieties, there is little relation (either ritual or statistical) between the number of relish types planted by the household and the number of swiddens made or the total number of hectares planted. There is one exception, involving a proscription against planting job's tears and Italian millet in the same swidden. The Kantu' say that if these two are planted together they will converse, and this conversation will frighten the rice, which will not grow well as a result (cf. Hill 1977:12). This belief appears to relate to the replacement of millet and job's tears by rice in the early agriculture of the ancestors of the Kantu' and other contemporary rice cultivators in Southeast Asia (see Chesnov 1973:4-5).
15. According to their Kantu' names, these are as follows: Dujung, Jerenang, Kunyit, Lulut, Manang Padi, Ncekur, Ngkerbai, Pencelap, Serai, Singkenyang, Tembako, Tubai (cf. Jensen 1974:182n).
16. As has been noted by all scholars of the subject, tubai is used in reference not only to the well known derris root - the full name for which is tubai urat 'root tuba', usually Derris elliptica Benth. - but also in reference to fish poisons taken from unrelated roots, creepers, shrubs and even trees. Thus, in addition to the just mentioned tubai urat, the Kantu' also frequently fish with tubai buah 'fruit tuba' (Croton tiglium L. ?). It is of interest in this regard that both Howell and Bailey (1900:177) and Richards (1981:395) note that a high quality tuba, produced from the fruit of a bush that is planted

in swiddens, is known among the Iban of Sarawak as tubai Kantu' 'the tuba of the Kantu'. This, coupled with the fact that the Kantu' themselves do not use the term tubai Kantu', suggests that they were the first to cultivate this particular plant for the use of its fruits for fish poison, or at least that they were the first to do so in Western Borneo.

17. This compares with Freeman's (1970:185) figure of 11.7 grains per hole for the Iban.
18. See Freeman (1970:186, 229-230).
19. ((49 holes sown/minute ÷ [0.85 x 80 holes dibbled/minute]) x 60 minutes) = 43 minutes dibbling/hour.
20. Planting a single rice variety in more than one microenvironment within a single swidden does not incur all of the risks attendant upon planting a single variety in differing microenvironments in different swiddens, as described in note #7 to this chapter, because the selection of seed can be monitored more easily and the characteristics selected for can be kept uniform more easily.
21. This assumes that the historic planting day resembled the contemporary one in all other respects, particularly with respect to the amount of time devoted to ritual. In fact, the Kantu' say that they did not use to make offerings on planting days as a matter of course, as they now do. In the absence of such offerings, the pre-1970 workday might have been just as long - in terms of actual work hours - as the current workday.
22. Schefold (1972-1973) has an apt term for the cessation of omen observance in circumstances such as this: he calls it 'religious devolution'.
23. N equals thirty-five plots, each one square meter in area.
24. The smaller the area of leaf surface, the slower the loss of the seedling's moisture - before replanting - through evapotranspiration. Cutting back the leaves of the seedling also helps to ensure that it will stand erect when transplanted. This is a matter for concern, given the loose, watery soil of the swamp swiddens, and also given that the transplant initially has only a minimal root system to support it (see Grist 1975:155).
25. See Grist (1975:149-151).
26. Given that in dryland swiddens the average planting density in dibbling-sowing equals $7.3/m^2$, the planting density in filling-in equals $4.6/m^2$, the sowing rate in dibbling-sowing equals 61.5 li/ha, the sowing rate in filling-in equals 7 li/ha, and the percentage of the average swidden in which filling-in is performed equals 18%; then the number of plants produced by each liter of seed equals ((7.3 x 10,000) + (4.6 x [.18 x 10,000])) ÷ (61.5 + 7) = 1187 plants/liter.

    Given that in swampland swiddens the average planting density equals $14.1/m^2$, the number of transplants taken from each successful seeding equals 2.6, and the sowing rate equals 146 li/ha; then the number of plants produced by each liter of seed equals ([14.1 x 10,000] ÷ 2.6) ÷ 146 = 371 plants/liter.

27. N equals 120 different pairs of households ([16-1]x[16÷2]).

28. (15.5 holes ÷ 80 holes dibbled/minute) + (15.5 holes ÷ 49 holes sown/minute) = 0.51 minute or 31 seconds.

29. Given that: (a) the labor input needed to completely fill-in one hectare equals 2,967 minutes, based on an average rate of 15.5 holes/minute and an average density of 4.6 holes/m^2; and (b) this input is the equivalent of 9.9 workdays, assuming a shorter workday of just 300 minutes in length due to the greater physical strain of filling-in, compared with ordinary dibbling-sowing; and (c) just 1.8 workdays is devoted in fact to filling-in the average hectare of swidden; then (d) the percentage of the average hectare or swidden that is filled-in equals (1.8 workdays/ha ÷ 9.9 workdays/ha) = 18 percent.

30. Based on data from swiddens #41, 42, 44, 67, and 68.

31. Based on data from swiddens #41, 42, 43, 44, 67, and 68.

32. Marginal workers sometimes are enabled to participate in reciprocal work groups during planting by an arrangement whereby the labor of one household's marginal worker is reciprocated by the labor of an equally marginal worker from the other household. These arrangements usually are made for adolescent workers who cannot yet match adult work rates.

33. The inverse association between the location of a swidden entirely in swampland and the use of reciprocal labor in its planting was highly significant (X_c^2=47.1, $P < .001$ for the sixty-nine 1975 and 1976 swiddens).

34. Assuming that: (a) the Kantu' planted their secondary forest swiddens before their primary forest swiddens, and that they began planting each swidden as soon as possible, meaning the day after the burn, then the 1976 planting season would have begun on the day after the first secondary forest swidden (#50) was burned (26 August), thus on 27 August; and (b) each swidden was planted using only intrahousehold labor, and the duration of planting equalled the constant labor input of 12.9 workdays per hectare multiplied by the number of hectares in each swidden and divided by the number of workers in the household (including both workers capable and incapable of reciprocal labor, the workday of the latter being equated to one-half of the workday of the former); and (c) the total labor expenditure of the longhouse on planting equalled the constant 12.9 workdays/hectare multiplied by the sum of the hectares in all twenty-seven swiddens that in fact were planted in 1976 under a reciprocal labor system (the eight swiddens that were not planted using reciprocal labor being excluded from these calculations), this sum equalling 65.9 hectares and the product thus equalling 850 workdays; and (d) the 1976 planting season ended on 2 October, based on the fact that the last burn (in swidden #45) was carried out on 21 September and the duration of planting in this swidden was eleven days; then (e) given that the longhouse planting season would have spanned thirty-six days (viz.,

27 August to 2 October), and given that the total planting work force of the longhouse equalled 67.5 persons (viz., 64 persons capable of reciprocal labor plus 7 persons [each valued at 0.5] incapable of such labor), the utilization of labor under this theoretical nonreciprocal system - expressed as a fraction of the labor available - would have equalled 850 workdays ÷ (36 days x 67.5 workers) = 35 percent.

35. Assuming that: (a) the 1976 planting season began on 27 August (for the same reasons cited in note #34); and (b) each swidden was planted using both intrahousehold labor and extrahousehold reciprocal labor, and a sufficient proportion of the longhouse's total labor resources was used in each swidden to enable its planting to be completed in one day; and (c) the total labor resources of the longhouse equalled 64 workers capable of reciprocal labor plus 0.4 worker incapable of reciprocal labor, the latter figure based on the assumption that an average of just one-sixteenth of the seven nonreciprocal workers in the longhouse (viz., 1/16 x 7 = 0.4) would have participated in the planting of any given swidden, given that each one could not plant in the swiddens of the fifteen other households in the longhouse but only in the swiddens of his/her own household; and (d) the total labor expenditure of the longhouse on planting equalled 850 workdays (for the same reasons cited in note #34); and (e) the 1976 planting season ended on the day after the date of the last burn among the twenty-seven swiddens (21 September in swidden #45), thus on September 22; then (f) given that the longhouse planting season would have spanned just twenty-six days (viz., 27 August to 22 September), and given that the total planting work force of the longhouse equalled 64.4 workers, the maximum theoretical utilization of labor under this reciprocal labor system - expressed as a fraction of the labor available - would have equalled 850 workers ÷ (26 days x 64.4 workers) = 51 percent.

36. This does not belie the statement in the Introduction that rubber tapping well complements swidden farming. It is complementary, but in a grove that has not been worked for a while it takes several days for the latex to start flowing strongly. Thus, a respite of a week or two from swidden labor can profitably be employed in the rubber groves, but a respite of just a day or two cannot be.

37. (1 - .35 [utilization of labor]) x 36 days (length of planting season) = 23 idle days under the theoretical nonreciprocal system. (1 - .51 [utilization of labor] x 26 days (length of planting season) = 13 idle days under the theoretical reciprocal system.

38. There is considerable variation in the type and amount of goods exchanged for planting labor under these several arrangements. In the rare cases in which the exchange is arranged beforehand, the standard wage for swidden labor is paid. This is two <u>kolak</u> of unhusked rice or 300 rupiah (to which another 100 rupiah

sometimes is added in lieu of meals). When workers come to the swidden on their own initiative and ask for rice in exchange for their planting labor, the owning household also is likely to pay them at the rate of two kolak/person/day. There is more variation when they ask for trade goods (or are given such goods by the preference of the swidden owners). For example, household #2 compensated two workers who assisted in planting its 1976 upstream Kenua' swidden with three sticks of tobacco plus one bottle (0.63 liter) of kerosene apiece, all of which could have been purchased from local traders for just 150 rupiah. That same year, household #9 compensated two workers who assisted in planting its upriver Kantu' swidden with four sticks of tobacco and one kilogram of rock salt apiece, all of which could have been purchased for 249 rupiah. In these and other cases the market value of compensation in trade goods always varied below the market value of compensation in rice, which reflects the differing access of the Kantu' to rice versus currency or trade goods.

39. Given that: (a) the 1976 swiddens at Tikul Batu comprised a total of 26.6 hectares of primary forest, 31.6 hectares of secondary forest, and 7.3 hectares of swampland; and (b) the labor input during rice planting is 12.9 days/hectare in primary forest, 14.7 days/hectare in secondary forest (including 1.8 days/hectare devoted to filling-in), and 16.7 days/hectare in swampland (not including transplanting, which is carried out after the planting season proper is finished); and (c) the labor input during relish planting is 1.6 days/hectare in primary and secondary forest swiddens (and negligible in swampland swiddens), and two-thirds of this (viz., 1.0 workday) is carried out during the rice planting season; and (d) the 1976 rice planting season was thirty-four days long (viz., 30 August through 2 October), after excluding the errant Malay household #17; and (e) the total planting work force of the longhouse equalled 67.5 persons; then (f) the intensity of utilization of labor during the 1976 rice planting season was equal to $(((26.6 \text{ ha} \times [12.9 \text{ days} + 1.0 \text{ day}]) + (31.6 \text{ ha} \times [14.7 \text{ days} + 1.0 \text{ day}]) + (7.3 \text{ ha} \times 16.7 \text{ days}) \div 67.5 \text{ persons}) \div 34 \text{ days} = .43$, or 43 percent.
Note the similarity between this empirical rate of utilization of labor and the maximum theoretical rate (viz., 51 percent) that was calculated earlier for a system of reciprocal labor.

40. The Kantu' also observe certain other phenomena to confirm their observations of the stellar cycles. For example, the Kantu' say that egg laying by crocodiles formerly marked the end of the fifth moon, and the incubation of these eggs marked the sixth moon.

41. The use here of the Malay term banyak 'many', as opposed to the equally adequate and indigenous Kantu' term mayoh 'many', probably can be attributed to the love of Kantu' bards for alliteration.

42. The Kantu' maintain that the 'good' period for planting formerly encompassed not only the seventh and eighth moons, but the ninth, tenth, and eleventh moons as well. This view may reflect not a diminishment in the lenth of the good period for planting, but a change in the position of this period from earlier to later within the stellar calendar. Such a change indeed has taken place, and continues to take place, as a result of the procession of the solar equinoxes. As Freeman (1970:172) has calculated, the important stellar positions occurred one month earlier, two thousand years ago, than they do today. At that time, however, the good planting period would have occupied the same relative position in the solar year that it does today, given that this period actually is a function of the movement of the earth vis-à-vis the sun and not a function of its movement vis-à-vis the stars. As a result, the stellar positions that today mark the end of the period in the solar year that is good for planting would have been followed, two thousand years ago, by one more month that was good for planting. This might explain why the Kantu' believe that the good planting period once extended past its contemporary limits, since once it actually did extend past the stellar dates or positions that are used to calculate it today. However, this does not explain why the Kantu' believe that the good planting period was once not just later but longer, overall, as well. That is, the good planting period extended one month later in the year, two thousand years ago, but it began one month later as well, so its overall length was two months, just as it is today. The fact that the Kantu' remember the later date of completion and not the later date of commencement as well, may be attributable to the rosy bias of cultural memory.

The length of time involved does not necessarily mitigate against this explanation of contemporary belief in a historically longer planting season. The proper time for planting is among the most important pieces of information in the possession of this culture, and hence it is the least likely to be lost over the generations. The ability of the Kantu' to have preserved knowledge of planting times, over a period of time sufficiently long to show the effects of the procession of the equinoxes, is attested to by the preservation in the oral traditions of most Ibanic peoples of the memory of the extinction of one of the seven Pleiades in the second millenium B.C. (Jensen 1974:155n).

43. See Dammerman (1929:225-226), Kalshoven (1981:106-108), and Rothschild (1970) on the population ecology of the rice bug.

44. Not included in this calculation of the 1976 planting season is the planting of the Sempinih swidden of household #17 on 24 August. This household was headed by a downriver, Islamicized Dayak, whose lack of knowledge of local swidden practices and constraints was reflected both in this relatively early planting date and in the terrible harvest that resulted.

45. ([18.5 secondary forest swiddens + 8.5 primary forest swiddens] x 5.5 primary forest swiddens) = 12.0 primary forest swiddens.

Chapter Six - Weeding

1. There also is one cultigen that is removed from the swidden during weeding. Maize plants that have grown poorly and show no promise of bearing fruit usually are weeded out to eliminate their competition for nutrients with other, more promising plants.
2. When a swidden is cut from a mixture of primary and secondary forest, however, the Kantu' say that the primary forest section will kena' kesah 'be infected' by the regrowth in the secondary forest section, forcing them to weed it as well.
3. Rarely, measures for providing water are more elaborate. In 1975 household #9 constructed several trenches in its Kantu' swidden. These trenches were unusually long, stretching for over one hundred meters across the swidden and terminating in an adjoining stream. Also unusual was the presence of rudimentary flood gates, constructed of woven mats and earth, at the points where the trenches connected with this stream. Whenever rainfall raised the level of the stream, the gates were opened in order to inundate the swidden. When the stream level began to subside, the gates were closed to retain sufficient water in the swidden to keep the land (but not the rice plants) covered. The value of this inundation, according to the members of household #9, lay in suppressing weed growth.
4. Cf. Freeman (1970:193).
5. Dove (1977b).
6. The records on total weeding inputs are from swiddens #41, 42, 43, 44, 67, and 68.
7. Cf. Freeman (1970:194-195).
8. The chronology of slashing bamboos and stump growth is determined by the same factors that determine the chronology of weeding. Indeed, when slashing is required in a swidden in which weeding is being performed, it is performed at the same time as or intermittent with the weeding.
9. Rarely, preburn regrowth is weeded as well. The Kantu' say that if a swidden (especially one in the floodzone) does not burn at all when it is fired, they will weed (and slash) any extant regrowth before carrying out redressive stack burning. Following this burning, regrowth eventually will spring up again in the swidden and this too will be weeded during the normal weeding season. In such cases, therefore, the entire swidden is weeded twice.

Chapter Seven - Guarding

1. The government issued this decree as part of its then campaign against insurgents operating along the Sarawak-Kalimantan border,

its intent being to restrict interaction between the insurgents and the local tribesmen.

2. The Kantu' believe that there is a general association between high productivity and high attractiveness to pests (cf. note #6, chapter 5).
3. This is in reciprocation of the perceived ills sent to the territory of the Kantu' by the national government on Java (corvée labor by Kantu' for military garrisons located near the border, etc.). See Hose and McDougall (1912, II : 89) for a similar Iban ritual.
4. Formerly this sacrifice always was performed collectively, the Kantu' say. Separate performances by individual households represents a recent development.
5. Cf. Hose and McDougall (1912, II:88-89) for a similar Iban ritual.
6. Cf. Medway (1977:147).
7. See note #40 to the Introduction.
8. These seasonal migrations are characteristic of the ecology of the bearded pig (Medway 1977:146-147; Pfeffer 1959).
9. The Kantu' say that deer hunting also is concentrated during the fruit season, not because more deer move into the Kantu' territory during this season, but rather because the indigenous population moves around the territory more during the fruit season. This, the Kantu' say, makes them easier to hunt.
10. Janzen (1974) suggests that the function of this 'mast fruiting' is to achieve satiation of seed predators (such as the pig), thus ensuring that at least a few seeds will survive to produce new trees.
11. But see Brooke cited in Roth (1896 I:421).
12. Given that: (a) the average labor cost of stretched lines and rail fences is 0.34 workday/hectare, and the average labor cost of brush fences and true fences is 2.4 workdays/hectare; and (b) there is a 70 percent chance that a stretched line or rail fence will have to be replaced with a brush fence or a rail fence and a 30 percent chance that it will not; then (c) the average labor cost of the strategy in which stretched lines or rail fences are erected first equals 0.34 day + (.70 x 2.4 days) = 2.0 days/hectare.
13. Cf. figures 10 and 11 in Roth (1896 I:440-441).
14. This swidden trapping resembles the Central American system of 'garden hunting' described by Linares (1976). However, whereas Linares suggests (1976:348) that such a system 'may even have substituted for actual animal domestication', the fact that the Kantu' also have domesticated pigs indicates that the two systems - exploitation of domesticated and nondomesticated animals - also may coexist.
15. The likelihood of encountering game at large in the forest, as opposed to at specific food sources, was higher in the past when the game populations were larger than they are today. This can be deduced from the historic use by the Kantu' of huge animal nets (called <u>badal</u>) in conjunction with a sweep

or drive through a large tract of forest by a massed group of hunters (cf. Nyandoh 1958). The use of a hunting technique such as this, both large-scale and undiscriminating as to locale, presupposed the existence of large animal populations distributed throughout the forest. The fact that the Kantu' no longer employ this technique is in accordance with other evidence of a historic decrease in the size of the game populations - especially the pig population - within the Kantu' territory. Smaller populations necessitate hunting techniques that are smaller in scale and more discriminatory in locale than the net and drive.

16. This too was not true in the past when game populations were larger. Formerly, the Kantu' commonly set special traps for deer and pig in the middle of the forest. These included baited *wa'* 'pitfalls' and *petok* 'sliding door traps', neither of which is in common use today.

17. Given (based on 1974-1976 data) that: (a) the average hunting trip was undertaken by 2.7 men and lasted three hours, thus consuming a total of 8.1 man hours; and (b) the probability of taking either one deer or one pig during the average hunting trip was .087/1.0; and (c) the average deer or pig weighed approximately seventy kilograms (undressed weight); then (d) the return on labor during hunting was equal to (.087 x 70 kg) ÷ 8.1 man hours = 0.75 kg/man hour. Given (again based on 1974-1976 data) that: (a) construction and tending of the average spring-spear (the use of which outnumbered use of the snare and the set-spear combined by at least ten-to-one) consumed a total of three man hours in the course of one swidden season; and (b) the probability of taking either one deer or one pig with a single trap during a single swidden season was .148/1.0; and (c) given as in [c] above; then (d) the return on labor during trapping was equal to (.148 x 70 kg) ÷ 3 man hours = 3.45 kg/man hour.

18. Cf. Aldrich-Blake (1980:155-157).

19. But see Chivers and Raemaekers (1980:251).

20. See figures 7-9 in Roth (1896 I:437-438).

21. See King (1976a).

22. This species of bamboo has many such ritual uses, perhaps due to the fact - known to the Kantu' - that some other Dayak groups make blowpipes from it. This would explain its use by the Kantu' for the *teruit* in particular, because this is in effect a weapon used to attack sorcerers.

Chapter Eight - Harvesting

1. This proscription conserves the thiamine in the rice (see Edmonson 1976:77; Whyte 1974:100).

2. Cf. Sather (1977) for comparative data from Saribas Iban.

3. The morphological explanation for this is that these grains look more like rice than do any of the other nonrice cultigens planted by the Kantu'. The historical explanation for this is that millet and job's tears likely preceded rice as the principal grains in the evolution of the Kantu' system of agriculture, meaning that what rice is to the Kantu' today, these two grains once were (see note #14, chapter 5). Cf. Chesnov (1973:4-5).

4. See Grist (1975:91) and Williams (1975:69-70).
5. See Miles (1979) and Dove (1980b) for a discussion of the broader role of the finger knife in the Iban and Kantu' swidden systems.
6. Wilkinson (1959:1238).
7. When harvesting immature rice, using a long cut, a deeper basket with an average volume of twenty liters is used. The Kantu' call this basket takin perupok 'pandanus basket', after the plant used in its construction.
8. (1 - [15.5 li + 88.5 li]) = .82, or 82 percent.
9. Glutinous and nonglutinous rice varieties may become mixed within a single stand due to mutation, cross-pollination between adjoining stands, or accident in the course of handling and processing the seed rice.
10. Based on data from households #3, 4,and 16.
11. But see Freeman (1970:190) on the wilful selection of poor grains for seed among the Iban.
12. In one instance observed, the grill was located outside the house, lashed atop posts so as to adjoin and be level with the drying platform. This location permitted the grain to be winnowed, by the action of the wind, as it fell through the grill to mats on the ground below.
13. Both measures of grain were corrected to take account of grain that was consumed by the household or paid out as wages before these measurements were made.
14. There also may be some empirical basis to this divination, since the rice beer could be expected to run out of the bottomless tubes quickest in porous, sandy soils, which happen to be very poor for cultivation.
15. The rice consumption requirements of each household were calculated as follows. First, measurements were made of the total daily consumption of rice in households #3, 4, and 16 over periods ranging from six to twelve months. These measurements then were utilized, with the assistance of a table on relative caloric requirements prepared by the Indonesian Institute of Sciences (reproduced in Whyte 1974:106), to produce the following table of actual rice requirements by age and gender. The figures in this table therefore reflect desired consumption levels, based on the Kantu's own consumption patterns. Actual consumption diverges from these estimates only when a household's rice stocks are nearing exhaustion. Using these estimates, then, and using also data on the age and gender composition of each Kantu' household, the daily and yearly rice requirements of each household in the longhouse during 1975 and 1976 were calculated.

Daily Requirements Of Threshed, Winnowed, Unhusked Rice

Age	Gender			
	Male	Female		
0 - 1	0.425 kg		0.425 kg	
1 - 3	0.565		0.565	
4 - 6	0.756		0.756	
7 - 9	0.896		0.896	
10 - 12	1.087		1.087	
13 - 15	1.368	1.134	1.274	1.512
16 - 19	1.414	0.99	1.134	1.368
20 - 39	1.228	0.943	1.087	1.321
40 - 59	1.134	0.896	1.036	1.274
60+	1.134		0.756	

NOTE: The second and third figures for four age categories of females apply to pregnant females and nursing females, respectively.

16. This ranking is based on a measurement of each household's harvests as a percentage of its minimal needs for seed and food, averaged over the two swidden years. The two households created in 1976, #16 and 17, were excluded from the calculations.
17. The 'Mean yield' and 'Mean intensification' in table 66 were calculated separately for each of the three main swidden types - primary forest, secondary forest, and swampland. The 'Extent of intensification' in each swidden was calculated using the following somewhat arbitrary values for each intensifying operation.

Intensification Of Swidden Cultivation

Swidden Phase	Intensifying Operation	Value if Performed
Felling	Trimming	1
Burning	Stack-burning	1
Planting	Redressive planting	1
Weeding	Before panicle emergence	½ (partial) 1 (complete)
	After panicle emergence	½
	Slashing of Stumps	1
Protecting	Fencing	½
	Trapping	½

18. In this analysis I have calculated swidden size and swidden

yield on the basis of all hectares cleared, as opposed to just hectares burned and planted. If the sizes of the 1975 and 1976 swiddens are recalculated based only upon the number of hectares that were burned/planted, and the swidden yields then are recalculated as well, the latter increase considerably, as shown in the following table. (Note: the calculation of the mean

Swidden Yields Calculated On The Basis Of Hectares Cleared Vs. Hectares Burned And Planted

Mean Yields		Primary Forest Swiddens	Secondary Forest Swiddens	Swamp Swiddens
1975	Based on hectares cleared	606 kg/ha	---------------	---------
	Based on hectares burned/ planted	735 kg/ha	---------------	---------
1976	Based on hectares cleared	293 kg/ha	492 kg/ha	1036 kg/ha
	Based on hectares burned/ planted	335 kg/ha	499 kg/ha	1246 kg/ha

yields in this table is restricted to those swiddens for which data are available on the percentage of cleared land that was burned and planted. This explains why the yields in this table based on the number of hectares cleared differ from those given earlier in the text.

19. The stage under discussion here corresponds to both Kanong laki and Kanong ino' in the list of stages in table 79. It is finished before the Terbit stage begins.
20. See Williams (1975:69, 74-75, 81-82) on the rice plant's sensitivity to environmental stress during its reproductive phase.
21. Given that: (a) the proportion of nonglutinous rice to glutinous rice in the average swidden is .76/.24; and (b) the proportion of eating rice to seed rice in the average harvest is .914/.086; then (c) the minimum proportion of the harvest threshed on a grill equals (.76 - [.086 x .76]), or 0.70; and the maximum proportion of the harvest threshed on mats equals (.24 + [.086 x .76]), or 0.30; and (d) given that the threshing rates on grills and mats are 0.63 minute/liter and 1.42 minutes/liter, respectively, then the maximum amount of labor devoted to thresh-

ing the average liter of grain equals ([.70 x .63 minute/liter] + [.30 x 1.42 minutes/liter]) = 0.87 minute/liter.

22. Using the large-scale technique for winnowing, grain is poured into the air at the rate of 0.83 liter/second, which is equal to a rate of 1.2 seconds/liter (of threshed but unwinnowed grain). Since there are not one but two people - the pourer and the fanner - working together during winnowing, this is in fact a rate of 2.4 seconds/liter. After the grain has been winnowed once by this technique, the approximately two-thirds of the grain that was <u>not</u> winnowed out is winnowed a second time, using the same technique. The total work rate for the two winnowings using the large-scale technique therefore equals 4.0 seconds/liter (viz., 2.4 seconds/liter + [2/3 x 2.4 seconds/liter]). Approximately one-third of the grain is winnowed out during the first of these winnowings and an additional one-sixth is winnowed out during the second. This one-half of the original grain then is winnowed twice more using the small-scale technique, at a rate of approximately 45.4 seconds/liter. During the first of these winnowings, two-thirds of the grain (viz., one-third of the original grain) is winnowed out. This fraction of the grain then is winnowed a second and final time by the small-scale technique. Therefore, the total work rate for the two winnowings using the small-scale technique equals 37.9 seconds/liter (viz., [½ x 45.5 seconds/liter] + [1/3 x 45.5 seconds/liter]). Added to the work rate for the large-scale technique, this yields an overall rate for the entire sequence - of two winnowings by the large-scale technique followed by two winnowings by the small-scale technique - of 41.9 seconds/liter or 0.7 minute/liter.

23. Given a work rate of 45.4 seconds/liter for the small-scale winnowing technique, and given that approximately one-third of the initial grain is winnowed out during the first winnowing (when not preceded by the large-scale technique) and then is winnowed a second time, the overall work rate for winnowing using the small-scale technique alone is 60.5 seconds/liter (viz., 45.4 seconds/liter + [1/3 x 45.4 seconds/liter]).

24. ([0.09 x 1.0 minute/liter] + [.91 x 0.7 minute/liter]) = 0.73 minute/liter.

25. Given that: (a) the work rate for threshing is 0.87 minute/liter of unthreshed and unwinnowed grain, and the work rate for drying is 0.08 minute/liter of threshed and unwinnowed grain, and the work rate for winnowing is 0.73 minute/liter of threshed and unwinnowed grain; and (b) since threshing results in a 25 percent reduction (approximately) in volume, and winnowing results in a further 25 percent reduction, these work rates can be corrected so that all are based on the same unit of measurement, namely a liter of threshed and winnowed grain, to yield rates of 1.55 minutes/liter (viz., 0.87 minute ÷ .56) for threshing, 0.11 minute/liter (viz., 0.08 minute ÷ .75) for drying, and 0.97 minute/liter (0.73 minute ÷ 0.75) for winnowing; then (c) the total work rate for the entire sequence of grain processing equals 2.63 minutes/liter of threshed and winnowed grain (viz., 1.55 minutes/liter + 0.11 minute/liter + 0.97 minute/liter).

26. Table 71 does not include all of the harvest dates in each swidden, but only those on which household #2 or 4 provided reciprocal labor to the other in one of the two swiddens. Household #4 gave a total of twenty-three reciprocal workdays to #2, while household #2 gave only twenty-one reciprocal workdays to #4. Household #2 made up the difference by performing two days of reciprocal labor in household #4's 1976 Serangir swidden.

27. For related reasons, gift labor also is little used during the harvest (viz., it was used during just four of the sixty-nine 1975 and 1976 swiddens). If a household's harvest is so good that it needs help to complete it, it is good enough to be used to pay for this help.

28. ([49.6 liters - 3 kolak] + 3.26 days) = 12.5 liters.

29. First, assume that: (a) 400 liters of ripe grain (as measured after threshing and winnowing) remain unreaped in a given household's swidden; and (b) there are four adult workers in the household, each of whom can reap a maximum of fifty liters of grain per day; and (c) the daily loss of ripe grain to pests also equals fifty liters (or one carrying basket of unthreshed grain, which is given by the Kantu' as a rule of thumb loss per day, per swidden, to deer and pigs combined); then (d) if the household completes reaping using only its own labor, it will require two more days, and the cost to the household in the interim will equal one hundred liters of grain (viz., 2 x 50 liters). Second, assume as in (a) and (c) above, except in this case the household hires four laborers, at a wage of 5.9 liters of grain per day, to assist in the completion of the harvest; then if the household completes the reaping using the four wage laborers plus its own labor, only one more day will be required, and the cost to the household will equal only 73.6 liters of grain (viz., [1 x 50 liters] + [4 x 5.9 liters]).

30. See Freeman (1970:196n) and Jensen (1974:184-186) for similar lists among the Iban.

31. If account also is taken of harvesting done by each household's workers outside of its own swiddens, for wages, the average work rate reached 14.5 kilograms/worker/day.

32. Viz., swiddens #41, 43, 67, and 68.

Chapter Nine - Carrying

1. Some Kantu' speak of carrying as a part of the overall stage of ngetau 'harvest' (e.g., just as threshing and winnowing are considered to be part of this stage). However, because opinion is divided on this, because the chronology (etc.)

of carrying is distinctive, and because it is an optional stage, I treat it in this study as a separate stage.

2. The custom of making an offering when carrying in the harvest is a new one, pioneered by household #4 in the early 1970s. The other households in the longhouse say that they began to follow suit after noting the subsequent harvest successes of household #4.
3. I use 'mantra' here in the sense of a spoken charm that is not directed towards any particular, specified supernatural agency.
4. See Sather (1980) for a detailed analysis of the ritual and symbolism related to rice storage among the Iban.
5. Given that in 1975 and 1976: (a) the average sizes of primary forest, secondary forest, and swampland swiddens were 3.4 hectares, 2.6 hectares, and 0.35 hectare, respectively; and (b) the average yields per hectare in primary forest, secondary forest, and swampland swiddens were 737 liters, 1213 liters, and 2132 liters, or 8.3, 13.7, and 24.1 carrying baskets of threshed grain, respectively; and (c) the ratios of longhouse storage to swidden storage among primary forest, secondary forest, and swampland swiddens were .56/.44, 70/.30, and .96/.04, respectively; and (d) the average distances to the longhouse among primary forest, secondary forest, and swampland swiddens were 2.7, 2.1, and 1.6 statute kilometers, respectively; and (e) the ratios of swiddens from which the harvest could be transported by water to swiddens from which it could be transported only by land were 0./1.0, .42/.58, and .65/.35 among the primary forest, secondary forest, and swampland swiddens, respectively; and (f) preparatory to transporting the rice, one workday had to be devoted to clearing each kilometer of trail between each swidden and the longhouse, and 0.07 workday had to be devoted to filling each carrying basket with grain; and (g) the work rates of transporting rice by land and by water were 0.21 workday/basket/kilometer and 0.15 workday/basket/kilometer, respectively; then (h) the work rate for carrying in the harvest from primary forest swiddens was equal to ((([.56 x 8.3 baskets] x .21 day) + (1.0 day ÷ 3.4 ha)) x 2.7 km) + ((.56 x 8.3 baskets) x .07 day) = 3.8 days/hectare; and in secondary forest swiddens it was equal to (((([.70 x 13.7 baskets] x .42) x .15 day) + (([.70 x 13.7 baskets] x .58) x .21 day) + ([1.0 day ÷ 2.6 ha] x .58)) x 2.1 km) + ([.70 x 13.7 baskets] x .07 day) = 4.9 days/hectare; and in swampland swiddens it was equal to (((([.96 x 24.1 baskets] x .65) x .15 day) + (([.96 x 24.1 baskets] x .35) x .21 day) + ([1.0 day ÷ .35 ha] x .35)) x 1.6 km) + ((.96 x 24.1 baskets) x .07 day) = 9.6 days/hectare.
6. There are occasional exceptions to this. In 1972-1973, when the government forbade the storage of any grain, no matter how briefly, in the swidden, the Kantu' carried their grain - still unthreshed - back to the longhouse on the same day that they harvested it.

7. Given that: (a) the total longhouse work forces available for cooperative carrying in 1975 and 1976 equalled twenty-four persons and twenty-five persons, respectively; and (b) totals of 174 workdays and 114 workdays were devoted to cooperative carrying in 1975 and 1976, respectively; and (c) the cooperative carrying season was fifty-seven days long in 1975 and nineteen days long in 1976; then (d) the intensity of utilization of the longhouse labor force equalled 13 percent in 1975 (174 days ÷ [57 days x 24 workers]) and 24 percent in 1976 (114 days ÷ [19 days x 25 workers]).

Chapter Ten - Harvesting Nonrice Cultigens

1. Table 85 examines 25 different types of relishes for which data on maturation and harvesting periods are available. For the purpose of constructing this table, it is assumed that all relish types are planted at the same time and that this is the same time at which the rice also is planted. For seven relish types (viz., spinach, sweet potatoe, pumpkin, taro, maize, cucumber and cassava), either more than one part of the plant is edible or else the plant is edible in both a mature and immature stage. In these cases the harvesting period is defined as the period between the earliest and latest harvestable dates.
2. Table 86 assumes that the 'Current Swidden' is planted at the same time of the year (viz., exactly twelve months after) the 'Previous Swidden'. It futher assumes that the maximum number of relish types under consideration, namely twenty-five types, is planted in both swiddens. The 'Total number of relish types simultaneously harvestable' represents the total number of different types, with one exception. Cassava yields edible leaves just three to four months after planting, and it continues to yield them - and eventually tubers as well - for sixteen to seventeen months thereafter; so it is harvestable simultaneously in both years' swiddens for a period of six months (viz., the third through ninth month in the current swidden, and the fifteenth through the twenty-first month in the previous year's swidden). Accordingly, one (1) relish type is subtracted from the total number of types harvestable for each of these six months.
3. The gathering of wild relishes in the forest (as described at the end of this chapter) is even more exclusively a female task. In contrast, the harvest of the cultivated but nonrelish utai, the plants with ritual importance that are grown in the 'place of the seed rice', is the province of males alone. The males in households #3, 4, and 16 (e.g.) harvested less than one-fifth of the relishes in their 1976 swiddens, on average, but they harvested virtually all of the ritual utai.
4. Table 87 presents the average number of days devoted to harvesting each type of relish between May 1975 and April 1976, among

households #3, 4, and 16. The summed total of days devoted to the subcategories of main side dishes exceeds the total given for the category because some relish types, which have both an edible leaf and an edible fruit or tuber, were counted in two different categories.

5. Table 88 is based on data from swiddens #41, 43, 44, 67.
6. Each case in table 89 represents a comparison of the incidence of harvesting of one relish type during the month(s) of the rice harvest in a given swidden, with the incidence of harvesting of that same type during the one month immediately preceding or immediately following the rice harvest. The data are based on the harvesting of five common relishes (viz., cucumber, pumpkin, taro, dish-cloth gourd, and long bean) in swiddens #41, 42, 43, 44, and 67.
7. In table 90, 'Request' denotes requesting and receiving permission to harvest relishes from the swiddens of another household, while 'Give' denotes giving permission to another household to do this. Each day on which one household gives relishes to or takes them from another household is calculated as one (1). Unsolicited gifts of relishes to or from any of the three households are excluded from this analysis since they have to do more with social obligation than material need.
8. On fungi see Chin (1981) and Sather (1978).
9. Table 91 is based on data from households #3, 4, and 16 during the 1976 swidden year.
10. The actual figures are 89 days for wild relishes and 153 days for cultivated relishes, based on data from households #3, 4, and 16 over the period from 1 May 1975 to 30 April 1976.

Chapter Eleven - House Making

1. See Frake (1955) for an early analysis of the problem of village-farm distance in a system of swidden agriculture. Ellen (1973) has made a more recent case study of the same topic.
2. This calculation is based on the fact that the average primary forest swidden is 3.4 hectares in size and requires labor inputs of twenty days per hectare during slashing and felling, while the respective figures for secondary forest swiddens are 2.6 hectares and thirteen workdays. This calculation is based on full workdays: if some slashing or felling is done on shortened or fractional workdays, the potential savings in travel time is higher still.
3. This is the floorplan of a <u>dampa</u>' under construction in 1975 and belonging to the nearby Kantu' longhouse called 'Kantu' Lalang'.
4. For further discussion of Kantu' warfare and settlement patterns, see Dove (1977b, 1982b).
5. Freeman (1970:161-169).
6. Given that: (a) the labor cost of constructing a nighttime swidden house is 19.5 workdays, while the cost of constructing

one household's section of a subsidiary longhouse is 32.5 workdays; and (b) the average nighttime swidden house is used for 1.4 years, while the average subsidiary longhouse is used for 3.0 years; then (c) the average yearly cost to a household of using a nighttime house versus a subsidiary longhouse is 13.9 workdays versus 10.8 workdays.

7. In those swiddens for which a day house is not regarded as sufficient by itself, one may be constructed nonetheless. If the night house being used for a given swidden is located outside that swidden's boundaries, then a day house may be constructed within the swidden by way of supplement. Day houses also are erected in swiddens whose owners are using a nearby dampa' as their nighttime shelter. (A dampa' always is located between, and hence outside of, the swiddens of its constituent households.) All such uses of supplementary day houses are a function of above average swidden size.
8. This labor input of 16 days does not include the 3.5 days that will be expended on adding a drying platform to the house later in the year.
9. The Kantu' have a term, durung, for a hut built (usually near the longhouse) especially for the storage of rice. It is unclear if this means that the Kantu' once made such huts themselves or that they know of this practise from neighboring groups. Certainly there are swidden groups in Kalimantan - such as the Ribun who live near Sanggau on the lower Kapuas river, among others throughout Indonesia - that store their rice or other crops in shelters built especially for this purpose.
10. Given that: (a) the labor costs of constructing day houses and night houses are 1½ workdays and 19½ workdays, respectively, and the cost of refurbishing an old night house is 2½ workdays; and (b) new swidden houses are constructed for 63 percent of the swiddens that are provided with nighttime shelter, while old houses are used in the remaining 37 percent; and (c) each newly constructed night house services an average of 1.2 swiddens, while each old house services an average of 1.3; and (d) the use of shelter in each swidden type is as follows:

The Use Of Shelter In The 1975 And 1976 Swiddens

	Nighttime Shelter Available	Daytime Shelter Alone Available	No Shelter Available	Totals
Primary forest swiddens	100%	0%	0%	100%
Secondary forest swiddens	95%	5%	0%	100%
Swampland swiddens	38%	27%	35%	100%

and (e) the average sizes of primary forest, secondary forest and swamp swiddens are 3.4, 2.6, and 0.35 hectares, respectively; then (f) the labor cost of shelter in primary forest swiddens equals ((.63 x [19.5 days ÷ 1.2 swiddens]) + (.37 x [2.5 days ÷ 1.3 swiddens])) ÷ 3.4 ha = 3.2 days/hectare; and in secondary forest swiddens equals ((.95 x ((.63 x [19.5 days ÷ 1.2 swiddens]) + (.37 x [2.5 days ÷ 1.3 swiddens]))) + (.05 x 1.5 days)) ÷ 2.6 ha = 4.0 days/hectare; and in swampland swiddens equals ((.38 x ((.63 x [19.5 days ÷ 1.2 swiddens]) + (.37 x [2.5 days ÷ 1.3 swiddens]))) + (.27 x 1.5 days)) ÷ .35 ha = 13.0 days/hectare.

11. While tensions may occur between a junior and a senior married couple living in the same household, they are not nearly as frequent or as severe as those that occur between two couples of the same generation living in the same household, as first discussed in the Introduction.
12. The other domestic animal of economic importance, the pig, usually is kept at the longhouse as opposed to the swiddens. In their case, however, disputes over ownership are minimized both by the lesser numbers involved and by use of a system of indentification of individual animals based on ear notching.

Chapter Twelve - Tool Making

1. This perception of danger is placed in greater perspective by the fact that plaiting by females may be equated with head-hunting by males (Haddon and Start 1936: 22; Pringle 1970: 24-25).
2. See the lelambak design in Haddon and Start (1936: 31, 104, 135).
3. Table 96 is based on observations in households #3, 4, and 16 during June-December 1975 and January-May 1976. For ease of reading, the latter period is placed before the former in the table.
4. Table 97 is based on observations in households #3 and 16 during June-December 1975 and January-May 1976.
5. These figures are based on detailed measurements in households #3, 4, and 16, over the period from 1 June 1975 to 31 May 1976, of all labor devoted to the gathering and preparation of plant materials, the plaiting itself, and all related activities such as lashing, post making, interlacing, threading, and mending. Given that: (a) these measurements yielded total labor inputs into plaiting of 609 hours, 468 hours, and 517 hours, for the three households, respectively (after discounting by the amounts of 31 percent, 41 percent, and 31 percent, respectively, to exclude labor devoted to plaited goods not used in or for the swiddens); and (b) the amounts of primary forest, secondary forest, and swampland cultivated in 1976 by households #3, 4, and 16 were as follows:

Forest And Land Types Cultivated By Households #3, 4, And 16 In 1976

Household	Primary Forest	Secondary Forest	Swampland
# 3	1.04 ha	3.30 ha	0.44 ha
4	2.97	2.69	0.30
16	0.	2.42	0.55

and (c) the average (hence expected) yields for primary forest, secondary forest, and swampland swiddens were 457, 752, and 1322 kilograms per hectare, respectively; and (d) the amount of labor devoted to plaiting per (expected) kilogram of harvest grain was equal in household #3 to 609 hours x 60 minutes ÷ ([1.04 ha x 457 kg] + [3.3 ha x 752 kg] + [.44 ha x 1322 kg]) = 10.33 minutes/kilogram; and in household #4 to 468 hours x 60 minutes ÷ ([2.97 ha x 457 kg] + [2.69 ha x 752 kg] + [.30 ha x 1322 kg]) = 7.43 minutes/kilogram; and in household #16 to 517 hours x 60 minutes ÷ ([2.42 ha x 752 kg] + [.55 ha x 1322 kg]) = 12.18 minutes/kilogram; then (e) taking the average of these three rates (viz., 9.98 min/kg), the average amounts of labor devoted to plaiting in primary forest, secondary forest, and swampland swiddens are estimated to equal, respectively, 13.0 days/hectare (viz., [457 kg x 9.98 min.] ÷ 350 min./day), 21.4 days/hectare (viz., [752 kg x 9.98 min.] ÷ 350 min./day), and 37.7 days/hectare (viz., [1322 kg x 9.98 min.] ÷ 350 min./day).

6. ([1.7 canoes x 2500 rupiah] ÷ 6 years) ÷ 400 rupiah/day = 1.8 workdays.

Conclusion

1. These calculations include material nonlabor resources - here expressed in workday equivalents using formulae based on local market values and the average daily wage for swidden labor - that are expended directly in the cultivation of the swiddens. These calculations exclude material resources expended on the feasting of interhousehold work parties, as well as the small amount of labor devoted specifically to the cultivation of the nonrice cultigens. They do not exclude labor devoted to the swidden as a whole that, while it primarily benefits the rice crop, also benefits the nonrice crops. These calculations include labor devoted to the threshing, drying, winnowing, and carrying of rice, but not labor devoted to its futher processing (viz., redrying, husking, and boiling in preparation for consumption).
2. See Freeman on the Iban (1970: 245, 248, 257, 266, 269), Conk-

lin on the Hanunóo (1957: 119, 145, 150, 152), Izikowitz on the Lamet (1951: 283-285, 287), Waddell on the Raiapu Enga (1972: 98), and Kunstadter on the Lua' and Karen (1978: 123).

3. This is a conservative estimate. When the labor devoted to nonrice cultigens is factored out in the case of the Hanunóo, the calculated return to labor rises from 4.5 kilograms per workday to 14.5 kilograms per workday (Conklin 1957:152).
4. Freeman (1970).
5. Conklin (1957: 17).
6. Freeman (1970: 64), Conklin (1957: 146).
7. (.75 x 1000 ha) + (.59 ha x [1 year + 7 years]) = 159 persons.
8. For further discussion of the primary determinants of Kantu' behavior, see Dove (1977b, 1979, 1980a, 1980d, 1981d, 1982b, 1983b, 1984).
9. Dove (1981c, 1983a, forthcoming b, Dove and Sumitro in press).
10. Dove (forthcoming b).
11. Dove (1983a, Dove and Sumitro in press).
12. This term is adapted from Appell (1977).

Glossary of Kantu' Terms for Flora and Fauna

This glossary presents the Kantu' term, the Linnaean designation, and an English gloss for all flora and fauna mentioned in the text. In some cases the identification is tentative, and is so designated by a question mark; and in a few cases no identification at all was possible. In transcribing the Kantu' terms, both here and in the text itself, I have followed Richards' (1981) example, except for those instances in which there is definite variation between the Kantu' and his Sarawak Iban. There are one or two rules of transcription that deserve special mention:

/ch/ is transcribed as /c/
/ŋ/ is transcribed as /ng/
Glottal stops are transcribed as /'/

In addition to Richards' dictionary, those prepared by Howell and Bailey (1900) and Scott (1956) may be consulted for comparative purposes. A longer, more general word list is in Dove (1981 a). Where this earlier list differs from the current list, the latter should be considered to be authoritative.

KANTU'	LINNAEAN DESIGNATION	ENGLISH GLOSS
Ama'	(various)	General term for crop pests.
Arom	Amaranthus sp.	A type of spinach, a swidden cultigen.
Bejampong	Platylophus galericulatus Cuvier.	The Crested Jay, one of the seven major omen birds.
Belaut	Ipomoea batatas (L.) Lam.	The sweet potatoe, a swidden cultigen.
Beman	Donax sp.?	A plant cultivated for use in plaiting.
Beragai	Harpactes duvauceli Temminck.	The Scarlet-rumped Trogon, one of the seven major omen birds.
Beruang	Helarctos malayanus euryspilus Horsfield.	The Malayan sun bear, an omen animal.
Betong	Bambusa sp.	A species of bamboo cultivated for use in flooring, etc.

Biuko'	Playsternum megacephalum Gray ?	A large, hard-shelled, edible turtle traded into the Kantu' territory from Malay fishing villages along the Kapuas river.
Buloh	(various)	General term for bamboos, both cultivated (in the swiddens) and uncultivated.
Buntak	Acrididae.	Grasshoppers, all of which are swidden pests, but some of which also are eaten.
Burong	Aves.	General term for birds.
Bwaya	Crocodilus siamensis Schneid.	A riverine crocodile formerly common in the Kantu' territory. Its image is used in ritual, and its form often is taken by evil spirits.
Bwaya ikan or Bwaya kenyulong	Tomistoma schlegelli S. Muell.	'The fish crocodile' or 'pointed crocodile', the false gavial.
Bwaya'	Varanus sp.	The rough-necked monitor, hunted for its flesh.
Cangkok	Sauropus androgynus (L.) Merr.	A swidden cultigen, whose leaves and fruit are eaten as relishes.
Cit	Rattus sp.	A type of rat, found both in the vicinity of the longhouse and in the swiddens, that preys on both crops and stored foodstuffs.
Dujung	(unidentified)	One of the ritually important plants in the lapik benih 'place of the seed rice', grown for its striking blossoms.
Garu	Gonystylus macrophyllus (Mig.) Arry Shaw.	A tree yielding scented wood credited with ritual properties, aloes wood.
Gemalai	Benincasa hispida (Thunb.) Cogn.	The wax gourd, a swidden cultigen.
Genok	Lagenaria siceraria (Molina) Standl.	The bottle gourd, cultivated in the swiddens for use as a relish when young and as

		a water vessel when older.
Gerenih	Licuala sp.?	A noncultigen gathered for use in plaiting sun hats. It also is part of the bungai jarau 'lopped blossom', formerly sent out as a summons to war.
Getah	Hevea brasiliensis (Willd ex A. Juss) Muell. Arg.	The Para rubber tree, cultivated by the Kantu' for its marketable latex.
Getah kawin & Getah lambau	,,	'Married' or grafted rubber.
Iba'	Ophidia.	General term for snakes.
Ikan	Osteichthyes.	General term for fish.
Ingkat	Tarsius bancanus borneanus Elliot.	The Western Tarsier, regarded as an omen animal and the ulun 'servant' of Pulang Gana, the deity of agriculture and the earth. This status appears to be related to the tarsier's perceived ability to bali' 'reverse' its eyes.
Jani'	Sus barbatus barbatus Müller	The bearded pig.
Jani' kampung	,,	The feral 'primary forest pig'.
Jani' laman	,,	The domesticated 'yard pig'.
Jani' apai Aloi	,,	'The pig of father Aloi', the clown figure of Kantu' folklore. This is the 'giant pig of Borneo', which Medway (1977:146) considers to be aberrant Sus barbatus.
Jani' ntangur	,,	A variety of wild pig with yellow fat and a cadaverous smell, said to be almost inedible.
Jawa'	Setaria italica (L.) Beauv.	Italian millet, a swidden cultigen.
Jejamu	(unidentified)	A forest plant.
Jejamu beras & Jejamu alang	Amaranthus caudatus L.	A swidden cultigen, having an edible fruit or grain but

		inedible leaf, grain amaranth.
Jejebung	Amaranthus hybridus L.	A type of spinach, a swidden cultigen.
Jelu	(various)	Feral animals.
Jelu laman	(various)	'Yard [domesticated] animals'.
Jerenang/Cerenga'	Daemonorhops sp.	A rattan yielding 'dragon's blood', the red dye. Planted in the lapik benih 'place of the seed rice'.
Kala	Palamneus sp.	A poisonous scorpion.
Kawa	Cucurbita moschata (Duch.) Poir.	The pumpkin, a swidden cultigen.
Keira'	Ficus benjamina L.?	A tree that usually is spared when clearing the forest for a swidden, on the grounds that it draws water to the surface of the soil and so stimulates the growth of rice planted near it.
Keladi	Colocasia esculenta (L.) Schott & Endl.	Taro, a swidden cultigen.
Kempang	Artocarpus elasticus Reinw. ex Bl.?	A tree, sometimes used instead of the omen birds in divination during the initial selection of a swidden site.
Kenas	Ananas comosus (L.) Merr.	The pineapple, a swidden cultigen.
Kenawang	Bungarus flaviceps Reinhardt.	The Red-headed Krait, regarded as an omen animal. Its coloring is very distinctive, consisting in longitudinal black stripes on a white body, with a bright red head and tail.
Kenyalang	Buceros rhinoceros borneoensis.	The Rhinocerous Hornbill, whose feathers are worn on ritual occasions by those who have taken heads. Its carved image also is used in ritual.
Kepayang	Pangium edule Reinw.	A forest tree yielding a fruit that can be used as a

		medicine after being boiled, peeled, and steeped in water.
Kera'	Macaca fascicularis mandibularis Elliot.	The long-tailed macaque, a swidden pest.
Keruai	(unidentified)	A greatly feared wasp that nests in secondary forest.
Kesa'	Oecophylla smaragdina Fabr.	The weaver ant, so named for its habit of making its nest from two leaves 'stitched' together. It is very aggressive and harries the Kantu' when they clear young secondary growth, which it favors.
Kesinap	(unidentified)	A species of forest tree.
Ketupong	Sasia abnormis Temminck.	The Rufous Piculet, one of the seven major omen birds.
Kijang	Muntiacus muntjak pleiharicus Kohlbrugge.	The barking deer, a small deer considered to be an omen animal.
Kijang kerangas	,,	'Heath forest barking deer', a variety distinguished by its smaller size and red coat.
Kijang taran	,,	'Indigo barking deer', a variety distinguished by its large size and black coat.
Kucai	Allium schoenoprasum L.	Chives, a swidden cultigen.
Kulat	(various)	General term for wild fungi, especially those collected for use as relishes.
Kunyit	Curcuma domestica Valeton.	Turmeric, planted in the lapik benih 'place of the seed rice' and used medicinally.
Kutok	Blythipicus rubiginosus.	The Maroon Woodpecker, one of the seven major omen birds.
Lada	Piper nigrum L.	Pepper, cultivated as a cash crop not in umai 'swiddens' but in kebun 'gardens'.
Lalang	Imperata cylindrica	Term for coarse grasses --

	(L.) Beauv., and others.	including but not limited to Imperata -- that occur as weeds in secondary forest swiddens but otherwise are scarce in the Kantu' territory.
Landak	Trichys fasciculata Shaw, and Hystrix brachyura longicauda Marsden.	The long-tailed and common porcupines, whose quills are credited with ritual powers.
Lang	Haliastur indus Boddaert.	The Brahminy Kite, a predator of chickens, also believed to be capable of becoming an antu 'spirit'.
Lenga'	Sesamun indicum L.	Sesame, cultivated in the swiddens as a condiment.
Lengkuas	Languas galanga (L.) Stuntz.	Greater galanggale, a swidden cultigen primarily used in the bumbu ciping 'yeast' used to brew rice beer.
Lepang	(unidentified)	A swidden cultigen with an edible fruit.
Lia'	Zingiber officinale Roxb.	Ginger, a swidden cultigen valued for its medicinal properties, especially in postpartum treatments.
Limih	(unidentified)	A swidden cultigen with edible leaves and fruit.
Lingkau	------------------	-------------------------
Lingkau amat or Lingkau keribang	Zea Mays (L.)	Maize, a swidden cultigen.
Lingkau lesit	Coix lachryma-jobi L.	Job's tears, a swidden cultigen primarily used in ritual offerings.
Lingkau tengkung or Lingkau gendum	(unidentified)	A grain resembling lingkau lesit but having smaller fruit.
Lulut	Symplocos sp.?	One of the plants in the lapik benih 'place of the seed rice', its leaves are made into a medicinal ointment. It is likened to kunyit 'turmeric'.

Madang	(unidentified)	A densely growing fernlike plant, common in younger secondary forest and difficult to clear. Also used in the construction of bubu 'fish traps'.
Belayan madang	,,	'Stand of madang'.
Manang padi	(unidentifed)	One of the plants in the lapik benih 'place of the seed rice', it keeps the rice spirits from sickness (literally, 'the shaman of the rice').
Manok	Gallus sp.	The domestic fowl, kept for use in sacrificial offerings and for cock fighting.
Mayas	Pongo pygmaeus pygmaeus L.	The orang-utan, hunted and eaten by some Kantu', by others considered to be kinsmen and antu 'spirits'.
Mayau	Felis domestica L.	The domestic cat, kept for rodent control and called the rajah rumah 'king of the longhouse'.
Melaban	Tristania sp.?	A large, hard-wooded tree found in primary forest.
Melipan	Scolopendra sp.	A large, reddish, poisonous centipede (having 42 feet).
Memuas	Lacedo pulchella Horsfield.	The Banded Kingfisher, one of the seven major omen birds.
Menturun	Arctictis binturong penicillata Temminck.	The bearcat, a predator of chickens.
Mpangau	Leptocorisa oratorius F.	The rice bug, a major swidden pest.
Mpartikar	(unidentified)	An insect that sings early in the morning, formerly regarded as an omen animal during the planting season.
Mpelawa'	Arachnida spp.	Generic term for spiders.
Mpelawa' mpeliau	Lampropelma sp.	A large, poisonous, burrowing spider - the notorious bird-eating spider - named

		after the mpeliau 'gibbon' whose fur its long body hair resembles.
Mpeliau	Hylobates muelleri Martin.	The Bornean gibbon, not a swidden pest.
Mpit	Gryllotalpa hirsuta Burm.	The mole cricket, a swidden pest that eats the roots of the rice plant.
Munsang	Paradoxurus hermaphroditus sabanus Thomas.	The common palm civet, a predator of chickens even in the vicinity of the longhouse.
Munti'	(unidentified)	A species of bamboo that occurs as a weed in secondary forest swiddens located within the riverine floodzone.
Mwanyi'	Apis sp.	A wild bee that furnishes the Kantu' with both honey and edible grubs. It has a painful sting and nests high up in giant forest trees.
Ncekur	Kaempfaria galanga L.?	An herb planted in the lapik benih 'place of the seed rice'.
Nenak	Copsychus malabaricus Scopoli.	The White-rumped Shama, one of the seven major omen birds.
Ngingit	Cicada sp.	The cicada, the late afternoon call of which formerly was regarded as an omen during the planting season.
Ngkabang	Isoptera borneensis Scheff ex Burck, and Shorea spp.	Candlenut trees, yielding a fruit every few years that is gathered and sold to be made into tallow.
Ngkayu	(various)	Anything eaten with the rice main dish.
Ngkerbai	Tarenna sp.?	One of the ritually important plants in the lapik benih 'place of the seed rice'.
Ngkulau	Neofelis nebulosa diardi Cuvier.	The clouded leopard, probably the largest cat in Borneo

		and a predator of chickens.
Nsabi	Brassica juncea (L.) Czern. & Coss.	Indian mustard, a swidden cultigen.
Ntamba'	Pteropus vampyrus natunae Andersen.	The 'flying fox', hunted for its flesh.
Ntimun	Cucumis sativus L.	The cucumber, a swidden cultigen.
Nyamok	Anopheles spp., and others.	Mosquitoes, some of which carry malaria, and all of which are diligently killed upon sight within the longhouse.
Nyumoh	Macaca nemestrina nemestrina L.	The pig-tailed macaque, a pest of maize, cucumbers and rice.
Padi	Oryza sativa L.	Rice, the principal swidden cultigen and food staple. More specifically, padi refers to unhusked rice as opposed to husked rice or beras, and it also refers to nonglutinous varieties as opposed to glutinous ones or puloi.
Pako'	(various)	Forest ferns that are gathered for their edible tips.
Pantak	Vespidae.	A wasp that is feared when clearing secondary forest, lives in large hives or nests.
Papau	Harpactes diardii Temminck.	Diard's Trogon, one of the seven major omen birds.
Pari'	Momordica charantia L.	The bitter gourd, a swidden cultigen.
Pelandok	Tragulus spp.	The mouse deer, hunted for its flesh and regarded as an omen animal.
Pencelap	(unidentified)	One of the ritually important plants in the lapik benih 'place of the seed rice'.
Penyaut	Shorea sp.	One of the most valued hardwood trees of the primary forest, used in making

		canoes.
Perupok	Pandanus sp.	A screw pine, planted for its long barbed leaves, which are used in making kajang 'awnings'.
Pipit	Erythrura spp. and Lonchura spp.	The rice sparrow, any one of several species of Munia that prey on the rice.
Pisang	Musa sapientum L.	The dessert banana, a swidden cultigen.
Puloi	Oryza sativa L.f. glutinosa Auct.	Glutinous rice varieties.
Purun	Fimbristylis globulosa (Retz.) Kunth ?	A sedge growing wild in swampland, commonly used in plaiting.
Pusut	Luffa cylindrica Auct. non Roem.	The dish-cloth gourd, a swidden cultigen.
Rabing	Crocodilus porosus Schneid.	The estuarine crocodile, formerly found occasionally in the Kantu' territory, distinguished from the bwaya by its short snout.
Rangki	Capsicum frutescens L.	Chili peppers, cultivated in the swiddens as a relish.
Rasong	Nasalis larvatus Wurmb.	The proboscis monkey, infrequently encountered in the Kantu' territory, hunted for its flesh.
Raung	Rana spp., and Bufo spp.	Frogs, some hunted for their flesh, others utilized for their poison sacks.
Remaung	Panthera tigris.	The tiger, seemingly known to the Kantu' although it is unclear whether it is native to Borneo.
Rengas	Gluta sp.	A tree whose toxic sap is burned in the swiddens to drive off insect pests.
Retak	Vigna unguiculata (L.) Walp.	The long bean, a swidden cultigen.
Rian	Durio zibethinus Murr.	The durian tree, the fruit of which is prized above

		all others by the Kantu'.
Rusa'	Cervus unicolor brookei Hose.	The sambar deer, a swidden pest but also a prized game animal.
Rusa' lalang	,,	'Grass deer', a short-horned, light-colored variety.
Rusa' balot	,,	'Covered deer', a short-horned, dark-colored and long-haired variety.
Sampok	Coptotermes sp.	A termite that preys on both standing and felled trees.
Sawa'	Python reticulatus.	The reticulated python, a predator of chickens, also hunted for its own flesh.
Semengkok	Citrullus lanatus (Thunb.) Mansf.	The watermelon, a swidden cultigen.
Serai	Cymbopogon citratus (DC) Stapf.	Lemon grass, a swidden cultigen used for flavoring in cooking fish, planted in the lapik benih 'place of the seed rice'.
Singan	(unidentified)	An uncultivated plant that is much used in making plaited goods. 'Good' people are reincarnated as this plant.
Singkenyang	Eurycles amboinensis (L.) Loud ?	Foremost of the ritually important plants in the lapik benih 'place of the seed rice'.
Tando'	(unidentified)	A variety of rattan growing wild in primary forest, valued for its great strength in making plaited goods.
Tebelian	Eusideroxylon zwageri Teijsm. & Binn.	The Bornean ironwood tree, providing the most valued of hardwoods for use in houses, canoes, etc.
Tebu	Saccharum officinarum L.	The sugar cane, a swidden cultigen.
Tedung	Naja naja L., and Naja hannah Cantor.	Hooded snakes, especially the Common Black Cobra and the King Cobra. Their form

		can be taken by evil spirits.
Tembako	Nicotiana tabacum L.	Tobacco, a swidden cultigen.
Temiang	Schizostachyum blumei Nees ?	A long-jointed bamboo used by some Dayaks groups for blow pipes, primarily used by the Kantu' in ritual defense of their swiddens.
Tengiling	Manis javanica Desmarest.	The pangolin or scaly anteater, hunted for the value of its scales (sold in the Chinese bazaars of Sarawak).
Tepus Sarawak	Zingiber sp.	A swidden cultigen, the leaves of which are used for seasoning in cooking.
Terong	Solanum melongena L.	The egg-plant, a swidden cultigen.
Tinting lema'	(unidentified)	A minor omen bird, said to be the ulun 'slave' of the papau and the beragai.
Tubai	(various)	A poison used to catch fish, obtained from various plants, some of which are planted in the lapik benih 'place of the seed rice'.
Tubai buah	Croton tiglium L. ?	'The fruit poison'.
Tubai urat	Derris sp.	'The root poison'.
Tubo'	(various)	Edible bamboo shoots, especially of the species munti'.
Tupai	Sciuridae.	Various species of squirrel, all swidden pests.
Ubi kayu	Manihot esculenta Crantz.	Cassava, a swidden cultigen.
Ubi pulai	,,	'Gone home cassava', remnants of a cultivated stand running wild in fallowed swidden land.
Ukoi	Canis familiaris L.	The domestic dog, kept primarily for hunting, but also eaten. The Kantu' seem to have knowledge of a wild dog as well, but the true

		dhole (Cuon alpinus [Pallas]) is supposed to be absent from Borneo (Harrison 1964:178; Medway 1977:131) although see Hose (1929:144).
Ulat	(various)	General term for caterpillar, grub, worm, etc.
Ulat jingkal	Spodoptora mauritia acronyctoides Gn.	'Looping caterpillar' or army worm, a rice pest.
Ulat tusok	Chilo spp. & Scirpophaga spp.	'Boring caterpillar' or stem borer, a rice pest.
Utai	(various)	General term for cultigens of ritual importance, planted in the 'place of the seed rice'.
Wi	(various)	General term for rattans, including both cultivated and uncultivated ones.

Bibliography

Aldrich-Blake, F.P.G.
1980 Long-Tailed Macaques. In Chivers, pp. 147-165.
Allan, W.
1965 The African Husbandman. Edinburgh: Oliver and Boyd.
Aminardi
1977 Personal communication.
Appell, George N.
1971 Systems of Land Tenure: A Problem in Ecological Determinism. Borneo Research Bulletin 3: 17-21.
1977 The Uses of Anthropological Research: Costing Social Change. Paper read at the Annual Meeting of the American Association for the Advancement of Science, 20-25 February, Denver.
n.d. Observational Procedures for Land Tenure and Kin Groupings in the Cognatic Societies of Borneo. Mimeographed. Walthum, Mass.: Brandeis University.
Appell, George N., ed.
1976a The Societies of Borneo: Explorations in the Theory of Cognatic Social Structure. Special Publication, no. 6. Washington (D.C.): American Anthropological Association.
1976b Studies in Borneo Society: Social Process and Anthropological Explanation. Center for Southeast Asian Studies, Northern Illinois University, Special Report no. 12. Detroit: Cellar Book Shop.
Avé, Jan; King, Victor; and de Wit, Joke
1983 West Kalimantan: A Bibliography. Koniklijk Instituut Voor Taal-, Land- En Volkenkunde, Bibliographical Series no. 13. Dordrecht: Foris Publications.
Barlett, Peggy F.
1980a Introduction: Development Issues and Economic Anthropology. In Barlett 1980, pp. 1-16.
1980b Adaptive Strategies in Peasant Agricultural Production. Annual Review of Anthropology 9: 545-573.
Barlett, Peggy F., ed.
1980 Agricultural Decision Making: Anthropological Contributions to Rural Development. New York: Academic Press.
Beckerman, Stephen
1983 Does the Swidden Ape the Jungle? Human Ecology 11 (1): 1-12.

Bennett, John W.
1973 Ecosystemic Effects of Extensive Agriculture. Annual Review of Anthropology 2: 36-45.

Boserup, Ester
1965 The Conditions of Agricultural Growth. Chicago: Aldine.

Bouman, M.A.
1924 Ethnografische Aanteekeningen omtrent de Governementslanden in de Boven-Kapoeas, Westerafdeeling van Borneo. Tijdschrift voor Indische Taal, Land en Volkenkunde 64 (185): 173-195.

Bucher, John Paul
1980 An Agricultural Development Program Plan for Slash and Burn Kantu Farmers in West Kalimantan. M.P.S. thesis, Department of Agriculture, Cornell University.

Burdon, T.W., and Parry, M.L.
1954 Malayan Fishing Methods. Monographs on Malay Subjects, no. 2. Journal of the Malay Branch of the Royal Asiatic Society 29 (2): 1-180.

Burkill, L.H.
1935 Dictionary of the Economic Products of the Malay Peninsula. London: Crown Agents for the Colonies.

Burling, Robbins
1965 Hill Farms and Padi Fields. Englewood Cliffs (N.J.): Prentice-Hall.

Cancian, Frank
1972 Change and Uncertainty in a Peasant Economy. Stanford: Stanford University Press.
1977 Can Anthropology Help Agricultural Development? Culture and Agriculture 2: 1-8.

Carneiro, R.L.
1960 Slash-and-Burn Agriculture: A Closer Look at its Implications for Settlement Patterns. In Men and Cultures; Selected Papers of the Fifth International Congress of Anthropological and Ethnological Sciences, ed. Anthony F.C. Wallace, pp. 229-234. Philadelphia: University of Pennsylvania Press.
1961 Slash-and-Burn Cultivation Among the Kuikuru and its Implications for Cultural Development in the Amazon Basin. Anthropological Supplement 2: 47-67.

Cense, A.A., and Uhlenbeck, E.M.
1958 Critical Survey of Studies on the Languages of Borneo. Koninklijk Instituut Voor Taal-, Land- En Volkenkunde, Bibliographical Series no. 2. The Hague: Martinus Nijhoff.

Chayanov, Alexander
1966 The Theory of Peasant Economy. Homewood (Illinois): Richard D. Irwin.

Chesnov, Ia.V.
1973 Domestication of Rice and the Origin of Peoples Inhabiting East and Southeast Asia. Paper read at the Ninth International Congress of Anthropological and Ethnological Sciences, August-September, Chicago.

Chin, Fook Hon

1981 Edible and Poisonous Fungi from the Forests of Sarawak. Sarawak Museum Journal 29 (50) n.s.: 211-225.

Chivers, David J., ed.

1980 Malayan Forest Primates: Ten Years' Study in Tropical Rain Forest. New York: Plenum.

Chivers, D.J., and Raemaekers, J.J.

1980 Long-Term Changes in Behavior. In Chivers pp. 209-260.

Clarke, William C.

1966 From Extensive to Intensive Shifting Cultivation: A Succession from New Guinea. Ethnology 5: 347-359.

1971 Place and People: The Ecology of a New Guinean Community. Berkeley: University of California Press.

Colfer, Carol J. Pierce

1981 Women, Man, and Time in the Forests of East Kalimantan. Borneo Research Bulletin 13 (2): 75-85.

1983 Change and Indigenous Agroforestry in East Kalimantan. Borneo Research Bulletin 15 (1): 3-21; 15 (2): 70-87.

Condominas, Georges

1977 We Have Eaten the Forest. New York: Farrar, Strauss, Giroux.

Conklin, Harold C.

1954 The Relation of Hanunóo Culture to the Plant World. Ph.D. dissertation, Department of Anthropology, Yale University. Ann Arbor: University Microfilms.

1957 Hanunóo Agriculture: A Report on an Integral System of Shifting Cultivation in the Philippines. Forestry Development Paper, no. 12. Rome: FAO.

1959 Shifting Cultivation and Succession to Grassland Climax. Proceedings of the Ninth Pacific Science Congress 7: 60-62.

1963 The Study of Shifting Cultivation. Studies and Monographs, no. 6. Washington (D.C.): Pan American Union.

1980 Ethnographic Atlas of Ifugao: A Study of Environment, Culture, and Society in Northern Luzon. New Haven: Yale University Press.

Cook, S.

1973 Production, Ecology and Economic Anthropology: Notes Toward an Integrated Frame of Reference. Social Science Information 12 (1): 25-52.

Cotter, Conrad P., and Saito, S.

1965 Bibliography of English Language Sources on Human Ecology, Eastern Malaysia and Brunei. Asian Studies at Hawaii, no. 1. Honolulu: University of Hawaii, Asian Studies Program.

Crain, Jay B.

1976 Ritual Process in a Bornean Rice Harvest. In Appell 1976 b.

Dammerman, K.W.

1929 The Agricultural Zoology of the Malay Archipelago: The Animals Injurious and Beneficial to Agriculture, Horticulture and Forestry in the Malay Peninsula, the Dutch East Indies and the Philippines. Amsterdam: J.H. de Bussy.

Delacour, Jean
1947 Birds of Malaysia. New York: Macmillan.
De Schlippe, Pierre
1956 Shifting Cultivation in Africa. London: Routledge and Kegan Paul.
Dixon, G.
1974 Dayak Land Tenure: An Alternative to Ecological Determinism. Borneo Research Bulletin 6 (1): 5-15.
Dove, Michael R.
1977a Points on the Collection of Ethnographic Data Among Tribal Peoples in Borneo. Borneo Research Bulletin 9 (2): 54-67.
1977b Pioneer Agriculture as a Wartime Adaptation. Paper read at the 76th Annual Meeting of the American Anthropological Association, December, Washington (D.C.).
1979 Pain and Sacrifice in Kantu' Drinking. Paper read at the 78th Annual Meeting of the American Anthropological Association, 28 November - 1 December, Los Angeles.
1980a Development of Tribal Land Rights in Borneo. Borneo Research Bulletin 12 (1): 3-11.
1980b The Use of the Finger Knife Among the Iban. American Ethnologist 1 (2): 371-373.
1980c The Swamp Rice Swiddens of the Kantu' of West Kalimantan. In Tropical Ecology and Development, ed. J.I. Furtado, pp. 953-956. Kuala Lumpur: International Society of Tropical Ecology.
1980d Sexual Equality and Economic Development Among Tropical Forest Horticulturalists. Paper read at the 79th Annual Meeting of the American Anthropological Association, 3-7 December, Washington D.C.
1981a Subsistence Strategies in Rainforest Swidden Agriculture: The Kantu' at Tikul Batu. Ph.D. dissertation, Department of Anthropology, Stanford University. Ann Arbor: University Microfilms.
1981b Household Composition and Intensity of Labour. Bulletin of Indonesian Economic Studies 17 (3): 86-93.
1981c Symbiotic Relationships Between Human Populations and Imperata Cylindrica. In Conservation Inputs from Life Sciences, eds. M. Nordin et al., pp. 187-200. Bangi: Universiti Kebangsaan Malaysia.
1981d A Tribal Pork Taboo in Local and Regional Perspectives. Paper read at the 80th Annual Meeting of the American Anthropological Association, December, Los Angeles.
1982a Epistemological Issues in the Analysis of Land Tenure in Borneo. Borneo Research Bulletin 14 (1): 27-34.
1982b The Myth of the Communal Longhouse in Rural Development. In Too Rapid Rural Development, eds. Colin MacAndrews and L.S. Chin, pp. 14-78. Athens: Ohio University Press.
1982c The Dual Economy of Rubber and Swiddens in Kalimantan, and Its Implications for Development. Paper read at the Pertemuan Ilmiah Se Kalimantan, 11-14 November, Palangka Raya.

1983a Theories of Swidden Agriculture and the Political Economy of Ignorance. Agroforestry Systems 1 (3): 85-99.

1983b Forest Preference in Swidden Agriculture. Tropical Ecology 24 (1): 122-142.

1983c Evidence from Contemporary Mythology Regarding the Origins of Agriculture in Southeast Asia. Paper read at the 82nd Annual Meeting of the American Anthropological Association, 16-20 November, Chicago.

1983d Review of H.C. Conklin's 'Ethnographic Atlas of Ifugao' and its Implications for Theories of Agricultural Evolution in Southeast Asia. Current Anthropology 24 (4): 516-519.

1984 The Chayanov Slope in a Swidden Economy. In Durrenberger, pp. 97-132.

Forthcoming a. The Chimaera of Unbalanced Reciprocity in Peasant Labor: A Dayak Example. In Cooperatives: Self-Help Organization in the Third World, eds. Michael D. Levin and Jacque Meullmeester. Amsterdam: Royal Tropical Institute.

Forthcoming b. Intensive Versus Extensive Systems of Agriculture in Inner and Outer Indonesia. In Prospects and Resources for Village-Level Modernization and Livelihood Enhancement in Southeast Asia, ed. G.B. Hainsworth. Vancouver: UBC Press.

n.d. Pig-Humans and Human-Pigs: Transformation and Sacrifice in Melaban Kantu' Ritual. Mimeographed. Stanford, Calif.: Stanford University.

Dove, Michael R., and Sumitro, Achmad

In Press. Local Versus National Perception and Use of Natural Resources: A Case Study of Banjarese Ecology and Indonesian Development in South Kalimantan. Journal of Southeast Asian Studies.

Drake, Allen

1983 The Material Provisioning of Mualang Society in Hinterland Kalimantan Barat, Indonesia. Ph.D. dissertation, Department of Anthropology, Michigan State University. Ann Arbor: University Microfilms.

Driessen, P.M.; Buurman, P.; and Permadhy

1976 The Influence of Shifting Cultivation on a 'Podzolic' Soil from Central Kalimantan. In Peat and Podzolic Soils and Their Potential For Agriculture in Indonesia. Bulletin no. 3. Bogor: Soil Research Institute.

Dunselman, Donatus

1955 Kana Sera, Zang der Zwangerschap. Koninklijk Instituut Voor Taal-, Land- En Volkenkunde, Verhandelingen no. 17. The Hague: Martinus Nijhoff.

Durrenberger, E. Paul, ed.

1984 Chayanov, Peasants and Economic Anthropology. New York: Academic Press.

Edmonson, Wade C.

1976 Land, Food and Work in East Java. Monograph Series, no. 4. Armidale (N.S.W): Department of Geography, University of New England.

Ellen, Roy F.
1973 Nualu Settlement and Ecology: An Approach to the Environmental Relations of an Eastern Indonesian Community. Koninklijk Instituut Voor Taal-, Land- En Volkenkunde, Verhandelingen no. 83. The Hague: Martinus Nijhoff.

Enthoven, J.J.K.
1903 Bijdragen tot de Geographie van Borneo's Wester-Afdeeling. Leiden: E.J. Brill.

Frake, Charles O.
1955 Social Organization and Shifting Cultivation Among the Sindangan Subanun. Ph.D. dissertation, Department of Anthropology, Yale University. Ann Arbor: University Microfilms.

Freeman, Derek
1955 Iban Agriculture: A Report on the Shifting Cultivation of Hill Rice by the Iban of Sarawak. Colonial Research Studies, no. 18. London: Her Majesty's Stationery Office.
1957 The Family System of the Iban of Borneo. In The Developmental Cycle in Domestic Groups, ed. Jack Goody, pp. 15-52. Cambridge: Cambridge University Press.
1960a The Iban of Western Borneo. In Social Structure in Southeast Asia, ed. George P. Murdock, pp. 65-87. Viking Fund Publications in Anthropology, no. 29. New York: Wenner-Gren Foundation.
1960b Iban Augury. In Smythies 1960, pp. 73-98.
1961 On the Concept of the Kindred. Journal of the Royal Anthropological Institute 91: 192-220.
1967 Shaman and Incubus. The Psychoanalytic Study of Society 4: 315-343.
1970 Report on the Iban. London School of Economics Monographs on Social Anthropology, no. 41. London: The Athlone Press.
1981 Some Reflections on the Nature of Iban Society. Occasional Paper. Canberra: Department of Anthropology, Research School of Pacific Studies, the Australian National University.

Friedman, Jonathan
1975 Tribes, States and Transformations. In Marxist Analyses and Social Anthropology, ed. M. Bloch, pp. 161-202. London: Malaby.

Geddes, Wilhelm R.
1954 The Land Dyaks of Sarawak: A Report on a Social Economic Survey of the Land Dyaks of Sarawak Presented to the Colonial Social Science Research Council. Colonial Research Studies, no. 14. London: Her Majesty's Stationery Office.

Geertz, Clifford
1963 Agricultural Involution: The Processes of Ecological Change in Indonesia. Association of Asian Studies Monographs and Papers, no. 11. Berkeley: University of California Press.

Glenister, A.G.
1951 The Birds of the Malay Peninsula, Singapore and Penang. Kuala Lumpur: Oxford University Press.

Gourou, Pierre
1953 The Tropical World: Its Social and Economic Condition and Future Status. London: Longmans.
Grijpstra, B.G.
1976 Common Efforts in the Development of Rural Sarawak, Malaysia. Assen: Van Gorcum.
Grist, Donald H.
1975 Rice. London: Longmans.
Gudeman, S.
1978 The Demise of a Rural Economy: From Subsistence to Capitalism in a Latin American Village. Boston: Routledge and Kegan Paul.
ter Haar, B.
1948 Adat Law in Indonesia. New York: Institute of Pacific Relations.
Haddon, A.C., and Start, L.E.
1936 Iban or Sea Dayak Fabrics and Their Patterns. Cambridge: Cambridge University Press.
Hanks, Lucien M.
1972 Rice and Man: Agricultural Ecology in Southeast Asia. Arlington Heights (Ill.): AHM.
Harner, Michael
1970 Population Pressure and the Social Evolution of Agriculturalists. Southwestern Journal of Anthropology 26: 67-86.
Harris, David R.
1972 Swidden Systems and Settlement. In Man, Settlement and Urbanism, eds. P.J. Ucko, R. Tringham, and G.W. Dimbleby, pp. 245-262. London: Duckworth.
Harris, Marvin
1966 The Cultural Ecology of India's Sacred Cattle. Current Anthropology 7: 51-59.
1979 Cultural Materialism: The Struggle for a Science of Culture. New York: Random House.
Harrison, John
1964 An Introduction to the Mammals of Sabah. Kota Kinabalu: The Sabah Society.
Harrisson, Tom
1960 Birds and Men in Borneo. In Smythies 1960, pp. 20-61.
Hempel, Carl
1958 The Logic of Functional Analysis. In Symposium on Sociological Theory, ed. Llewellyn Gross, pp. 271-310. Evanston: Row, Peterson.
Hill, R.D.
1977 Rice in Malaya: A Study in Historical Geography. Kuala Lumpur: Oxford University Press.
Hoben, Allan
1982 Anthropologists and Development. Annual Review of Anthropology 11: 349-375.
Hoffman, Carl L.
1983 Punan. Ph.D dissertation, Department of Anthropology, Uni-

versity of Pennsylvania. Ann Arbor: University Microfilms.
Holleman, J.F., ed.
1981 Van Vollenhoven on Indonesian Adat Law. KITLV Translation Series, no. 20. The Hague: Martinus Nijhoff.
Hooker, M.B.
1978 Adat Law in Indonesia. Kuala Lumpur: Oxford University Press.
Hose, Charles
1929 The Field-Book of a Jungle-Wallah. London: H.F. and G. Witherby.
Hose, Charles, and McDougall, William
1912 The Pagan Tribes of Borneo. London: MacMillan & Co.
Howard, Allan
1963 Land, Activity Systems, and Decision-Making Models in Rotuma. Ethnology 2: 407-440.
Howell, Rev. William, and Bailey, D.J.S.
1900 A Sea Dayak Dictionary. Singapore: American Mission Press.
Hudson, Alfred E.
1970 A Note on Selako: Malayic and Land Dyak Languages in Western Borneo. Sarawak Museum Journal 18 (36-37) n.s.: 301-318.
Izikowitz, Karl G.
1951 Lamet: Hill Peasants in French Indochina. Etnografiska Studier, no. 17. Gothenburg: Etnografiska Museet.
Janzen, Daniel H.
1974 Tropical Blackwater Rivers, Animals and the Mast Fruiting of the Dipterocarpaceae. Biotropica 6 (2): 69-103.
Jensen, Erik
1974 The Iban and Their Religion. Oxford Monographs on Social Anthropology. Englewood Cliffs (N.J.): Oxford University Press.
Jessup, Timothy C.
1981 Why Do Apo Kayan Shifting Cultivators Move? Borneo Research Bulletin 13 (1): 16-32.
Johnson, Allan
1971 Sharecroppers of the Sertão. Stanford: Stanford University Press.
Kalshoven, L.G.E.
1981 Pests of Crops in Indonesia. Translated by P.A. van der Laan and G.H.L. Rothschild. Jakarta: P.T. Ichtiar Baru/ Van Hoeve. (Original: De Plagen van de Cultuurgewassen in Indonesië. The Hague/Bandung, 1950.)
Kielstra, E.G.
1893 Bijdragen tot de Geschiedenis van Borneo's Westerafdeeling. De Indische Gids 15 (2): 2091-2104.
King, Victor
1972 Additional Notes on the Malohs and Related Peoples of Kalimantan Barat. Sarawak Museum Journal 20 (40-41) n.s.: 83-105.
1973 Kantus of the Kapuas River. Sarawak Gazette 1389: 254-257.
1975a Further Problems in Bornean Land Tenure Systems: Comments

on an Argument. Borneo Research Bulletin 7 (1): 12-16.
1975b Stones and the Maloh of Indonesian West Borneo. Journal of the Malaysian Branch of the Royal Asiatic Society 48 (1): 104-119.
1976a Cursing, Special Death and Spirits in Embaloh Society. Bijdragen 132 (1): 124-145.
1976b The Peoples of the Middle and Upper Kapuas: Possible Research Projects in West Kalimantan. Borneo Research Bulletin 8 (2): 87-105.
1977 Unity, Formalism and Structure: Comments on Iban Augury and Related Problems. Bijdragen 133 (1): 63-87.
1979 Ethnic Classification and Ethnic Relations: A Borneo Case Study. Occasional Papers, no. 2. Hull: The University of Hull, Centre for South-East Asian Studies.

King, Victor, ed.
1978 Essays on Borneo Societies. Hull Monographs on South-East Asia, no. 7. Oxford: Oxford University Press.

Kunstadter, Peter
1978 Subsistence Agricultural Economies of Lua' and Karen Hill Farmers, Mae Sariang District, Northwestern Thailand. In Kunstadter et al., pp. 74-133.

Kunstadter, Peter; Chapman, E.C.; and Sabhasri, Sanga
1978 Farmers In The Forest. Honolulu: East-West Center.

Lawless, Robert
1975 Effects of Population Growth and Environment Changes on Divination Practises in Northern Luzon. Journal of Anthropological Research 31: 18-33.

Linares, Olga F.
1976 "Garden Hunting" in the American Tropics. Human Ecology 4 (4): 331-349.

Loh, Chee Yin
1980 An Index to All the Issues Published from 1911 to 1979. Sarawak Museum Journal 28 (49) n.s.: 1-198.

Low, Hugh
1848 Sarawak: Its Inhabitants and Productions. London: Richard Bentley.

Lumholtz, Carl
1920 Through Central Borneo: An Account of Two Years' Travel in the Land of the Head-Hunters Between the Years 1913 and 1917. New York: Charles Scribner's Sons.

MacDonald, Malcolm
1956 Borneo People. London: Jonathan Cape.

Maxwell, Allen
1978 Generalization about Place Names: Some Borneo Examples. Borneo Research Bulletin 10 (2): 75-82.
1979 Local Classification and the Organization of Rice Subsistence Production in a Kadayan Malay Village. Paper read at the Fifth International Symposium of Tropical Ecology, 16-21 April, Kuala Lumpur.

Medway, Lord
1977 Mammals of Borneo. Monograph no. 7. Kuala Lumpur: The Malaysian Branch of the Royal Asiatic Society.
Metcalf, Peter
1976 Birds and Dieties in Borneo. Bijdragen 132 (1): 96-123.
Miles, Douglas
1972 Land, Labour and Kin Groups Among Southeast Asian Shifting Cultivators. Mankind 8: 185-197.
1979 The Finger Knife and Ockham's Razor: A Problem in Asian Culture History and Economic Anthropology. American Ethnologist 6 (2): 223-234.
Moerman, M.
1968 Agricultural Change and Peasant Choice in a Thai Village. Berkeley: University of California Press.
Moore, Omar Khayam
1957 Divination - A New Perspective. American Anthropologist 59: 69-74.
van Naerssen, F.H.
1951-1952 Een Streekenderzoek in West Borneo. Indonesie 5: 145-156.
Nagel, Ernest
1961 The Structure of Science. New York: Harcourt, Brace and World.
Neill, Wilfred T.
1973 Twentieth Century Indonesia. New York: Columbia University Press.
Netting, Robert McC.
1965 Household Organization and Intensive Agriculture: The Kofyar Case. Africa 35: 422-429.
1968 Hill Farmers of Nigeria: Cultural Ecology of the Kofyar of the Jos Plateau. Seattle: University of Washington Press.
1974 Agrarian Ecology. Annual Review of Anthropology 3: 21-56.
Nicholson, D.I.
1965 A Study of Virgin Forest Near Sandakan North Borneo. Proceeding of the Symposium on Ecological Research Into Humid Tropics Vegetation. Kuching: UNESCO.
Nyandoh, R.
1958 How Land Dayaks Net Wild Boar. Sarawak Museum Journal 8 (11) n.s.: 374-382.
Nye, D.H., and Greenland, D.J.
1960 The Soil Under Shifting Cultivation. Commonwealth Bureau of Soils, Technical Communication no. 15. Harpenden (England): Commonwealth Agricultural Bureaux.
Ochse, J.J.
1977 Vegetables of the Dutch East Indies: Survey of the Indigenous and Foreign Plants Serving As Pot-Herbs and Side-Dishes. In collaboration with R.C. Bakhuizen van den Brink. Canberra: Australian National University Press. (Original: Indische Groenten, Bogor 1931.)

Orlove, B.S.
1977 Integration through Production: The Use of Zonation in Espinar. American Ethnologist 4 (1): 84-101.
Ormeling, F.J.
1956 The Timor Problem: Geographical Interpretation of an Underdeveloped Island. The Hague/Jakarta: Groningen.
Padoch, Christine
1978 Migration and its Alternatives Among the Iban of Sarawak. Ph.D. dissertation, Faculty of Political Science, Columbia University. Ann Arbor: University Microfilms.
1982 Land Use in New and Old Areas of Iban Settlement. Borneo Research Bulletin 14 (1): 3-14.
1983 Agricultural Practices of the Kerayan Lun Dayeh. Borneo Research Bulletin 15 (1): 33-38.
Padwick, G. Watts
1950 Manual of Rice Diseases. Kew: Commonwealth Mycological Institute.
Pelzer, Karl J.
1948 Pioneer Settlement in the Asiatic Tropics. Special Publication, no. 29. New York: American Geographical Society.
Pfeffer, P.
1959 Biologie et Migrations du Sanglier de Borneo. Mammalia 23: 227-303.
Pringle, Robert
1970 Rajahs and Rebels: The Ibans of Sarawak Under Brooke Rule, 1841-1941. Ithaca: Cornell University Press.
Provinse, J.H.
1937 Cooperative Ricefield Cultivation Among the Siang Dyaks of Central Borneo. American Anthropologist 39: 77-102.
Rambo, A. Terry
1981 Fire and the Energy Efficiency of Swidden Agriculture. EAPI Preprint. Honolulu: East-West Center.
Rappaport, Roy A.
1968 Pigs for the Ancestors: Ritual in the Ecology of a New Guinea People. Hew Haven: Yale University Press.
1977 Ecology, Adaptation and the Ills of Functionalism. Michigan Discussions in Anthropology 2: 139-190.
1979 Ecology, Meaning and Religion. Richmond (California): North Atlantic Books.
Richards, Anthony
1981 An Iban-English Dictionary. Oxford: Clarendon Press.
Richards, Paul W.
1952 The Tropical Rain Forest. Cambridge: Cambridge University Press.
Robequain, Charles
1954 Malaya, Indonesia, Borneo and the Philippines. London: Longmans, Green and Company.
Rosaldo, Michelle Z.
1980 Knowledge and Passion: Ilongot Notions of Self and Social Life. Cambridge: Cambridge University Press.

Ross, Eric B., ed.

1980 Beyond the Myths of Culture: Essays in Cultural Materialism. New York: Academic Press.

Roth, Henry Ling

1896 The Natives of Sarawak and British North Borneo. London: Truslove and Hanson.

Rothschild, G.H.L.

1970 Observations on the Ecology of the Rice-Ear Bug (Leptocorisa oratorius (F.) (Hemiptera: Alydidae) in Sarawak (Malaysian Borneo). Journal of Applied Ecology 7: 147-167.

Sahlins, Marshall

1957 Land Use and the Extended Family in Moala, Fiji. American Anthropologist 59: 449-462.

1971 The Intensity of Domestic Production in Primitive Societies: Social Inflections of the Chayanov Slope. In Studies in Economic Anthropology, ed. George Dalton, pp. 30-51. Anthropological Studies, no. 7. Washington (D.C.): American Anthropological Association.

1972 Stone Age Economics. Chicago: Aldine.

St. John, Spenser B.

1862 Life in the Forests of the Far East. London: Smith, Elder.

Sanchez, P.A.

1976 Properties and Management of Soils in the Tropics. New York: John Wiley and Sons.

Sandin, Benedict

1967 The Sea Dayaks of Borneo Before White Rajah Rule. London: MacMillan.

1980 Iban Adat and Augury. Penang: Penerbit University Sains Malaysia.

Sather, Clifford

1977 Nanchang Padi: Symbolism of Saribas Iban First Rites of Harvest. Journal of the Malaysian Branch of the Royal Asiatic Society 50 (2): 150-170.

1978 Iban Folk Mycology. Sarawak Museum Journal 26 (47) n.s.: 81-102.

1980 Symbolic Elements in Saribas Iban Rites of Padi Storage. Journal of the Malaysian Branch of the Royal Asiatic Society 53 (2): 67-95.

Schefold, Reimar

1972-1973 Religious Involution: Internal Change, and its Consequences, in the Taboo System of the Mentawaians. Tropical Man 5: 46-81.

Schieffelin, Edward L.

1975 Felling the Trees On Top of the Crop. Oceania 46 (1): 25-39.

Schlegel, Stuart A.

1979 Tiruray Subsistence: From Shifting Cultivation to Plow Agriculture. Quezon City: Ateneo de Manila University Press.

Scott, James C.

1976 The Moral Economy of the Peasant: Rebellion and Subsistence in Southeast Asia. New Haven: Yale University Press.

Scott, N.C.
1956 A Dictionary of Sea Dyak. London: School of Oriental and African Studies, University of London.

Seavoy, Ronald E.
1973 The Shading Cycle in Shifting Cultivation. Annals of the Association of American Geographers 63 (4): 522-528.
1975 The Origin of Tropical Grasslands in Kalimantan. Journal of Tropical Geography 40: 48-52.
1977 Social Contraints on Food Production in Indonesian Subsistence Culture. Journal of Southeast Asian Studies 8 (1): 15-30.

Shelford, Robert W.C.
1917 A Naturalist in Borneo. New York: E.P. Dutton.

Smythies, Bertram E.
1981 The Birds of Borneo. 3rd rev. ed. Kota Kinabalu/Kuala Lumpur: The Sabah Society with The Malayan Nature Society.

Smythies, Bertram E., ed.
1960 The Birds of Borneo. Edinburgh: Oliver and Boyd.

Spencer, J.E.
1966 Shifting Cultivation in Southeastern Asia. University of California Publications in Geography, vol. 19. Berkeley: University of California Press.

Spooner, Brian, ed.
1972 Population Growth: Anthropological Implications. Cambridge: MIT Press.

Street, J.
1969 An Evaluation of the Concept of Carrying Capacity. Professional Geographer 21 (2): 104-107.

Sutlive, Vinson H.
1976 The Iban Manang: An Alternate Route to Normality. In Appell 1976b, pp. 64-71.

Vayda, Andrew P.
1961 Expansion and Warfare Among Swidden Agriculturalists. American Anthropologist 63: 346-358.
1981 Research in East Kalimantan on Interactions Between People and Forests: A Preliminary Report. Borneo Research Bulletin 13 (1): 3-15.

Vayda, Andrew P., and McCay, Bonnie J.
1975 New Directions in Ecology and Ecological Anthropology. Annual Review of Anthropology 4: 293-306.

Vayda, Andrew P., and Jessup, Timothy C.
1980 Shifting Cultivation and Patch Dynamics in an Upland Forest in East Kalimantan, Indonesia. Research proposal submitted to the 'Consortium for the Study of Man's Relationship with the Global Environment'. Typescript.

Waddell, Eric
1972 Agricultural Practises, Environment and Society in the Central Highlands of New Guinea. Seattle: University of Washington Press.

Wallace, Benjamin J.
1970 Hill and Valley Farmers. Cambridge: Schenkman.
Watters, R.F.
1971 Shifting Cultivation in Latin America. Forestry Development Paper, no. 17. Rome: FAO.
Weinstock, Joseph A.
1979 Land Tenure Practises of the Swidden Cultivators of Borneo. M.A. thesis, Department of Rural Sociology, Cornell University.
White, Gilbert
1789 The Natural History of Selborne. Reprint. Oxford: Oxford University Press, 1902.
Whitmore, Timothy C.
1975 The Tropical Rain Forests of the Far East. London: Oxford University Press.
Whyte, Robert O.
1974 Rural Nutrition in Monsoon Asia. Kuala Lumpur: Oxford University Press.
Wilkinson, R.J.
1959 A Malay-English Dictionary. London: MacMillan.
Williams, C.N.
1975 The Agronomy of the Major Tropical Crops. Kuala Lumpur: Oxford University Press.
Wittfogel, Karl
1957 Oriental Despotism. New Haven: Yale University.
Worsley, P.
1956 The Kinship System of the Tallensi: A Reevaluation. Journal of the Royal Anthropology Institute 86: 37-77.

Indexes

Following Condominas (1977: xi-xii) and Conklin (1980: 108), I have made the indexes an integral part of the analysis that is contained in this volume. I have attempted to make not just a concordance of names or terms found in the text, but an analytic index containing all concepts, constructs, associations, and arguments occurring in the text. This index is intended to help readers investigate both the main topics of the book, and topics that were discussed only in passing, but for which relevant data is nevertheless present in the text.

The material that is indexed here includes not only the written text, but also all tables and figures, as well as the section of notes and the glossary at the end of the book, although not the bibliography. The resulting index is subdivided for ease of use into six categories, and hence six separate subindexes, as follows: (I) authors, (II) ethnic and geographic names, (III) flora and fauna, (IV) Tikul Batu households, (V) Tikul Batu swiddens, and (VI) the main, general index. Entries are extensively cross-referenced both within and among the separate indexes. When a cross-reference to another index is made, the latter is designated by its roman numeral (I-VI, as listed above). For example, the entry 'Gardens' in the General Index (VI) is followed by 'SEE ALSO Pepper (III)', which indicates that further relevant references can be found in the Flora and Fauna Index (III), under the heading 'Pepper'.

In each entry, the material cited is ordered as follows: (1) page number, for all material in the text and glossary; (2) page number plus endnote number (indicated by subscript 'n' plus numeral), for all material in the section of notes (reference to which, therefore, is based on the number of the page to which the note refers, not the page on which it is printed); (3) table number; (4) figure number; and (5) cross references. Subheadings are listed in alphabetical order, with two exceptions. First, when the subheadings refer to specific stages of the swidden cycle (e.g., selecting, slashing, felling, etc.), they are usually listed in chronological order, just as they are in the text itself. Second, the subheading 'other' is usually listed after all other subheadings, even where this violates alphabetical order.

I. AUTHORS

II. ETHNIC AND GEOGRAPHIC NAMES

III. FLORA AND FAUNA

All references to specific flora and fauna are indexed in this section. This includes references in Kantu', Latin, and English. The last page reference in each case refers to the appropriate entry in the Glossary (viz., pages 419-431), which can be consulted for the identification of each plant or animal. References to certain general or analytic categories of flora and fauna, such as 'Forest' or 'Pests', are indexed in section VI.

IV. TIKUL BATU HOUSEHOLDS

This index contains all references to individual households within the longhouse Tikul Batu. These include the fourteen households that were present during the 1974-1975 farm year, and the sixteen households that were present during the 1975-1976 farm year. This index also includes all references to individual named members of each household. It does not include references to discussions, tables (etc.) that are based on data from a number of households but do not specifically mention the individual households involved.

V. TIKUL BATU SWIDDENS

This index contains all references to individual swiddens cultivated by the households of Tikul Batu. These include the thirty-four swiddens cultivated during the 1974-1975 farm year, numbered 1-34, and the thirty-five swiddens cultivated during the 1975-1976 farm year, numbered 35-69. Not included here are references to discussions, tables (etc.) that are based on data from a number of swiddens but do not specifically mention the individual swiddens involved.

VI. GENERAL

F

Y

Bei Fragen zur Produktsicherheit wenden Sie sich bitte an:
If you have any questions regarding product safety, please contact:

Walter de Gruyter GmbH
Genthiner Straße 13
10785 Berlin
productsafety@degruyterbrill.com